I0044411

Asymmetric Synthesis: Catalysis, Methods and Applications

Asymmetric Synthesis: Catalysis, Methods and Applications

Edited by
Jack Riley

Larsen & Keller
www.larsen-keller.com

Asymmetric Synthesis: Catalysis, Methods and Applications
Edited by Jack Riley
ISBN: 978-1-63549-673-4 (Hardback)

© 2018 Larsen & Keller

Larsen & Keller
Published by Larsen and Keller Education,
5 Penn Plaza,
19th Floor,
New York, NY 10001, USA

Cataloging-in-Publication Data

Asymmetric synthesis : catalysis, methods and applications / edited by Jack Riley.
p. cm.
Includes bibliographical references and index.
ISBN 978-1-63549-673-4
1. Asymmetric synthesis. 2. Organic compounds--Synthesis. I. Riley, Jack.
QD262 .A89 2018
547.2--dc23

This book contains information obtained from authentic and highly regarded sources. All chapters are published with permission under the Creative Commons Attribution Share Alike License or equivalent. A wide variety of references are listed. Permissions and sources are indicated; for detailed attributions, please refer to the permissions page. Reasonable efforts have been made to publish reliable data and information, but the authors, editors and publisher cannot assume any responsibility for the vailidity of all materials or the consequences of their use.

Trademark Notice: All trademarks used herein are the property of their respective owners. The use of any trademark in this text does not vest in the author or publisher any trademark ownership rights in such trademarks, nor does the use of such trademarks imply any affiliation with or endorsement of this book by such owners.

For more information regarding Larsen and Keller Education and its products, please visit the publisher's website www.larsen-keller.com

Table of Contents

Preface

As a branch of chemistry, asymmetric synthesis refers to that form of chemical synthesis, which produces stereoisomeric products and chirality in large unequal amounts. This synthesis is used in pharmaceutics because it produces diastereomers and enantiomers which are very helpful in drug discovery. This book presents the complex subject of asymmetric synthesis in the most comprehensible and easy to understand language. Different approaches, evaluations and methodologies have been included in it. This textbook will serve as a reference to a broad spectrum of readers.

A detailed account of the significant topics covered in this book is provided below:

Chapter 1- Asymmetric synthesis is an important topic in modern chemistry. Enantioselective catalysis is a commonly used method for producing chiralic elements and stereoisomeric products in unequal quantities. The chapter on asymmetric synthesis offers an insightful focus, keeping in mind the complex subject matter.

Chapter 2- An Alder-ene reaction is a chemical reaction between an allylic hydrogen and an alkene and an enophile, so that σ-bond is formed. These reactions are especially favored when the enophile is electrophilic. This chapter elucidates the crucial theories and principles of asymmetric synthesis.

Chapter 3- The Overman rearrangement converts readily available allylic alcohols to allylic amines by a two-step process. Carbon-heteroatom formation is an important part of asymmetric synthesis. Asymmetric synthesis is best understood in confluence with the major topics listed in the following chapter.

Chapter 4- The addition of Si-H bonds across unsaturated bonds is called hydrosilation. Alkyl and vinyl silanes are the products of alkenes and alkynes while silyl ethers are produced by aldehydes and ketones. The topics discussed in the chapter are of great importance to broaden the existing knowledge on asymmetric synthesis.

Chapter 5- When a chemical reaction occurs between molecular hydrogen and another element, usually in the presence of palladium or nickel as a catalyst, the resulting reaction is known as hydrogenation. Catalysts make the reactions feasible as non-catalytic reactions only take place in high temperatures. The aspects elucidated in this chapter are of vital importance, and provide a better understanding of hydrogenation.

I would like to make a special mention of my publisher who considered me worthy of this opportunity and also supported me throughout the process. I would also like to thank the editing team at the back-end who extended their help whenever required.

Editor

1

Understanding Asymmetric Synthesis

Asymmetric synthesis is an important topic in modern chemistry. Enantioselective catalysis is a commonly used method for producing chiralic elements and stereoisomeric products in unequal quantities. The chapter on asymmetric synthesis offers an insightful focus, keeping in mind the complex subject matter.

Enantioselective Synthesis

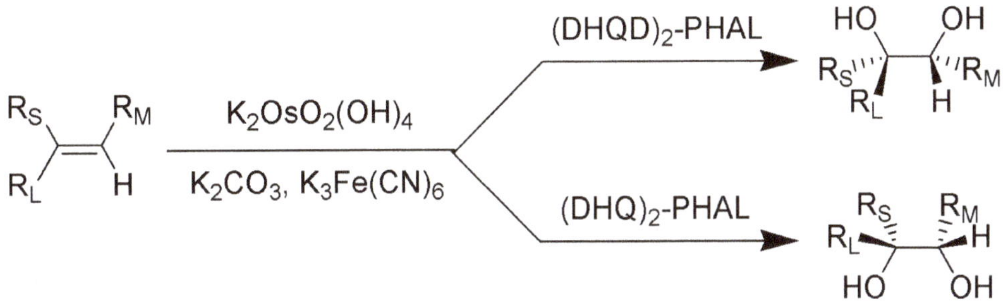

In the Sharpless dihydroxylation reaction the chirality of the product can be controlled by the "AD-mix" used. This is an example of enantioselective synthesis using asymmetric induction

Key: R_L = Largest substituent; R_M = Medium-sized substituent; R_S = Smallest substituent

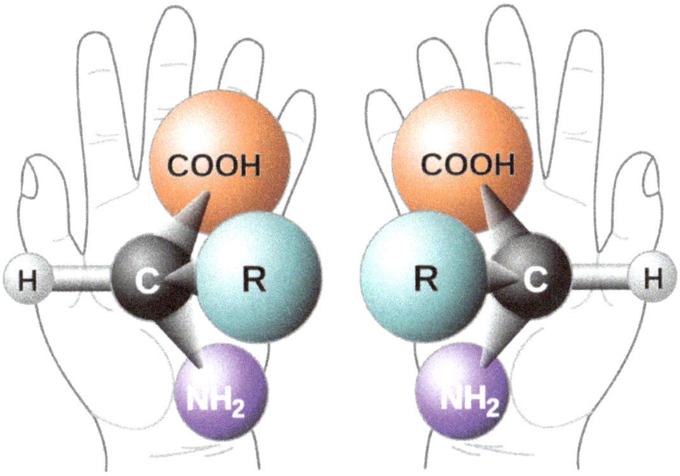

Two enantiomers of a generic alpha amino acid

Enantioselective synthesis, also called chiral synthesis or asymmetric synthesis, is a form of chemical synthesis. It is defined by IUPAC as: a chemical reaction (or reaction sequence) in which one or more new elements of chirality are formed in a substrate molecule and which produces the stereoisomeric (enantiomeric or diastereoisomeric) products in unequal amounts.

Put more simply: it is the synthesis of a compound by a method that favors the formation of a specific enantiomer or diastereomer.

Enantioselective synthesis is a key process in modern chemistry and is particularly important in the field of pharmaceuticals, as the different enantiomers or diastereomers of a molecule often have different biological activity.

Macroscopic Manifestations of Enantioselectivity

Many of the building blocks of biological systems, such as sugars and amino acids, are produced exclusively as one enantiomer. As a result, living systems possess a high degree of chemical chirality and will often react differently with the various enantiomers of a given compound. Examples of this selectivity include:

- Flavour: the artificial sweetener aspartame has two enantiomers. L-aspartame tastes sweet, yet D-aspartame is tasteless.
- Odor: *R*-(−)-carvone smells like spearmint yet *S*-(+)-carvone, smells like caraway.
- Drug effectiveness: the antidepressant drug Citalopram is sold as a racemic mixture. However, studies have shown that only the (*S*)-(+) enantiomer is responsible for the drug's beneficial effects.
- Drug safety: *D*penicillamine is used in chelation therapy and for the treatment of rheumatoid arthritis. However *L*penicillamine is toxic as it inhibits the action of pyridoxine, an essential B vitamin.

As such, enantioselective synthesis is of great importance; but it can also be difficult to achieve. Asymmetric induction can occur intramolecularly when given a chiral starting material. This behaviour can be exploited, especially when the goal is to make several consecutive chiral centres to give a specific enantiomer of a specific diastereomer. An aldol reaction, for example, is inherently diastereoselective; if the aldehyde is enantiopure, the resulting aldol adduct is diastereomerically and enantiomerically pure.

Approaches

Enantioselective Catalysis

Enantioselective catalysis is a widely practiced method for generating chiral compounds.

Principles

Enantiomers possess identical enthalpies and entropies and hence should be produced in equal amounts by an undirected process – leading to a racemic mixture. The nonselectivity can be biased using a chiral feature that favors the formation of one enantiomer over another via interactions at the transition state. This biasing is known as asymmetric induction and can involve chiral features in the substrate, reagent, catalyst, or environment and works by making the activation energy required to form one enantiomer lower than that of the opposing enantiomer.

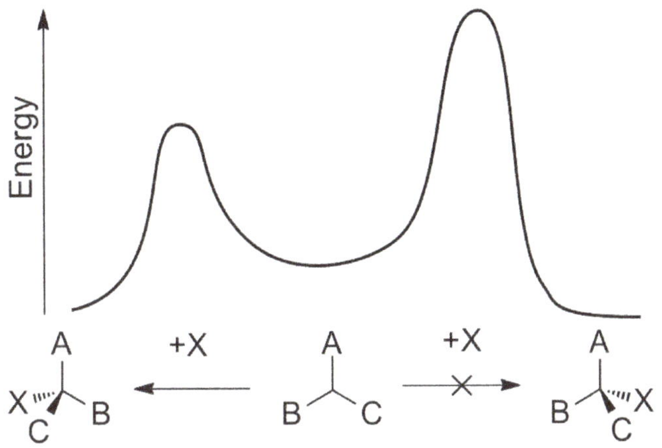

A energy profile of an enantioselective addition reaction.

Enantioselectivity is usually determined by the relative rates, k_1/k_2, of an enantiodifferentiating step. The differing barrier heights, ΔΔG*, for these divergent steps determines the relative rates:

k_1/k_2 = 10^(ΔΔG*/(temperature*1.98*2.3))

k_1/k_2 is also sensitive to temperature, which is more noticeable for modest values of ΔΔG*.

ΔΔG* (kcal)	k1/k2 (273 K)	k1/k2 (298 K)	k1/k2 (323 K)
1.0	6.37	5.46	4.78
2.0	40.6	29.8	22.9
3.0	259	162	109
4.0	1650	886	524
5.0	10500	4830	2510

Chiral Catalysis

In general, enantioselective catalysis (known traditionally as asymmetric catalysis) are chiral co-ordination complexes. Catalysis is effective for a broader range of transformations than any other method of enantioselective synthesis. The catalysts are almost invariably rendered chiral by using chiral ligands. Most enantioselective catalysts are effective at low substrate/catalyst ratios. Given their high efficiencies, they are often suitable for industrial scale synthesis, even with expensive catalysts. A versatile example of enantioselective synthesis is asymmetric hydrogenation, which is used to reduce a wide variety of functional groups.

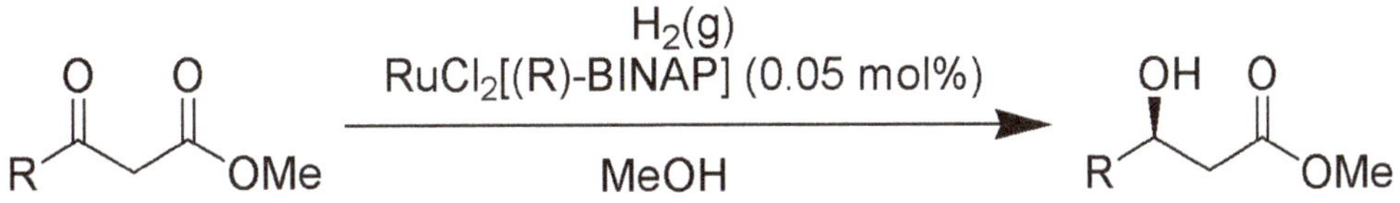

The design of new catalysts is very much dominated by the development of new classes of ligands. Certain ligands, often referred to as 'privileged ligands', have been found to be effective in a wide range of reactions; examples include BINOL, Salen, and BOX. Few catalysts are however applicable to more than one type of asymmetric reaction. For example, Noyori asymmetric hydrogenation with BINAP/Ru requires a β-ketone, although another catalyst, BINAP/diamine-Ru, widens the scope to α,β-olefins and aromatics.

Chiral Auxiliaries

A chiral auxiliary is an organic compound which couples to the starting material to form new compound which can then undergo enantioselective reactions via intramolecular asymmetric induction. At the end of the reaction the auxiliary is removed, under conditions that will not cause racemization of the product. It is typically then recovered for future use.

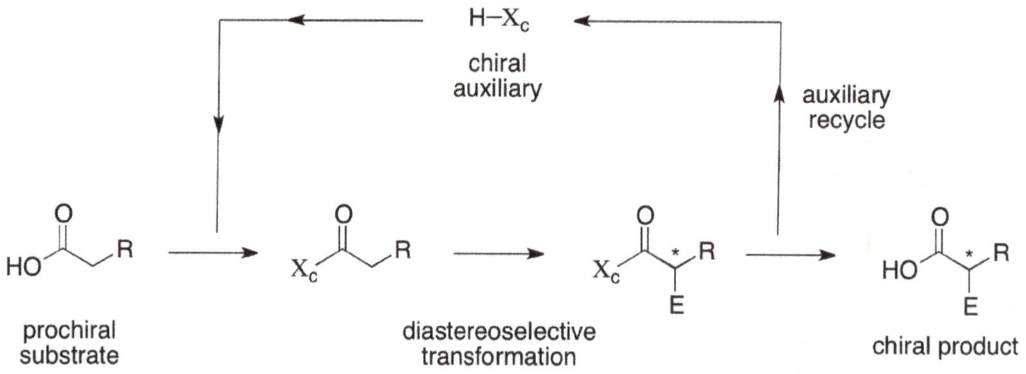

Chiral auxiliaries must be used in stoichiometric amounts to be effective and require additional synthetic steps to append and remove the auxiliary. However, in some cases the only available stereoselective methodology relies on chiral auxiliaries and these reactions tend to be versatile and very well-studied, allowing the most time-efficient access to enantiomerically pure products. Additionally, the products of auxiliary-directed reactions are diastereomers, which enables their facile separation by methods such as column chromatography or crystallization.

Biocatalysis

Biocatalysis makes use of biological compounds, ranging from isolated enzymes to living cells, to perform chemical transformations. The advantages of these reagents include very high ee's and reagent specificity, as well as mild operating conditions and low environmental impact. Biocatalysts are more commonly used in industry than in academic research; for example in the production of statins. The high reagent specificity can be a problem, however, as it often requires that a wide range of biocatalysts be screened before an effective reagent is found.

Enantioselective Organocatalysis

Organocatalysis refers to a form of catalysis, where the rate of a chemical reaction is increased by an organic compound consisting of carbon, hydrogen, sulfur and other non-metal elements. When the organocatalyst is chiral enantioselective synthesis can be achieved; for example a number of carbon–carbon bond forming reactions become enantioselective in the presence of proline with the aldol reaction being a prime example. Organocatalysis often employs natural compounds and secondary amines as chiral catalysts; these are inexpensive and environmentally friendly, as no metals are involved.

Chiral Pool Synthesis

Chiral pool synthesis is one of the simplest and oldest approaches for enantioselective synthesis. A readily available chiral starting material is manipulated through successive reactions, often using achiral reagents, to obtain the desired target molecule. This can meet the criteria for enantioselective synthesis when a new chiral species is created, such as in an S_N2 reaction.

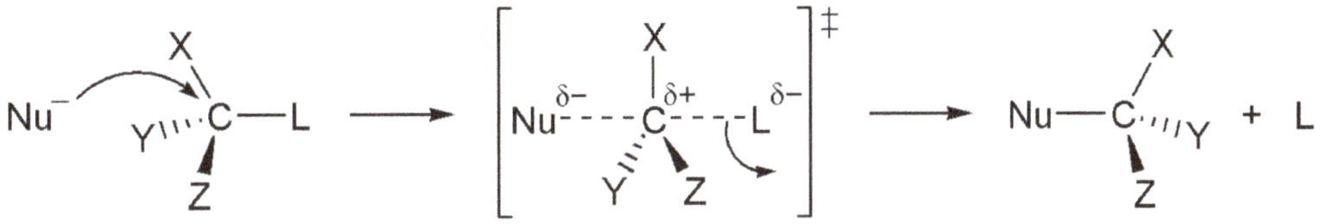

Chiral pool synthesis is especially attractive for target molecules having similar chirality to a relatively inexpensive naturally occurring building-block such as a sugar or amino acid. However, the number of possible reactions the molecule can undergo is restricted and tortuous synthetic routes may be required (e.g. Oseltamivir total synthesis). This approach also requires a stoichiometric amount of the enantiopure starting material, which can be expensive if it is not naturally occurring.

Alternative Approaches

Alternatives to enantioselective synthesis usually involve the isolation of one enantiomer from a racemic mixture by any of a number of methods. If the cost in time and money of making such racemic mixtures is low (or if both enantiomers may find use) then this approach may remain cost-effective. Common methods of separation are based around chiral resolution or kinetic resolution.

Separation and Analysis of Enantiomers

The two enantiomers of a molecule possess the same physical properties (e.g. melting point, boiling point, polarity etc.) and so behave identically to each other. As a result, they will migrate with an identical R_f in thin layer chromatography and have identical retention times in HPLC and GC. Their NMR and IR spectra are identical.

This can make it very difficult to determine whether a process has produced a single enantiomer (and crucially which enantiomer it is) as well as making it hard to separate enantiomers from a reaction which has not been 100% enantioselective. Fortunately, enantiomers behave differently in the presence of other chiral materials and this can be exploited to allow their separation and analysis.

Enantiomers do not migrate identically on chiral chromatographic media, such as quartz or standard media that has been chirally modified. This forms the basis of chiral column chromatography, which can be used on a small scale to allow analysis via GC and HPLC, or on a large scale to separate chirally impure materials. However this process can require large amount of chiral packing material which can be expensive. A common alternative is to use a chiral derivatizing agent to convert the enantiomers into a diastereomers, in much the same way as chiral auxiliaries. These have different physical properties and hence can be separated and analysed using conventional methods. Special chiral derivitizing agents known as 'chiral resolution agents' are used in the NMR spectroscopy of stereoisomers, these typically involve coordination to chiral europium complexes such as $Eu(fod)_3$ and $Eu(hfc)_3$.

The enantiomeric excess of a substance can also be determined using certain optical methods. The oldest method for doing this is to use a polarimeter to compare the level of optical rotation in the product against a 'standard' of known composition. It is also possible to perform ultraviolet-visible spectroscopy of stereoisomers by exploiting the Cotton effect.

One of the most accurate ways of determining the chirality of compound is to determine its absolute configuration by Xray Crystallography. However this is a labour-intensive process which requires that a suitable single crystal be grown.

History

Inception (1815–1905)

In 1815 the French physicist Jean-Baptiste Biot showed that certain chemicals could rotate the plane of a beam of polarised light, a property called optical activity. The nature of this property remained a mystery until 1848, when Louis Pasteur proposed that it had a molecular basis originating from some form of *dissymmetry*, with the term *chirality* being coined by Lord Kelvin a year later. The origin of chirality itself was finally described in 1874, when Jacobus Henricus van't Hoff and Joseph Le Bel independently proposed the tetrahedral geometry of carbon; structural models prior to this work had been two-dimensional, and van't Hoff Le Bel theorized that the arrangement of groups around this tetrahedron could dictate the optical activity of the resulting compound.

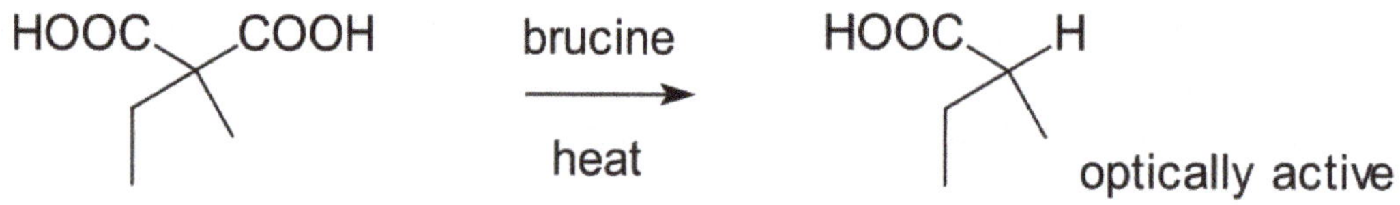

Marckwald's brucine-catalyzed enantioselective decarboxylation of 2-ethyl-2-methylmalonic acid, resulting in a slight excess of the levorotary form of the 2-methylbutyric acid product.

In 1894 Hermann Emil Fischer outlined the concept of asymmetric induction; in which he correctly ascribed selective the formation of D-glucose by plants to be due to the influence of optically active substances within chlorophyll. Fischer also successfully performed what would now be regarded as the first example of enantioselective synthesis, by enantioselectively elongating sugars via a process which would eventually become the Kiliani–Fischer synthesis.

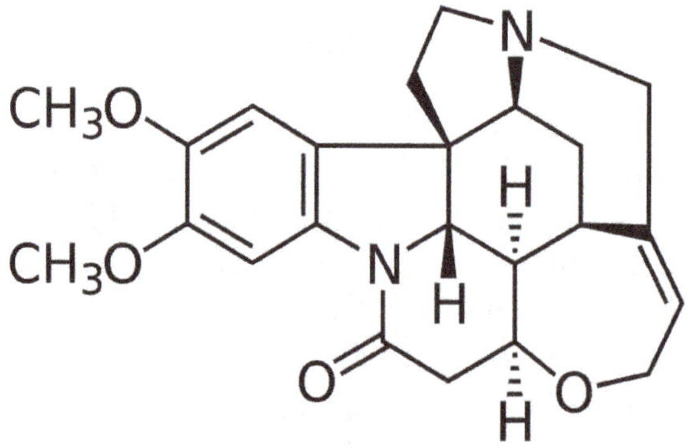

Brucine, an alkaloid natural product related to strychnine, used successfully as an organocatalyst by Marckwald in 1904.

The first enantioselective chemical synthesis is most often attributed to Willy Marckwald, Universität zu Berlin, for a brucine-catalyzed enantioselective decarboxylation of *2-ethyl-2-methylmalonic acid* reported in 1904. A slight excess of the levorotary form of the product of the reaction, 2-methylbutyric acid, was produced; as this product is also a natural product—e.g., as a side chain of lovastatin formed by its diketide synthase (LovF) during its biosynthesis—this result constitutes the first recorded total synthesis with enantioselectivity, as well other firsts (as Koskinen notes, first "example of asymmetric catalysis, enantiotopic selection, and organocatalysis"). This observation is also of historical significance, as at the time enantioselective synthesis could only be understood in terms of vitalism. Natural and artificial compounds were fundamentally different, it was argued, and chirality could only exist in natural compounds. Unlike Fischer, Marckwald had performed an enantioselective reaction upon an achiral, *un-natural* starting material, albeit with a chiral organocatalyst (as we now understand this chemistry).

Early Work (1905–1965)

The development of enantioselective synthesis was initially slow, largely due to the limited range of techniques available for their separation and analysis. Diastereomers possess different physical properties, allowing separation by conventional means, however at the time enantiomers could only be separated by spontaneous resolution (where enantiomers separate upon crystallisation) or kinetic resolution (where one enantiomer is selectively destroyed). The only tool for analysing enantiomers was optical activity using a polarimeter, a method which provides no structural data.

It was not until the 1950s that major progress really began. Driven in part by chemists such as R. B. Woodward and Vladimir Prelog but also by the development of new techniques. The first of these was Xray Crystallography, which was used to determine the absolute configuration of an organic compound by Johannes Bijvoet in 1951. Chiral chromatography was introduced a year later by Dalgliesh, who used paper chromatography to separate chiral amino acids. Although Dalgliesh was not the first to observe such separations, he correctly attributed the separation of enantiomers to differential retention by the chiral cellulose. This was expanded upon in 1960, when Klem and Reed first reported the use of chirally-modified silica gel for chiral HPLC chromatographic separation.

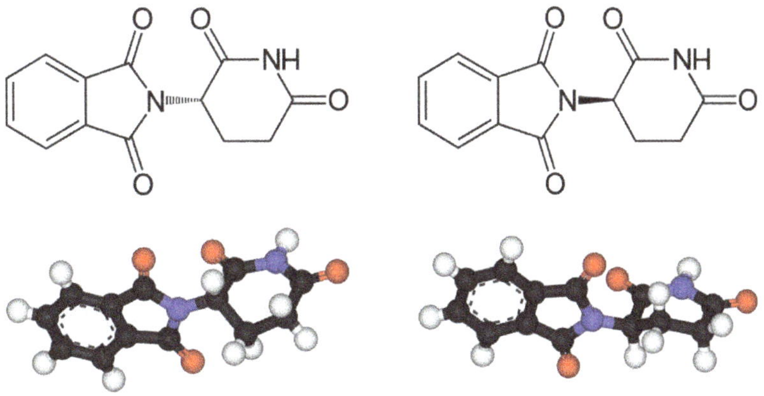

The two enantiomers of thalidomide: Left: (*S*)-thalidomide Right: (*R*)-thalidomide

Thalidomide

While it had long been known that the different enantiomers of a drug could have different activities, this was not accounted for in early drug design and testing. However, following the thalidomide disaster the development and licensing of drugs changed dramatically.

First synthesized in 1953, thalidomide was widely prescribed for morning sickness from 1957 to 1962, but was soon found to be seriously teratogenic, eventually causing birth defects in more than 10,000 babies. The disaster prompted many countries to introduce tougher rules for the testing and licensing of drugs, such as the Kefauver-Harris Amendment (U.S.) and Directive 65/65/EEC1 (E.U.).

Early research into the teratogenic mechanism, using mice, suggested that one enantiomer of thalidomide was teratogenic while the other possessed all the therapeutic activity. This theory was later shown to be incorrect and has now been superseded by a body of research. However it raised the importance of chirality in drug design, leading to increased research into enantioselective synthesis.

Modern Age (Since 1965)

The Cahn–Ingold–Prelog priority rules (often abbreviated as the CIP system) were first published in 1966; allowing enantiomers to be more easily and accurately described. The same year saw first successful enantiomeric separation by gas chromatography an important development as the technology was in common use at the time.

Metal catalysed enantioselective synthesis was pioneered by William S. Knowles, Ryōji Noyori and K. Barry Sharpless; for which they would receive the 2001 Nobel Prize in Chemistry. Knowles and Noyori began with the development of asymmetric hydrogenation, which they developed independently in 1968. Knowles replaced the achiral triphenylphosphine ligands in Wilkinson's catalyst with chiral phosphine ligands. This experimental catalyst was employed in an asymmetric hydrogenation with a modest 15% enantiomeric excess. Knowles was also the first to apply enantioselective metal catalysis to industrial-scale synthesis; while working for the Monsanto Company he developed an enantioselective hydrogenation step for the production of L-DOPA, utilising the DIPAMP ligand.

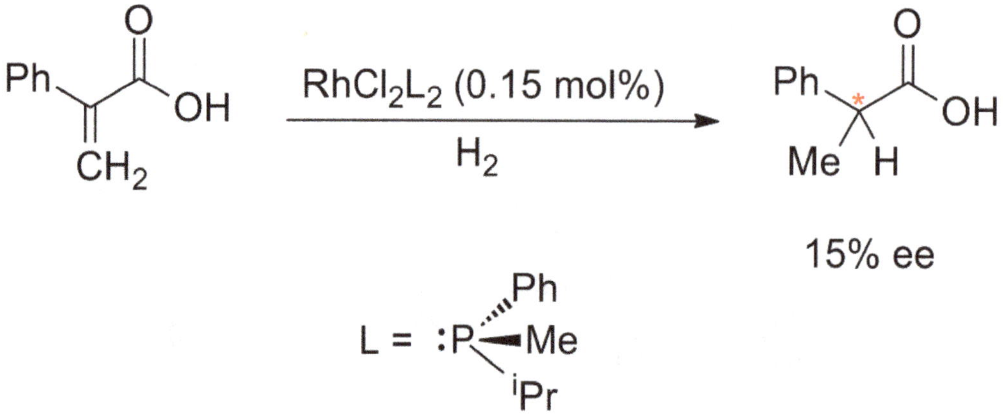

Knowles: Asymmetric hydrogenation (1968)

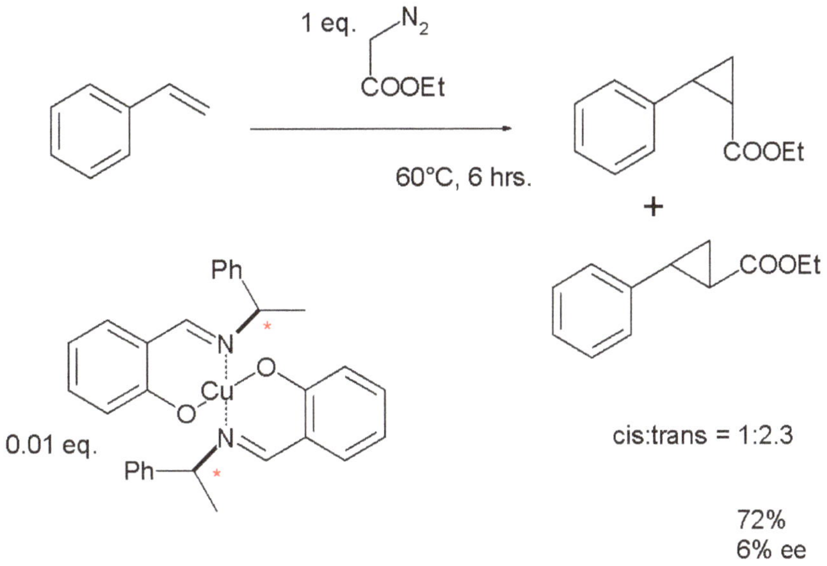

Noyori: Enantioselective cyclopropanation (1968)

Noyori devised a copper complex using a chiral Schiff base ligand, which he used for the metal-carbenoid cyclopropanation of styrene. In common with Knowles' findings, Noyori's results for the enantiomeric excess for this first-generation ligand were disappointingly low: 6%. However continued research eventually led to the development of the Noyori asymmetric hydrogenation reaction.

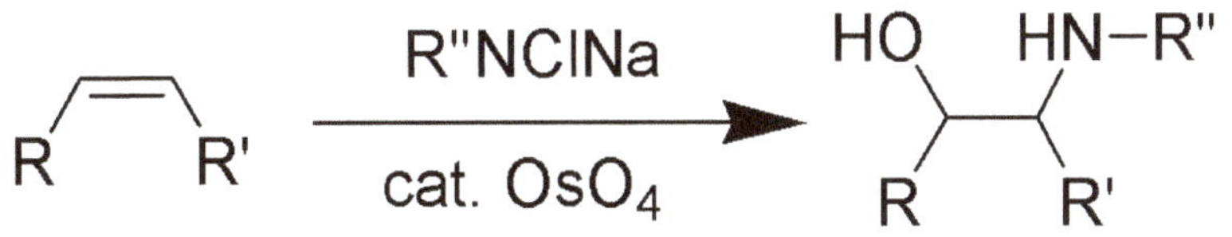

The Sharpless oxyamination

Sharpless complemented these reduction reactions by developing a range of asymmetric oxidations (Sharpless epoxidation, Sharpless asymmetric dihydroxylation, Sharpless oxyamination) during the 1970s to 1980's. With the asymmetric oxyamination reaction, using osmium tetroxide, being the earliest.

During the same period, methods were developed to allow the analysis of chiral compounds by NMR; either using chiral derivatizing agents, such as Mosher's acid, or europium based shift reagents, of which $Eu(DPM)_3$ was the earliest.

Chiral auxiliaries were introduced by E.J. Corey in 1978 and featured prominently in the work of Dieter Enders. Around the same time enantioselective organocatalysis was developed, with pioneering work including the Hajos–Parrish–Eder–Sauer–Wiechert reaction. Enzyme-catalyzed enantioselective reactions became more and more common during the 1980s, particularly in industry, with their applications including asymmetric ester hydrolysis with pig-liver esterase. The emerging technology of genetic engineering has allowed the tailoring of enzymes to specific processes, permitting an increased range of selective transformations. For example, in the asymmetric hydrogenation of statin precursors.

Chiral Lewis Acid

Chiral Lewis acids (CLAs) are a type of Lewis acid catalyst that effects the chirality of the substrate as it reacts with it. In such reactions the synthesis favors the formation of a specific enantiomer or diastereomer. The method then is an enantioselective asymmetric synthesis reaction. Since they affect chirality, they produce optically active products from optically inactive or mixed starting materials. This type of preferential formation of one enantiomer or diastereomer over the other is formally known as an asymmetric induction. In this kind of Lewis acid. the electron-accepting atom is typically a metal, such as indium, zinc, lithium, aluminium, titanium, or boron. The chiral-altering ligands employed for synthesizing these acids most often have multiple Lewis basic sites (often a diol or a dinitrogen structure) that allow the formation of a ring structure involving the metal atom.

Achiral Lewis acids have been used for decades to promote the synthesis of racemic mixtures in a myriad different reactions. Starting in the 1960s chemists have use the chiral acids to induce the enantioselective reactions. Common reaction types include Diels-Alder reactions, the ene reaction, [2+2] cycloaddition reactions, hydrocyanation of aldehydes, and most notably, Sharpless expoxidations.

Theory

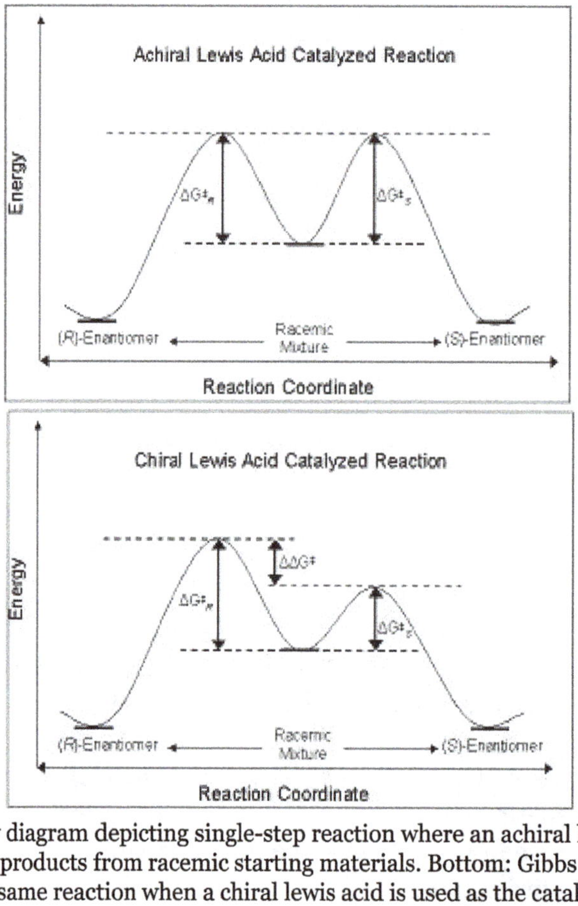

Figure: Top: Gibbs Free Energy diagram depicting single-step reaction where an achiral lewis acid is catalyzing the formation of a racemic mixture of products from racemic starting materials. Bottom: Gibbs free energy diagram depicting the same reaction when a chiral lewis acid is used as the catalyst

The enantioselectivity of CLAs derives from their ability to perturb the free energy barrier along the reaction coordinate pathway that leads to either the *R*- or *S*- enantiomer. Ground state diastereomers and enantiomers are of equal energy in the ground state, and when reacted with an achiral lewis acid, their diastereomeric intermediates, transition states, and products are also of equal energy. This leads to the production of racemic mixtures of products. However, when a CLA is utilized in the same reaction, the energetic barrier of formation of one diastereomer is less than that of another – the reaction is under kinetic control. If the difference in the energy barriers between the diastereomeric transition states are of sufficient magnitude, and high enantiomeric excess of one isomer should be observed.

Applications of CLAs in Asymmetric Synthesis

Diels-Alder Reaction

Diels-Alder reactions occur between a conjugated diene and an alkene (commonly known as the dienophile). This cycloaddition process allows for the stereoselective formation of cyclohexene rings capable of possessing as many as four contiguous stereogenic centers.

Diels-Alder reactions can lead to formation of a variety of structural isomers and stereoisomers. The molecular orbital theory considers that endo transition state, instead of the exo transition state, is favored (endo addition rule). Also, augmented secondary orbital interactions have been postulated as the source of enhanced endo diastereoselection.

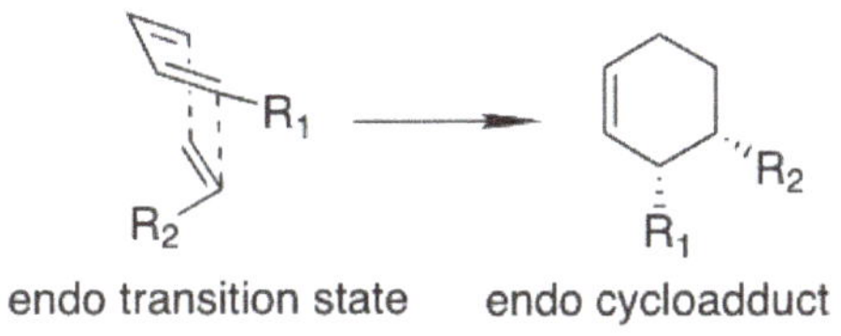

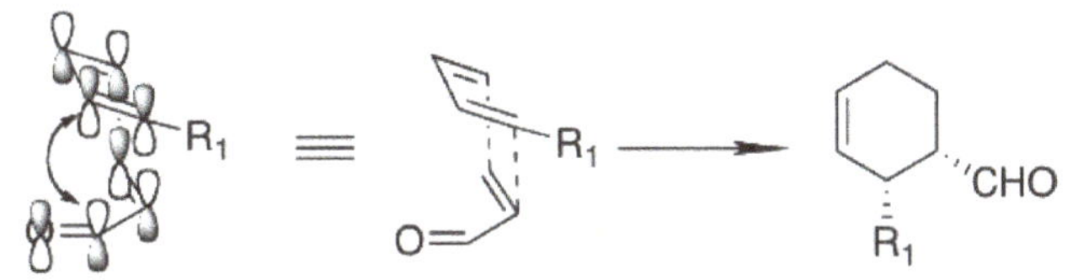

Secondary orbital interaction

The addition of a CLA selectively activates one component of the reaction (the diene or dienophile) while providing a stereodefined environment that permits unique enantioselectivity.

Koga and coworkers disclosed the first practical example of a catalytic enantioselective Diels-Alder reaction promoted by a CLA - menthoxyaluminum dichloride - derived from menthol and ethylaluminum dichloride.

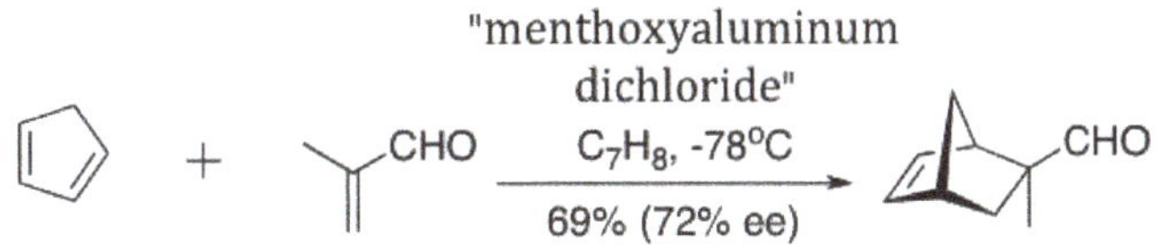

A decade later, Elias James Corey introduced an effective aluminum-diamine controller for

Diels-Alder reaction. Formation of the active catalyst is achieved by treatment of the bis(sulfonamide) with trimethylaluminum; recovery of the ligand was essentially quantitative. The proposed tetracoordinate aluminum prevent the imide acting as a chelating Lewis base, while enhance the α-vinyl proton of the dienphile and the benzylic proton of the catalyst.

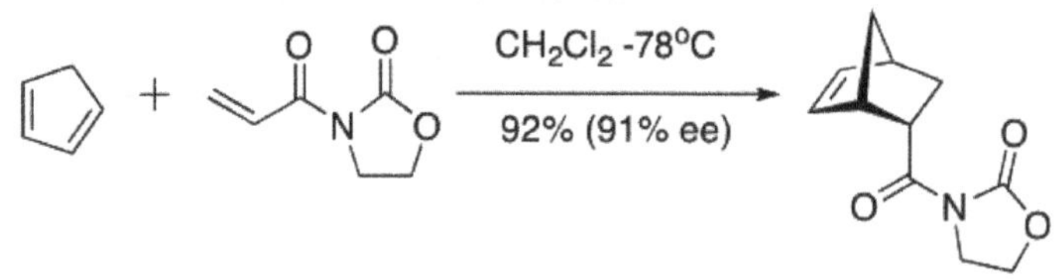

The X-ray structure of the catalyst showed a stereodefined environment.

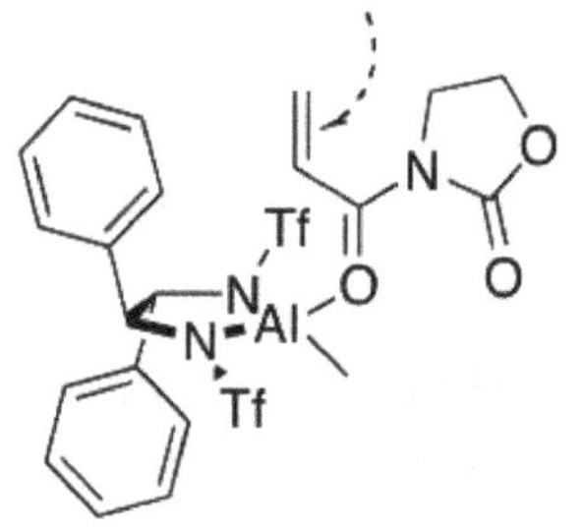

In 1993, Wulff and coworkers found a complex derived from diethylaluminium chloride and a "vaulted" biaryl ligand below catalyzed the enantioselective Diels-Alder reaction between cyclopentadiene and methacrolein. The chiral ligand is recovered quantitatively by silica gel chromatography.

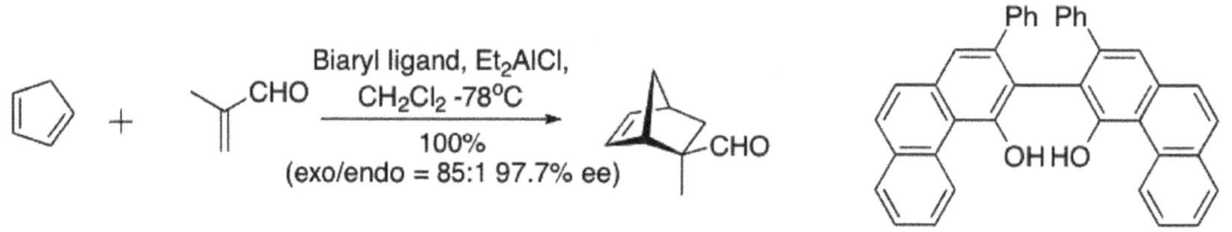

Hisashi Yamamoto and coworkers have developed a practical Diels-Alder catalyst for aldehyde dienophiles. The chiral (acyloxy)borane (CAB) complex is effective in catalyzing a number of aldehyde Diels-Alder reactions. NMR spectroscopic experiments indicated close proximity of the aldehyde and the aryl ring. Also, Pi stacking between the aryl group and aldehyde was suggested as an organizational feature which imparted high enantioselectivity to the cycloaddition.

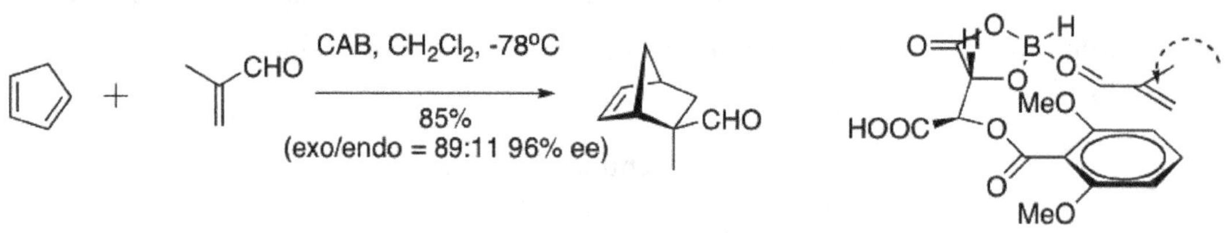

Yamamoto and co-wokers have introduced a conceptually interesting series of catalysts that incorporate an acidic proton into the active catalyst. This kind of what so called Bronsted acid-assisted chiral Lewis acid (BLA) catalyzes a number of diene-aldehyde cycloaddition reactions.

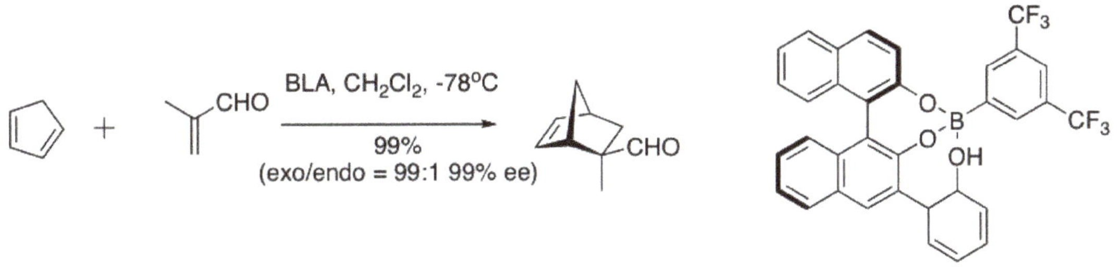

Aldol Reaction

In the aldol reaction, the diastereoselectivity of the product is often dictated by the geometry of the enolate according to the Zimmerman-Traxler model. The model predicts that the Z enolate will give syn products and that E enolates will give anti products. Chiral Lewis acids allow products that defy the Zimmerman-Traxler model, and allows for control of absolute stereochemistry. Kobayashi and Horibe demonstrated this in the synthesis of dihydroxythioester derivatives, using a tin-based chiral Lewis acid.

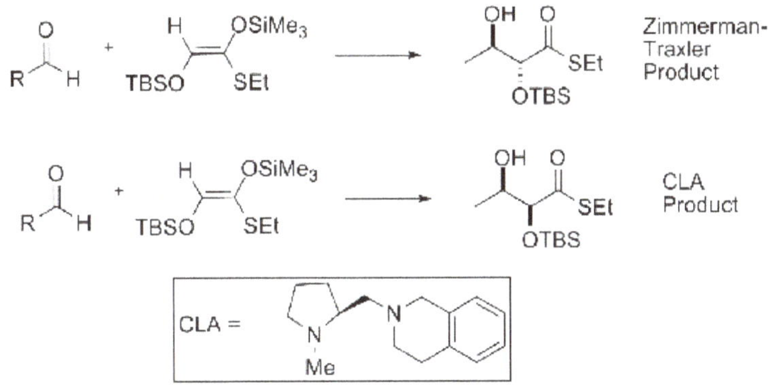

The transition structures for reactions with both the R and S catalyst enantiomers are shown below.

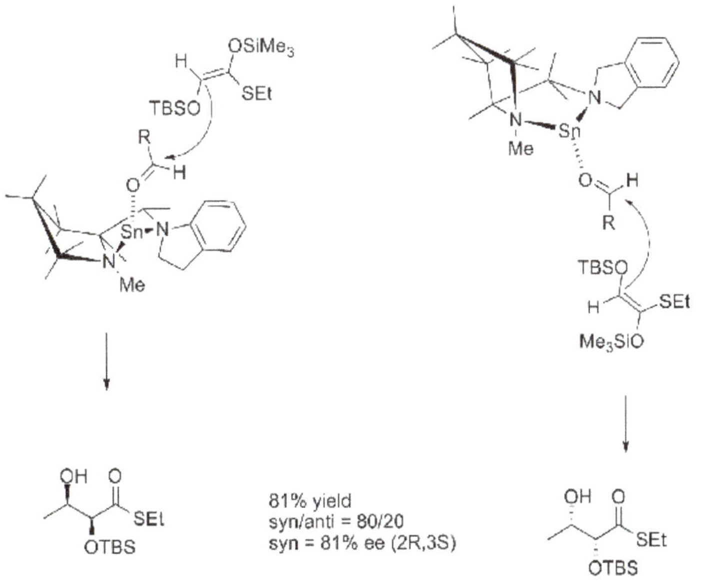

Baylis-Hillman Reaction

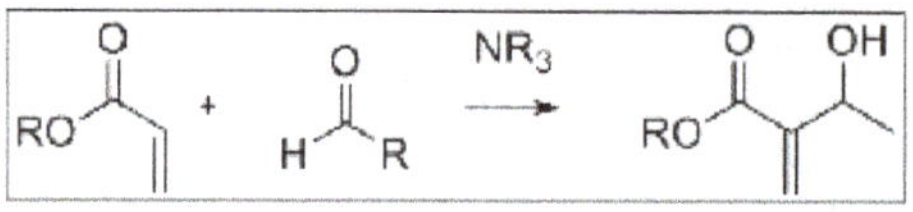

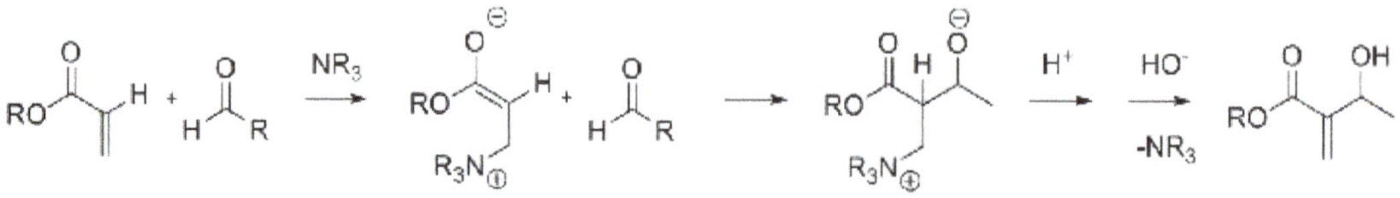

The Baylis-Hillman reaction is a route for C-C bond formation between an alpha, beta-unsaturated carbonyl and an aldehyde, which requires a nucleophilic catalyst, usually a tertiary amine, for a Michael-type addition and elimination. The stereoselectivity of these reactions is usually poor. Chen et al. demonstrated an enantioselective chiral Lewis acid-catalyzed reaction. Lanthanum was used in this case. Similarly a chiral amine may also be used to achieve stereoselectivity.

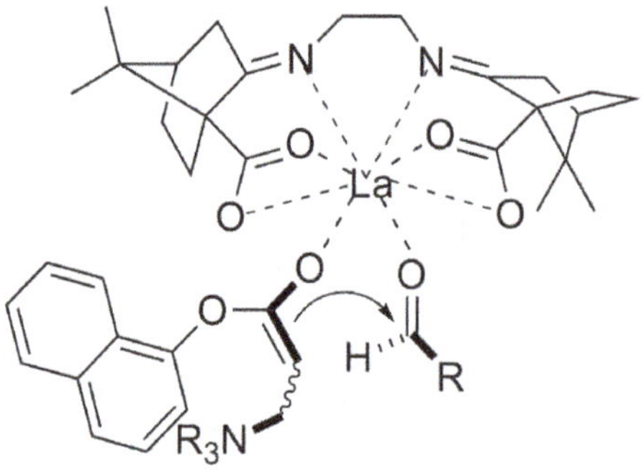

The product obtained by the reaction using the chiral catalyst was obtained in good yield with excellent enantioselectivity.

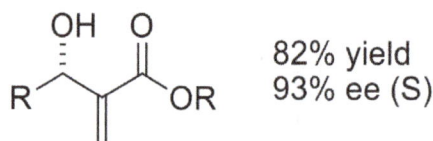

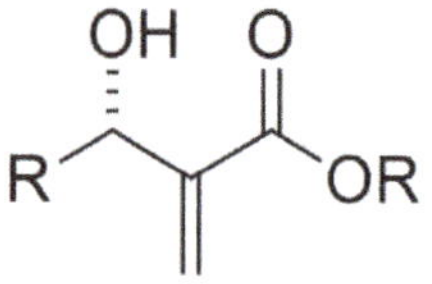

Ene Reaction

Chiral lewis acids have also proven useful in the ene reaction. When catalyzed by an achiral lewis acid the reaction normally provides good diastereoselectivity.

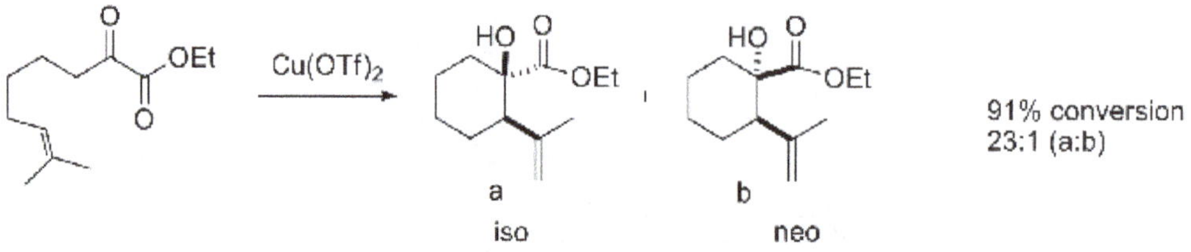

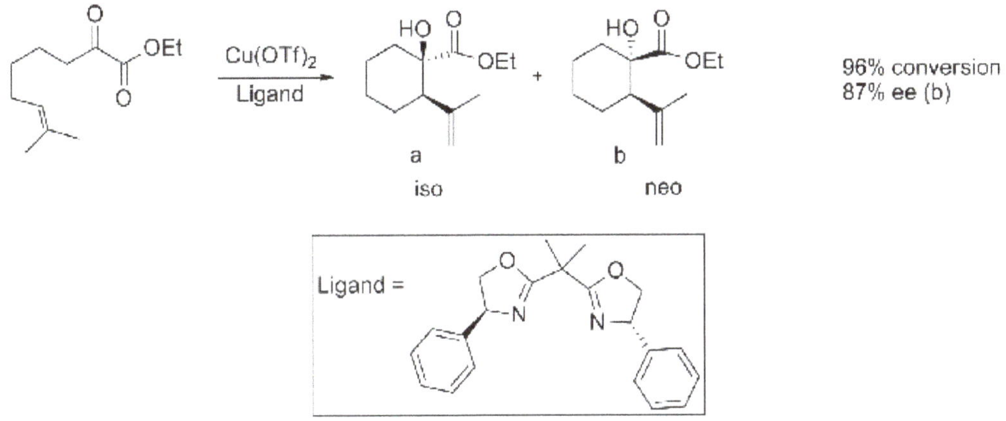

When a chiral lewis acid catalyst was used good enantioselectivity was observed.

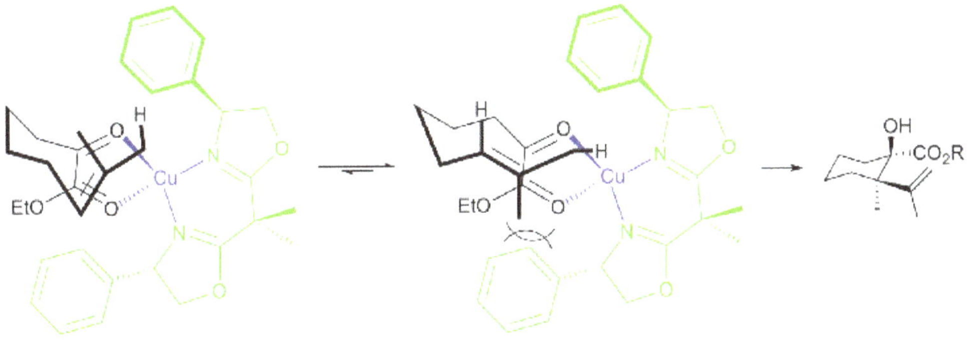

The enantioselectivity is believed to be due to the steric interactions between the methyl and phenyl group, which makes the transition structure of the iso product considerably more favorable.

Examples of Achiral Lewis acids in Stereoselective Synthesis

Nickel catalyzed coupling of 1,3-dienes with aldehydes In some cases an achiral Lewis acid may provide good stereoselectivity. Kimura et al. demonstrated the regio- and diastereoselective coupling of 1,3-dienes with aldehydes.

Utility of Chiral Lewis Acids

Asymmetric synthesis and production of enantiomerically pure substances through the use of CLAs is of particular interest to organic chemists and pharmaceutical corporations. Because many pharmaceuticals target enzymes which are specific for a particular enantiomer, compounds intended for patient administration must be of a high optical purity. Furthermore, resolution of a particular enantiomer from a racemic mixture is both costly and wasteful.

Brønsted Acid-Assisted Lewis Acid (BLA)

Chiral Brønsted acid-assisted Lewis acids (BLAs) are efficient and versatile chiral Lewis acids for a wide range of catalytic asymmetric cycloaddition reactions. Some of the representative examples follow:

Diels–Alder Reaction

The Diels–Alder reaction is an organic chemical reaction (specifically, a [4+2] cycloaddition) between a conjugated diene and a substituted alkene, commonly termed the dienophile, to form a substituted cyclohexene system. It was first described by Otto Diels and Kurt Alder in 1928, for which work they were awarded the Nobel Prize in Chemistry in 1950. The Diels–Alder reaction is particularly useful in synthetic organic chemistry as a reliable method for forming 6-membered systems with good control over regio- and stereochemical properties. The underlying concept has also been applied to other π-systems, such as carbonyls and imines, to furnish the corresponding heterocycles, known as the hetero-Diels–Alder reaction. Diels–Alder reactions can be reversible under certain conditions; the reverse reaction is known as the retro-Diels–Alder reaction.

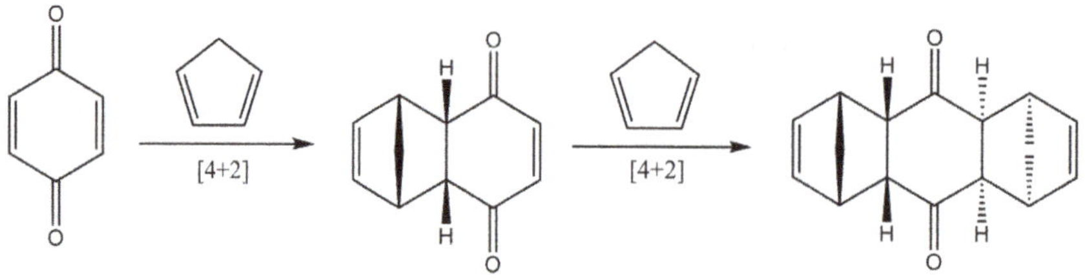

Reaction Mechanism

The reaction is an example of a concerted pericyclic reaction. It is believed to occur via a single, cyclic transition state, with no intermediates generated during the course of the reaction. As such, the Diels–Alder reaction is governed by orbital symmetry considerations: it is classified as a [4πS+2πS] cycloaddition, indicating that it proceeds through the suprafacial/suprafacial interaction of a 4π electron system (the diene structure) with a 2π electron system (the dienophile structure), an interaction that is thermally allowed as a 4n+2 cycloaddition.

A consideration of the reactants' frontier molecular orbitals (FMO) makes plain why this is so. We note that for a 'normal' electron demand Diels–Alder reaction, the electron-rich diene's Ψ2 is the highest occupied molecular orbital (HOMO) while the electron-deficient dienophile's π* is the lowest unoccupied molecular orbital (LUMO). However, the HOMO-LUMO energy gap is such that the roles can be reversed by switching the substitution pattern: i.e. the diene's Ψ3 might be considered the LUMO if electron withdrawing group (EWG) substituents make it sufficiently electron-deficient and electron donating groups (EDGs) raise the dienophile's filled π orbital's energy sufficiently to make it the HOMO. Such a scenario is termed an inverse electron demand Diels–Alder reaction. Regardless of which situation pertains, the HOMO and LUMO of the components are in phase and a bonding interaction results as can be seen in the diagram below. Since the reactants are in their ground state, the reaction is initiated thermally and does not require activation by light.

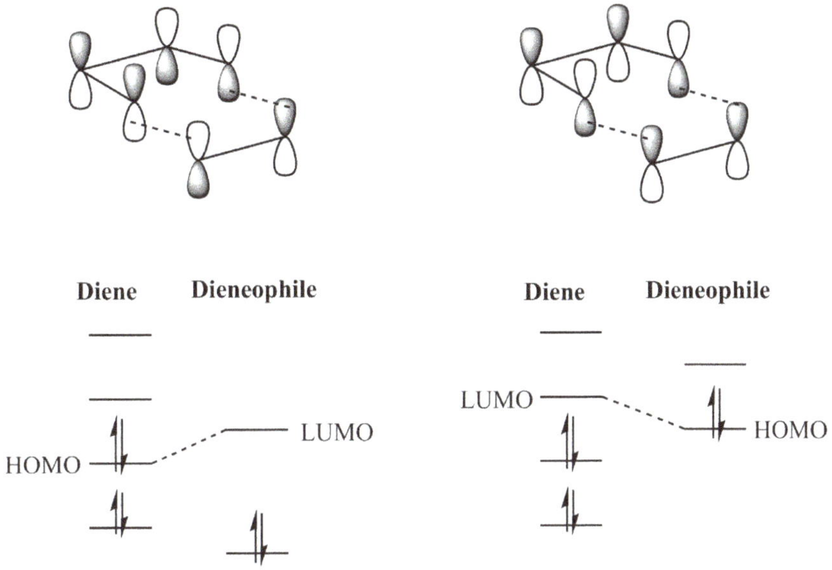

The "prevailing opinion" is that most Diels–Alder reactions proceed through a concerted mechanism; the issue, however, has been thoroughly contested. Despite the fact that the vast majority of Diels–Alder reactions exhibit stereospecific, syn addition of the two components, a diradical intermediate has been postulated (and supported with computational evidence) on the grounds that the observed stereospecificity does not rule out a two-step addition involving an intermediate that collapses to product faster than it can rotate to allow for inversion of stereochemistry.

There is a notable rate enhancement when certain Diels–Alder reactions are carried out in polar organic solvents such as dimethylformamide and ethylene glycol. and even in water. The reaction of cyclopentadiene and butenone for example is 700 times faster in water relative to 2,2,4-trimethylpentane as solvent. Several explanations for this effect have been proposed, such as an increase in effective concentration due to hydrophobic packing or hydrogen-bond stabilization of the transition state.

Regioselectivity

FMO analysis has also been used to explain the regioselectivity patterns observed in Diels–Alder reactions of substituted systems. Calculation of the energy and orbital coefficients of the components' frontier orbitals provides a picture that is in good accord with the more straightforward analysis of the substituents' resonance effects. If we assume that the centers with the largest frontier orbital coefficients will react more readily when matched than mismatched (i.e. largest HOMO coefficient will react more readily with the largest LUMO coefficient), we can easily predict the major regioisomer that will result from a given diene-dienophile pair.

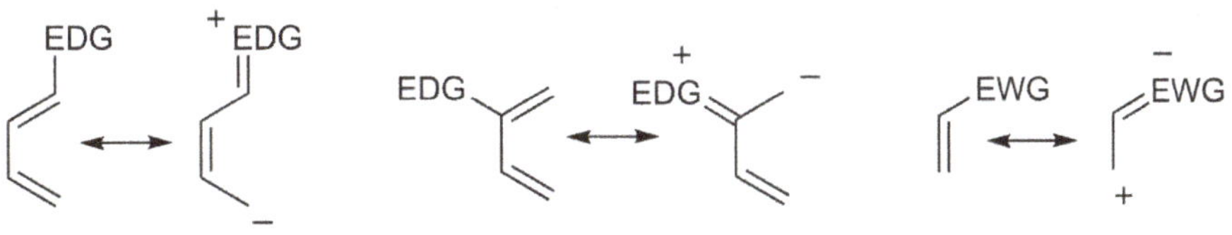

For example, in a normal-demand scenario, a diene bearing an EDG at C1 has its largest HOMO coefficient at C4, while the dienophile has the largest LUMO coefficient at C2. Pairing these two coefficients gives the "ortho" product as seen in case 1 in the figure below. A diene substituted at C2 as in case 2 below has the largest HOMO coefficient at C1, giving rise to the "para" product. Similar analyses for the corresponding inverse-demand scenarios gives rise to the analogous products as seen in cases 3 and 4.

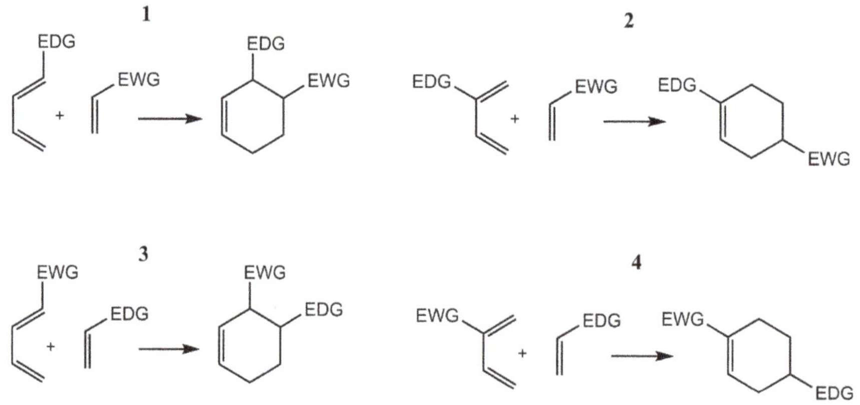

Stereoselectivity

Diels–Alder reactions, as concerted cycloadditions, are stereospecific, i.e. stereochemical information in the reactants is retained in the products. *E*- and *Z*-dienophiles, for example, give rise to the adducts with corresponding *syn* or *anti*-stereochemistry:

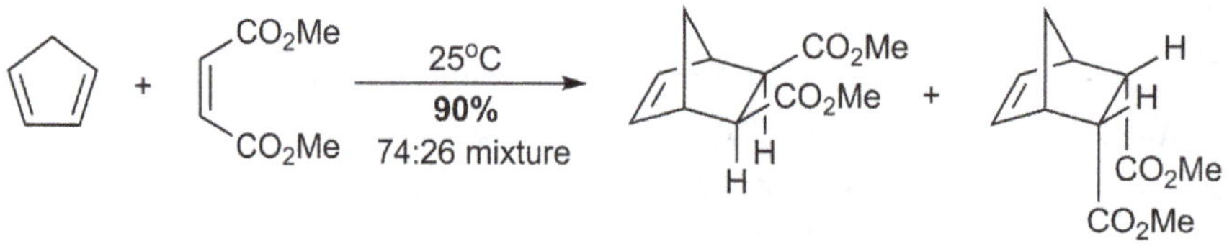

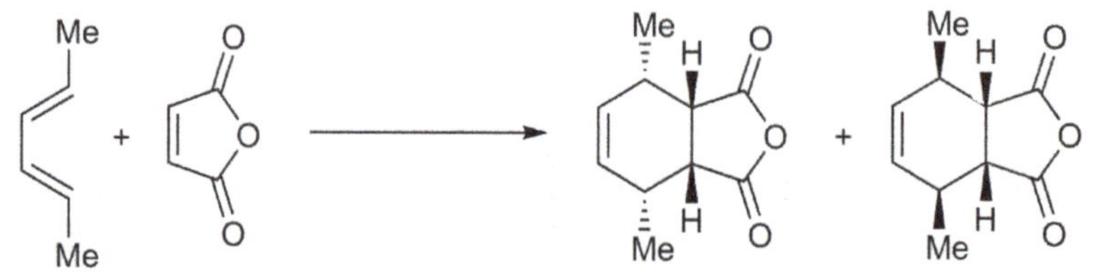

Unsymmetrical dienophiles imply two different possible transition states, which are called the *endo* and *exo* transition states, each leading to adducts of different stereochemistry. In the *endo* transition state, the substituent on the dienophile is oriented towards the diene π system, while in the *exo* it is oriented away from it. For normal demand Diels-Alder scenarios, with electron-withdrawing substituents such as carbonyls attached to the dienophile, the *endo* transition state is typically preferred, despite often being more sterically congested. This preference is known as the *Alder rule*. *Endo* selectivity is typically higher for rigid dienophiles such as maleic anhydride and benzoquinone; for others, such as acrylates and crotonates, selectivity is not very pronounced The most widely accepted explanation for the origin of this effect is a favorable interaction between the dienophile substituent's π system and the diene's (termed *secondary orbital effects*), though dipolar and van der Waals attractions may play a part as well.

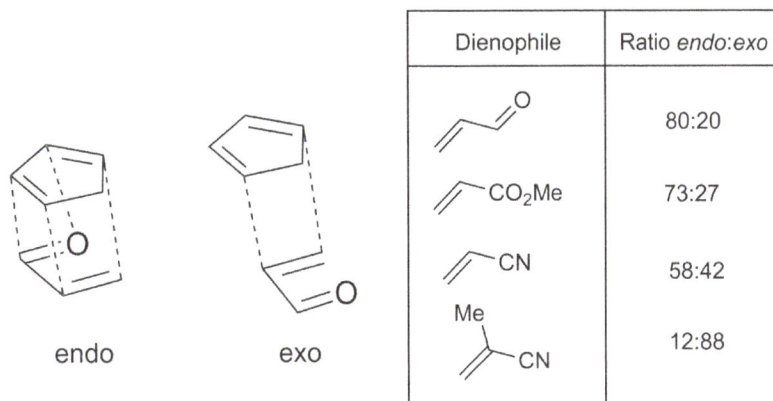

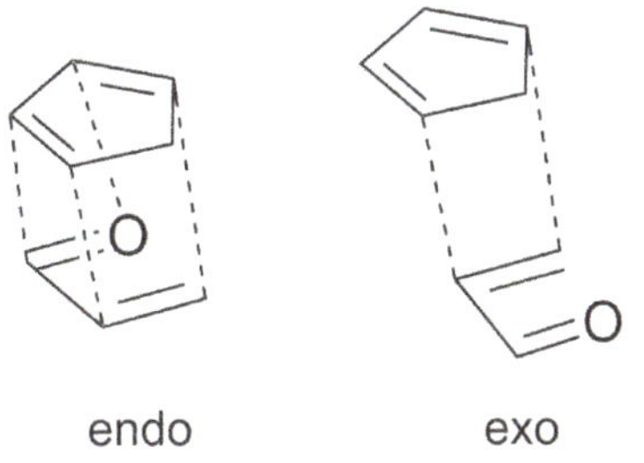

Dienophile	Ratio *endo*:*exo*
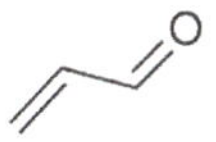	80:20
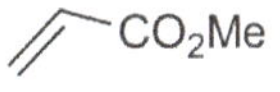	73:27
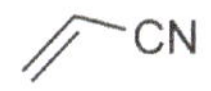	58:42
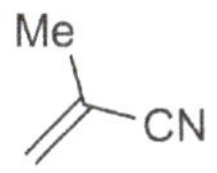	12:88

Oftentimes, as with highly substituted dienes or very bulky dienophiles, steric effects can greatly influence *endo*/*exo* selectivity. Dienes with bulky terminal substituents (C1 and C4) decrease the rate of reaction, presumably by impeding the approach of the diene and dienophile; however, bulky substituents at the C2 or C3 position actually increase reaction rate by destabilizing the s-*trans* conformation and forcing the diene into the reactive s-*cis* conformation. 2-*tert*-butyl-1,3-butadiene, for example, is 27 times more reactive than simple butadiene.

The Diene

The diene component of the Diels–Alder reaction can be either open-chain or cyclic, and it can host many different types of substituents; it must, however, be able to exist in the s-*cis* conformation, since this is the only conformer that can participate in the reaction. Though butadienes typically prefer the s-trans conformation, for most cases the barrier to rotation is small (~2–5 kcal/mol).

An especially reactive diene is 1-methoxy-3-trimethylsiloxy-1,3-butadiene, otherwise known as Danishefsky's diene. It has particular synthetic utility as means of furnishing α,β–unsaturated cyclohexenone systems by elimination of the 1-methoxy substituent after deprotection of the enol silyl ether. Other synthetically useful derivatives of Danishefsky's diene include 1,3-alkoxy-1-trimethylsiloxy-1,3-butadienes (Brassard dienes) and 1-dialkylamino-3-trimethylsiloxy-1,3-butadienes (Rawal dienes). The increased reactivity of these and similar dienes is a result of synergistic contributions from donor groups at C1 and C3, raising the HOMO significantly above that of a comparable monosubstituted diene.

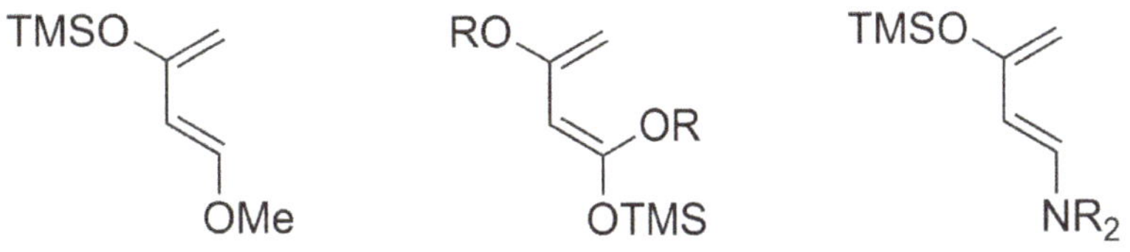

Danishefsky's diene Brassard diene Rawal diene

Unstable (and thus highly reactive) dienes, of which perhaps the most synthetically useful are o-quinodimethanes, can be generated in situ. A strong driving force for the [4+2] cycloaddition of such species is a result of the establishment (or reestablishment) of aromaticity. Common methods for generating o-quinodimethanes include pyrolysis of benzocyclobutenes or the corresponding sulfone, 1,4-elimination of ortho benzylic silanes or stannanes, and reduction of α,α'-ortho benzylic dibromides.

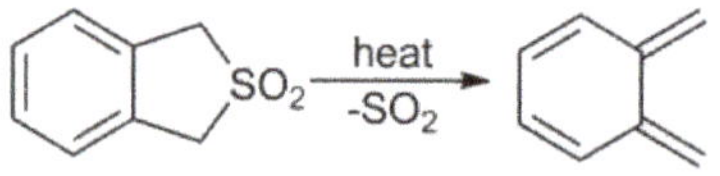

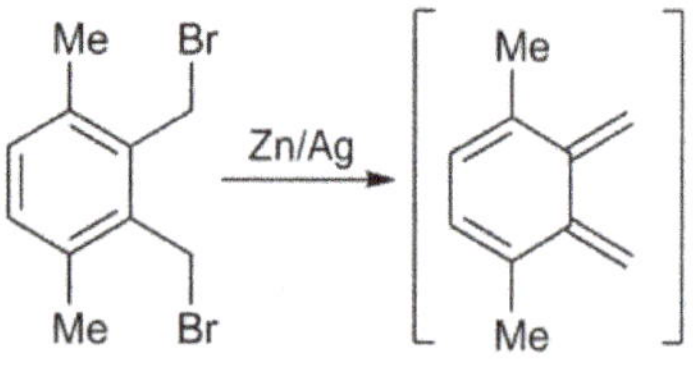

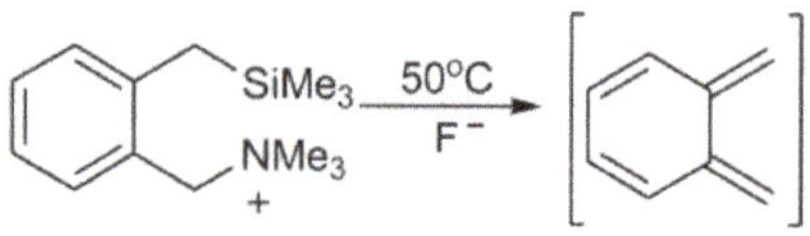

On the contrary, stable dienes are rather unreactive and undergo Diels–Alder reactions only at elevated temperatures: for example, naphthalene can function as a diene, leading to adducts only with highly reactive dienophiles, such as N-phenyl-maleimide. Anthracene, being less aromatic (and therefore more reactive for Diels-Alder syntheses) in its central ring can form a 9,10 adduct with maleic anhydride at 80 degrees Celsius and even with acetylene, a weak dienophile, at 250 degrees.

The Dienophile

In a normal demand Diels–Alder reaction, the dienophile has an electron-withdrawing group in conjugation with the alkene; in an inverse-demand scenario, the dienophile is conjugated with an electron-donating group. Dienophiles can be chosen to contain a "masked functionality". The dienophile undergoes Diels–Alder reaction with a diene introducing such a functionality onto the product molecule. A series of reactions then follow to transform the functionality into a desirable group. The end product cannot not be made in a single DA step because equivalent dienophile is either unreactive or inaccessible. An example of such approach is the use of α-chloroacrylonitrile (CH_2=CClCN). When reacted with a diene, this dienophile will introduce α-chloronitrile functionality onto the product molecule. This is a "masked functionality" which can be then hydrolyzed to form a ketone. α-Chloroacrylonitrile dienophile is an equivalent of ketene dienophile (CH_2=C=O), which would produce same product in one DA step. The problem is that ketene itself cannot be used in Diels–Alder reactions because it reacts with dienes in unwanted manner (by [2+2] cycloaddition), and therefore "masked functionality" approach has to be used. Other such functionalities are phosphonium substituents (yielding exocyclic double bonds after Wittig reaction), various sulfoxide and sulfonyl functionalities (both are acetylene equivalents), and nitro groups (ketene equivalents).

Hetero-Diels–Alder

Diels–Alder reactions involving at least one heteroatom are also known and are collectively called hetero-Diels–Alder reactions Carbonyl groups, for example, can successfully react with dienes to yield dihydropyran rings, a reaction known as the oxo-Diels–Alder reaction, and imines can be used, either as the dienophile or at various sites in the diene, to form various *N*-heterocyclic compounds via the aza-Diels–Alder reaction. Nitroso compounds (R-N=O) can react with dienes to form oxazines. Chlorosulfonyl isocyanate can be utilized as a dienophile to prepare the Vince lactam.

Lewis Acid Activation

Lewis acids such as zinc chloride, boron trifluoride, tin tetrachloride, aluminum chloride, etc. can act as catalysts of normal-demand Diels–Alder reactions by coordination to the dienophile. The complexed dienophile becomes more electrophilic and more reactive toward the diene, increasing the reaction rate and often improving the regio- and stereoselectivity as well. Lewis acid catalysis also enables Diels–Alder reactions to proceed at low temperatures, i.e. without thermal activation.

Asymmetric Diels–Alder

Many methods have been developed for influencing the stereoselectivity of the Diels–Alder reaction, such as the use of chiral auxiliaries, catalysis by chiral Lewis acids, and small organic molecule catalysts. Evans' oxazolidinones, oxazaborolidines, *bis*-oxazoline–copper chelates, imidazoline catalysis, and many other methodologies exist for effecting diastereo- and enantioselective Diels-Alder Reactions.

Synthetic Applications

One of the earliest and most important examples of the Diels–Alder reaction in total synthesis was in R. B. Woodward's landmark 1952 syntheses of the steroids cortisone and cholesterol. The reaction of butadiene with the quinone below quickly furnished the C and D rings of the steroid skeleton with the desired regiochemistry. Syn addition of butadiene gave rise to the desired stereochemistry of the final target's methyl group, and thereafter a straightforward epimerization could be effected selectively to give the requisite trans-decalin system.

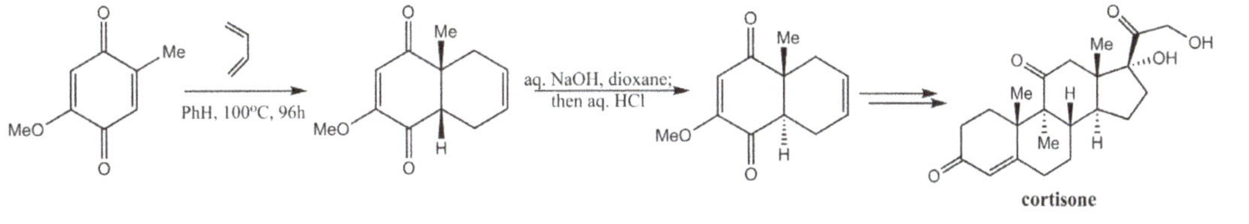

E. J. Corey, in his original 1969 synthesis of prostaglandins F2α and E2, utilized a Diels–Alder reaction early in the synthesis to establish the relative stereochemistry of three contiguous stereocenters on the prostaglandin cyclopentane core. To mitigate isomerization of the substituted cyclopentadiene via 1,5-hydride shift, it was found necessary to keep this intermediate below 0 °C until the Diels-Alder could take place. Thus activation by strongly Lewis acidic cupric tetrafluoroborate was required to allow for the reaction to take place. The use of 2-chloroacrylonitrile as dienophile is a viable synthetic equivalent for ketene, a structure that typically underdoes a [2+2] cycloaddition to give a cyclobutanone dimer rather than participating in Diels–Alder reactions with 1,3-dienes. Hydrolysis of the epimeric mixture of chloronitrile adducts revealed the desired bicycloheptanone in high yield.

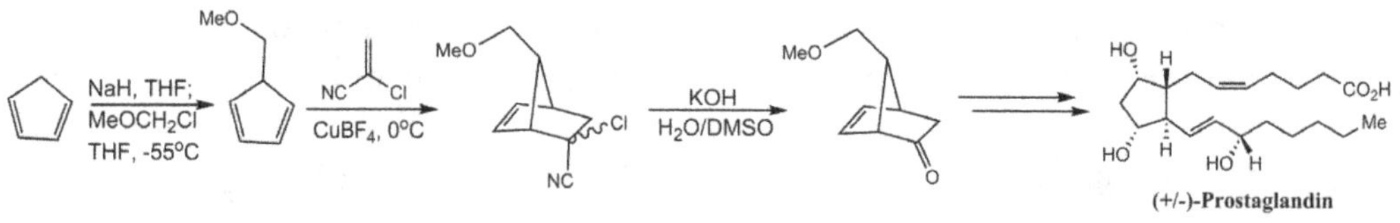

Samuel J. Danishefsky used a Diels–Alder reaction to synthesize disodium prephenate, a biosynthetic precursor of the amino acids phenylalanine and tyrosine, in 1979. This sequence is notable as one of the earliest to feature 1-methoxy-3-siloxybutadiene, the so-called Danishefsky diene, in total synthesis. Its utility is apparent below, namely, the ready furnishing of α,β–unsaturated cyclohexenone systems.

In their 1980 synthesis of reserpine, Paul Wender and coworkers used a Diels–Alder reaction to set the cis-decalin framework of the D and E rings of the natural product. The initial Diels-Alder between 2-acetoxyacrylic acid and the 1,2-dihydropyridine-1-carboxylate shown below put the newly installed carboxyl group in a position to rearrange exclusively to the cis-fused rings after conversion to the isoquinuclidene shown below. The cis-fusion allowed for the establishment of the stereochemistry at C17 and C18: first by cleavage of the acetate group at C18 to give a ketone that can modulate the stereochemistry of the methoxy group C17, and then by reduction of the ketone at C18 from the exo face to achieve the stereochemistry of the final product.

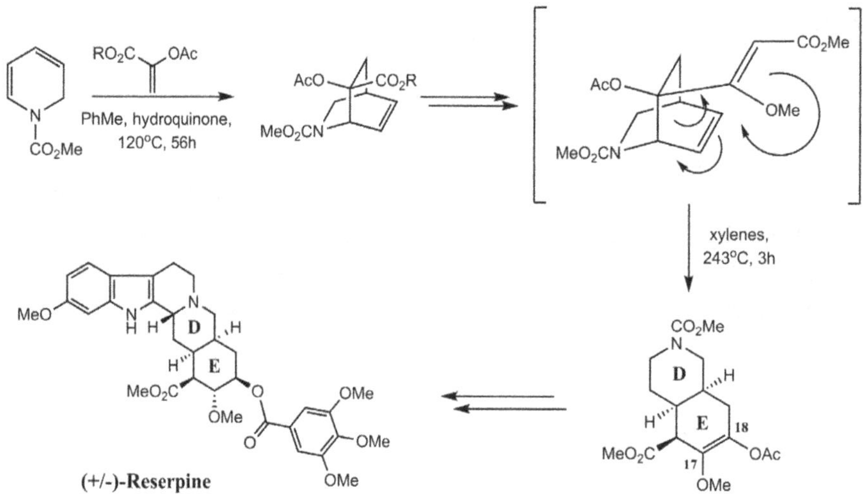

In Stephen F. Martin's synthesis of reserpine, the cis-fused D and E rings were also formed by a Diels–Alder reaction. Intramolecular Diels-Alder of the pyranone below with subsequent extrusion of carbon dioxide via a retro [4+2] afforded the bicyclic lactam. Epoxidation from the less hindered α-face, followed by epoxide opening at the less hindered C18 afforded the desired stereochemistry at these positions, while the cis-fusion was achieved with hydrogenation, again proceeding primarily from the less hindered face.

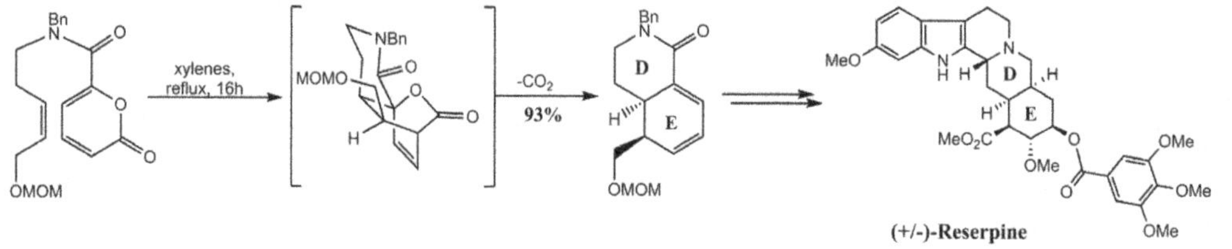

A pyranone was similarly used as the dienophile by K. C. Nicolaou's group in the total synthesis of taxol. The intermolecular reaction of the hydroxy-pyrone and α,β–unsaturated ester shown below suffered from poor yield and regioselectivity; however, when directed by phenylboronic acid the desired adduct could be obtained in 61% yield after cleavage of the boronate with 2,2-dimethyl-1,3-propanediol. The stereospecificity of the Diels–Alder reaction in this instance allowed for the definition of four stereocenters that were carried on to the final product.

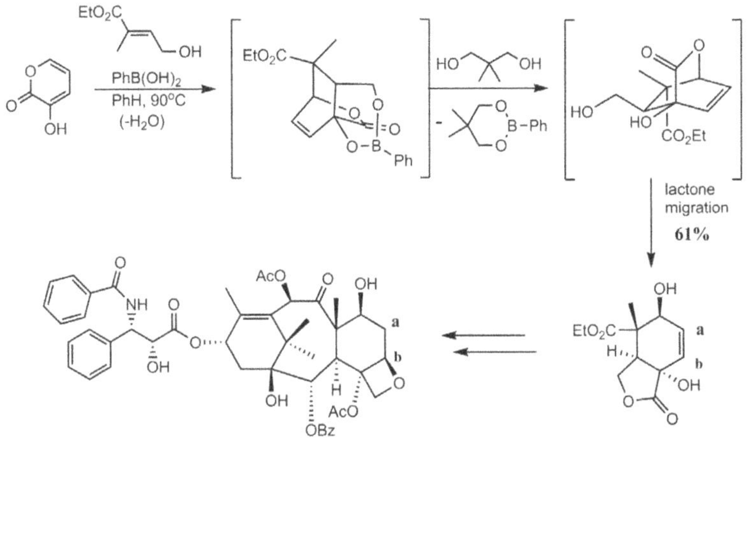

(-)-taxol

A Diels–Alder reaction was the key step in Amos Smith's synthesis of (-)-furaquinocin C. Lactone 1 was converted to the requisite diene by two successive silylations with TMSCl, and reaction with the bromoquinone as seen below furnished the final target upon aromatization with good overall yield. The diene in this instance is notable as a rare example of a cyclic derivative of Danishefsky's diene.

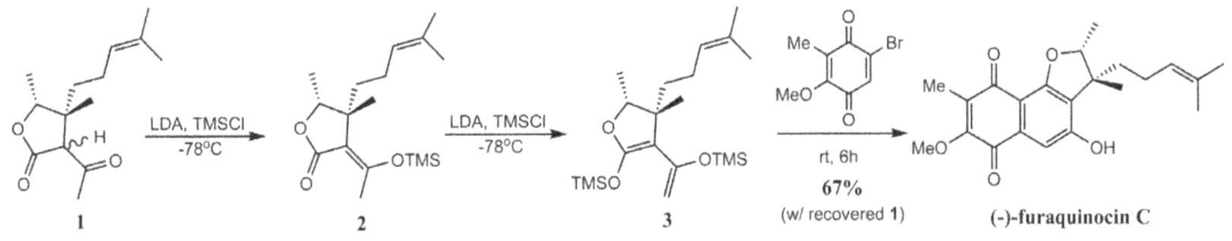

Virish Rawal and Sergey Kozmin, in their 1998 synthesis of tabersonine, used a Diels-Alder to establish cis relative stereochemistry of the alkaloid core. Conversion of the cis-aldehyde to its corresponding alkene by Wittig olefination and subsequent ring-closing metathesis with a Schrock catalyst gave the second ring of the alkaloid core. The diene in this instance is notable as an example of a 1-amino-3-siloxybutadiene, otherwise known as a Rawal diene.

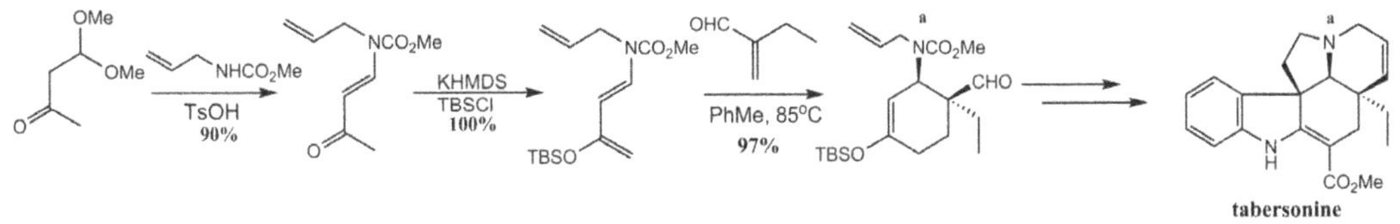

In 1988, William Okamura and Richard Gibbs reported an enantioselective synthesis of (+)-sterpurene that featured a remarkable intramolecular Diels–Alder reaction of an allene. The [2,3]-sigmatropic rearrangement of the thiophenyl group to give the sulfoxide as below proceeded enantiospecifically due to the predefined stereochemistry of the propargylic alcohol. In this way, the single allene isomer formed could direct the Diels-Alder to occur on only one face of the generated 'diene'.

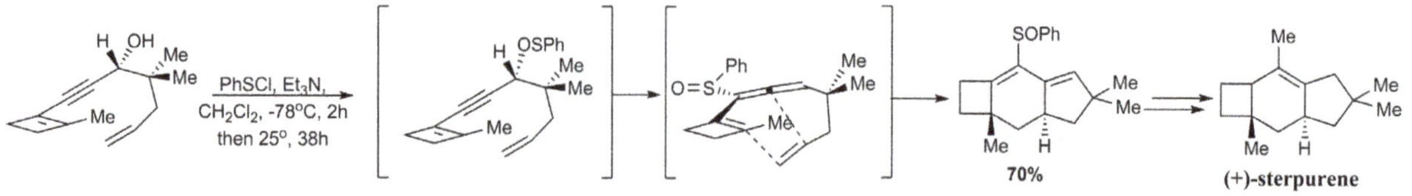

Andrew Myers' 2005 synthesis of (-)-tetracycline achieved the linear tetracyclic core of the antibiotic with a Diels–Alder reaction. Thermally initiated, conrotatory opening of the benzocyclobutene generated the o-quinodimethane, which reacted intermolecularly to give the tetracycline skeleton; the diastereomer shown was then crystallized from methanol after purification by column chromatography. The authors note that the dienophile's free hydroxyl group was integral to the success of the reaction, as hydroxyl-protected variants did not react under several different reaction conditions.

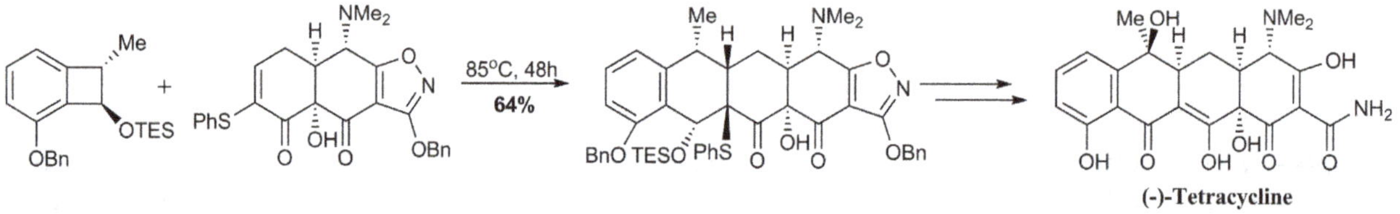

Takemura et al. synthesized cantharadrin in 1980 by Diels-Alder, utilizing high pressure in the reaction vessel.

Michel Addition

Michel addition of silyl ketene acetals to cyclic and acyclic α,β -unsaturated ketones has been studied. In these reactions, the addition of catalytic amount of Ph_3PO increases the enantioselectivity because it could trap Me_3Si species that could form during the reaction. For example, BLA 1b has been used for the Michel addition of cyclo hexenone with silyl ketene acetal to afford key intermediate for the enantioselective synthesis of caryophyllene. The absolute stereochemical course of the reaction can be rationalized by the above proposed transition states.

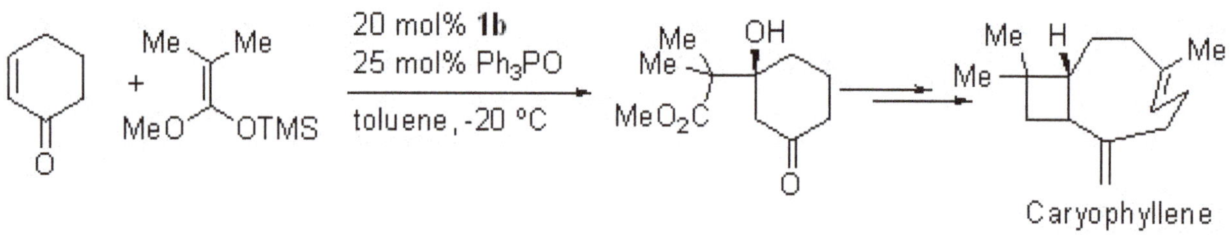

Examples

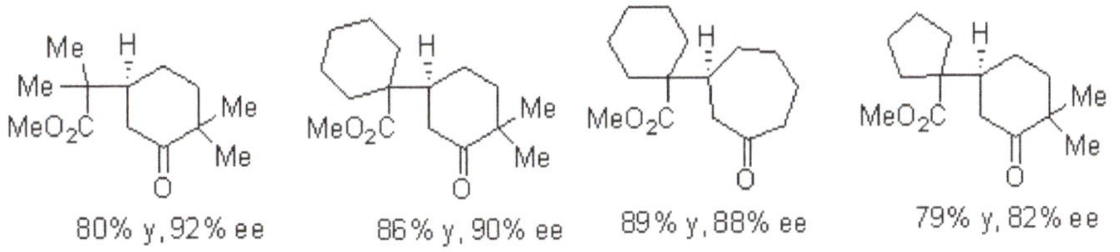

[3+2] Cycloaddition

Several benzoquinones proceed reactions with 2,3-dihydrofuran in the presence of BLA 1b to afford a variety of chiral phenolic tricycles with high enantioselectivities. The application of this reaction has been demonstrated in the total synthesis of aflatoxin B2. The reaction pathway has been elucidated by performing the reaction in the presence of excess of 2,3-dihydrofuran.

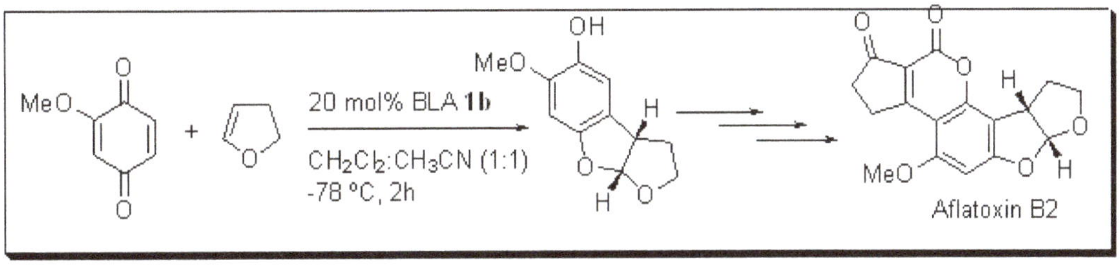

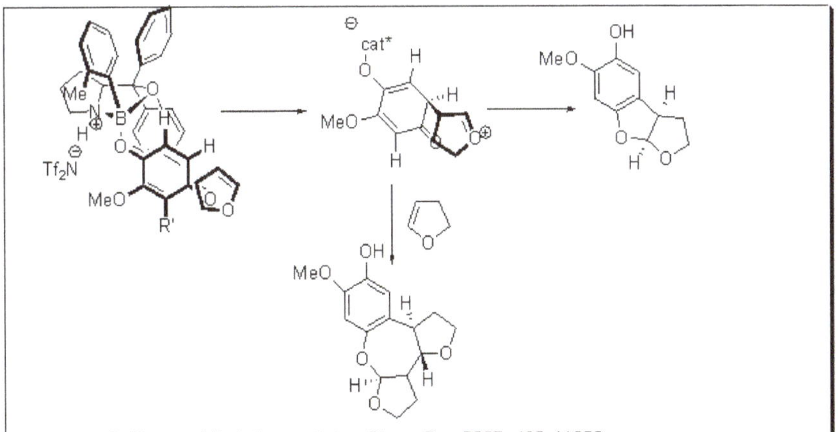

G. Zhou and E. J. Corey, *J. Am. Chem. Soc.* **2005**, *127*, 11958.

β-Lactone Synthesis

Chiral BLA 1c , derived from precatalyst zwitterions and tributyltintriflate, has been investigated for the reaction of aldehydes with ketene to afford β -lactones.

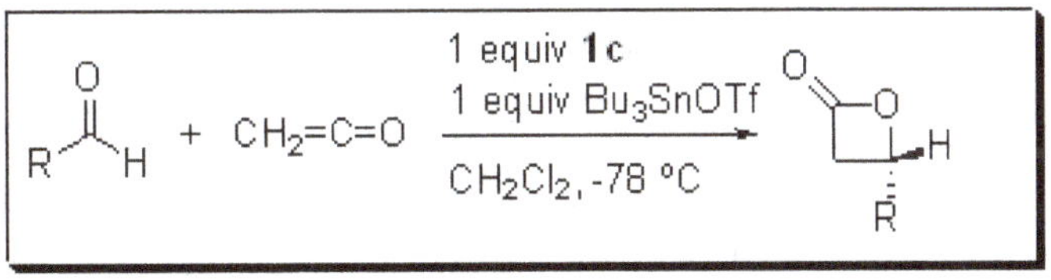

Examples

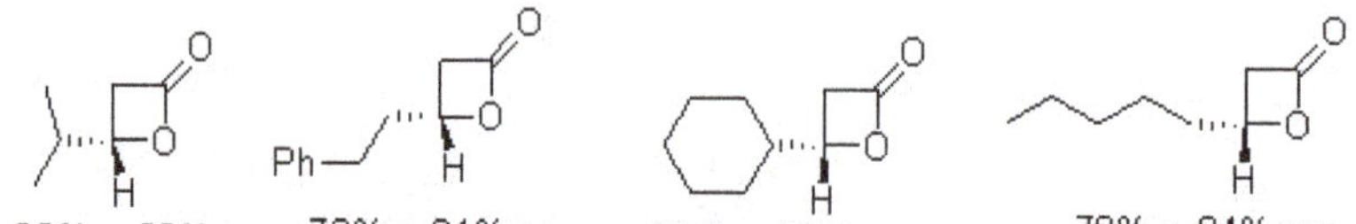

62% y, 68% ee   73% y, 81% ee   78% y, 70% ee   78% y, 84% ee

Proposed Mechanism

Reaction of the precatalyst 1c with tri- n -butyltintriflate may give an ion pair that could react with ketene to give sufficiently strong Lewis acid intermediate to make chelation with aldehydes. It is important to note that the formation β -lactone from α -branched aldehydes has been demonstrated for the first time.

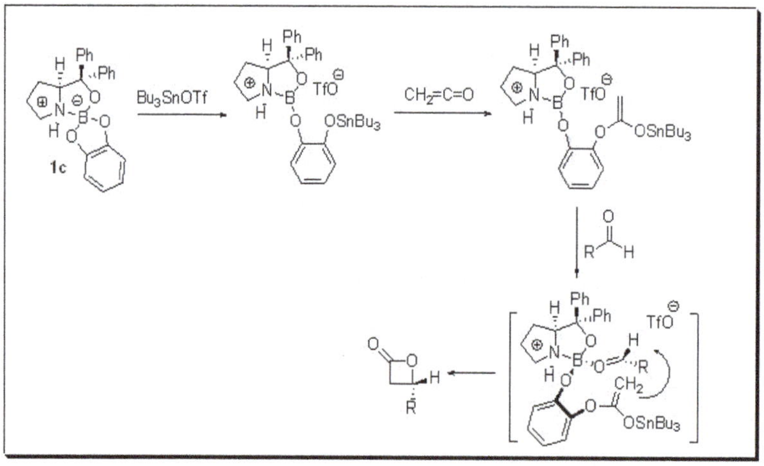

Modified BLA Catalysts

The following modified BLA catalysts 1d-e has been subsequently developed. These catalysts have also been demonstrated as powerful catalysts for Diels-Alder reactions.

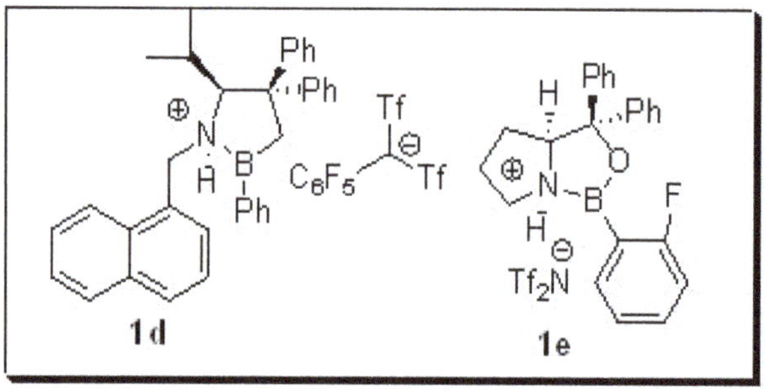

Lewis Acid-Assisted Lewis Acid (LLA)

In Lewis acid assisted chiral Lewis acids (LLAs), achiral Lewis acid is added to activate chiral Lewis acid via complex formation. The reactivity of LLA is much greater compared to that of achiral Lewis acid, and thus, the latter's presence does not affect the selectivity of the reaction.

Diels-Alder Reaction

The LLA 2a , derived from chiral valine-based oxazaborolidine and $SnCl_4$ as an activator, has been utilized as an efficient catalyst the for Diels-Alder reaction of wide range of substrates. In this system, the LLA 2a is more reactive compared to $SnCl_4$ and the ee is not affected because of the addition of excess $SnCl_4$.

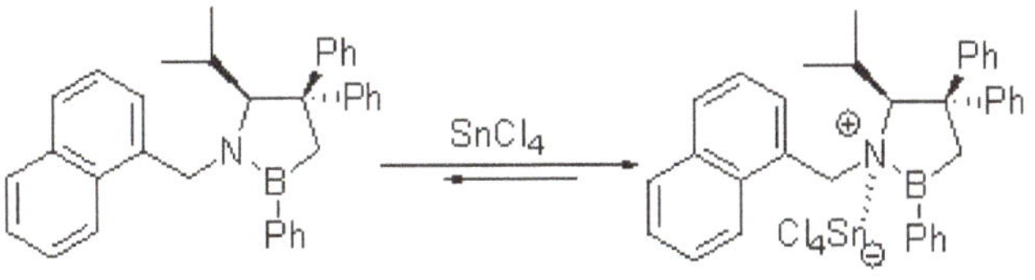

2a

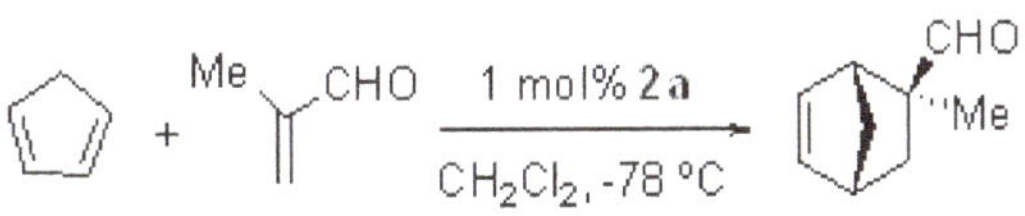

Yield: 99%
68:32 exo selectivity
95% ee (exo)
98% ee (endo)

Additional Examples

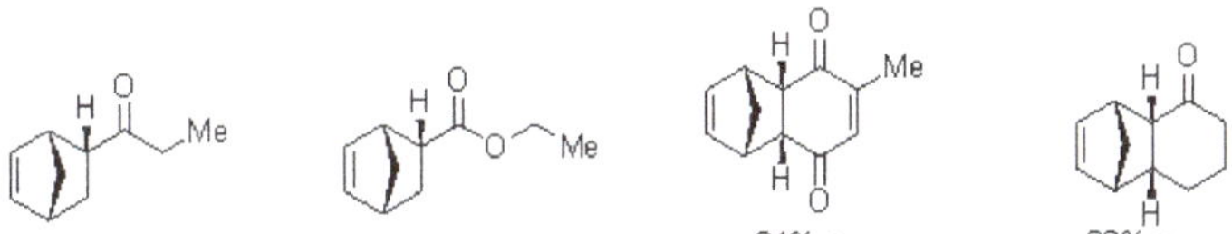

90% y, 99:1 endo selective 96% ee

96% y, 99:1 endo selective 95% ee

94% y, 99:1 endo selective 99% ee

93% y, 92:8 endo selective 95% ee

The LLA 2b , derived from the complexation of AlBr3 with chiral oxazaborolidine, has been shown as useful catalyst for Diels-Alder reaction. The observed results suggest that LLA 2b is considerably is more efficient catalyst than the corresponding BLA 1a or 1b since 10-20 mol% of BLA is usually needed for the optimum results.

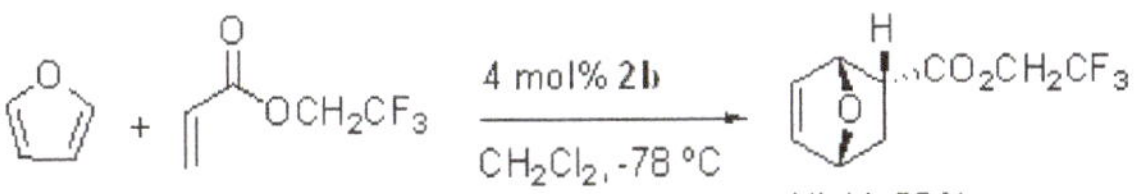

Yield: 99%
88:12 exo selectivity
99% ee (endo)

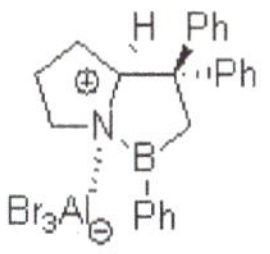

2b

Additional Examples

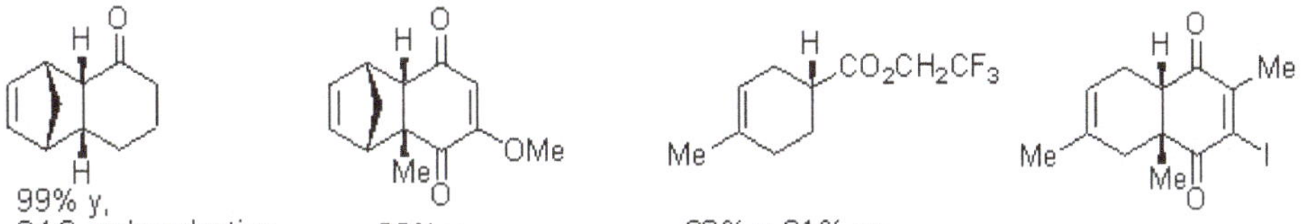

99% y, 94:6 endo selective 95% ee

99% y, 99:1 endo selective 88% ee

98% y, 91% ee

99% y 99% ee

Chiral Phosphoric Acids

Nucleophilic Additions of Aldimines

Chiral phosphoric acids (PAs) have been investigated as effective catalysts for Mannich type reactions. For examples, the reaction of imines with ketene silyl acetals has been studied using PA 1 in which introduction of 4-nitrophenyl substituents at 3,3'-positions has a beneficial effect on obtaining the high enantioselectivity. Based on DFT calculations a nine-membered zwitterionic transition state has been proposed to explain the stereoinduction.

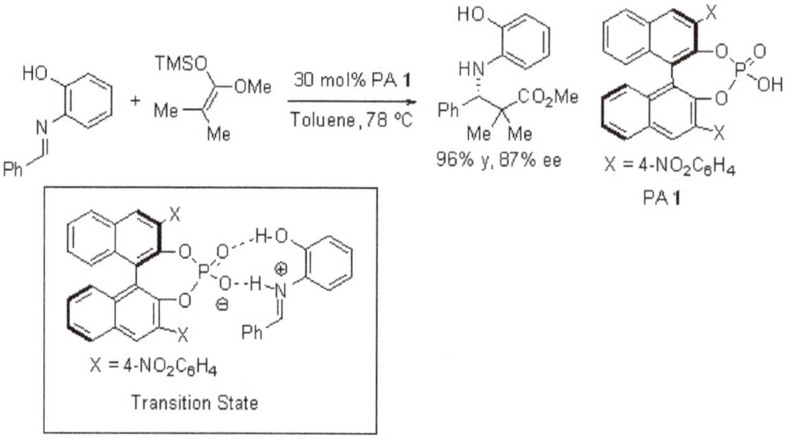

The reaction of acetylacetone with *N* -boc-protected imines has been subsequently reported employing 2 mol% PA 2 with excellent yield and enantioselectivites. The procedure is compatible with a series of substrates to afford target products in high enantioselectivities.

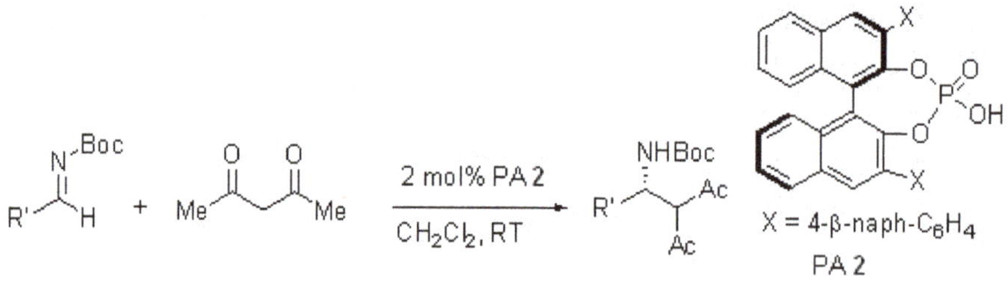

Examples

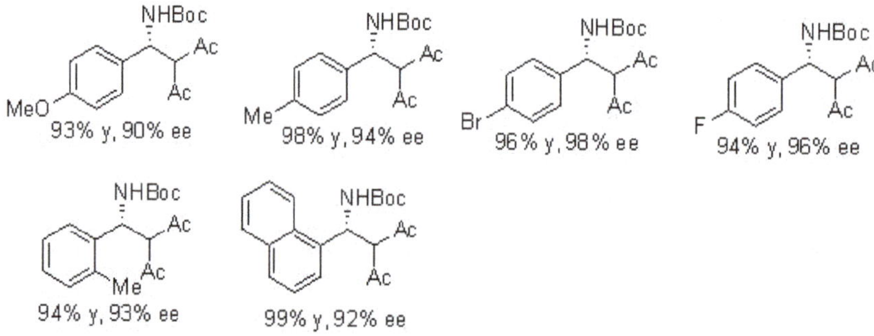

Phosphoric acid PA 3 derived from H_8-BINOL derivative has been further studied for the direct Mannich reactions between in situ generated N -aryl imines and ketones. The authors have proposed TS-1 for the acid-promoted enolization of the ketone and its addition to the protonated aldimine.

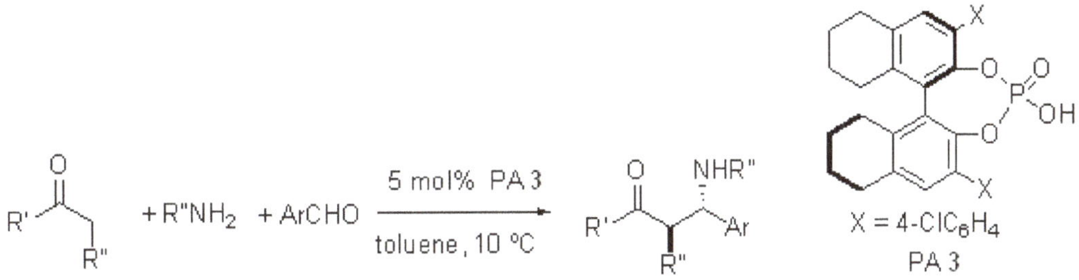

Examples

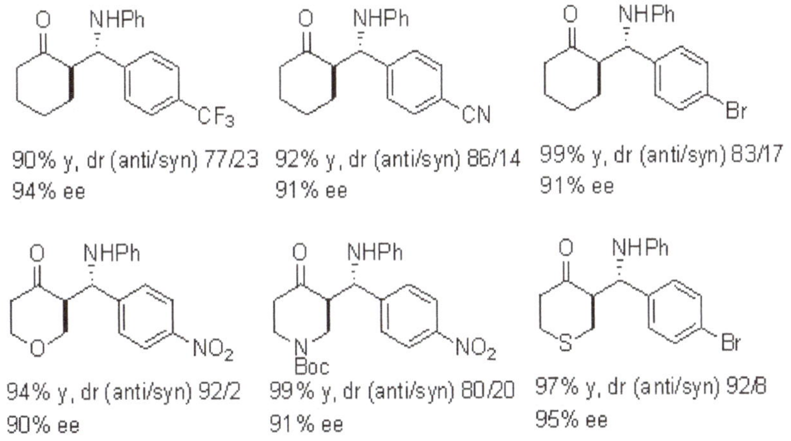

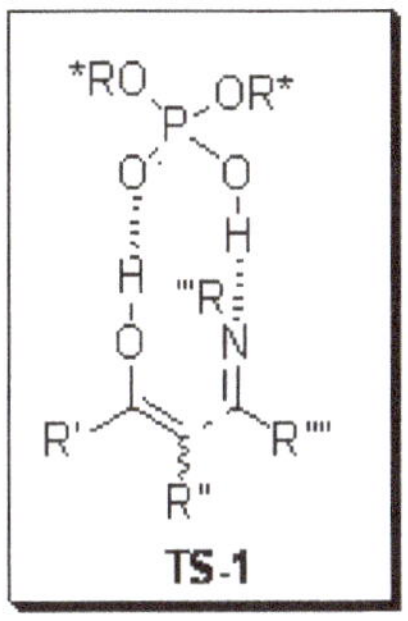

Hydrophosphorylation of aldimines with dialkyl phosphate has been studied using PA 4 to afford optically active α -amino phosphonates in good to high yields and enantioselectivities. The proposed transition state is shown in TS-2, where PA 4 acts as a bifunctional catalyst: the OH in phosphoric acid activates the aldimine as Brønsted acid and the phosphoryl oxygen activates the nucleophile as a Lewis base, thereby orienting both nucleophile and electrophile.

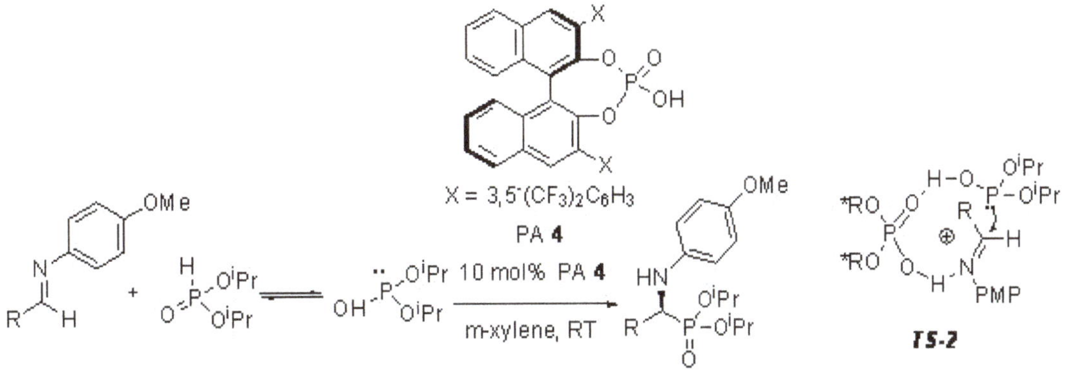

Examples

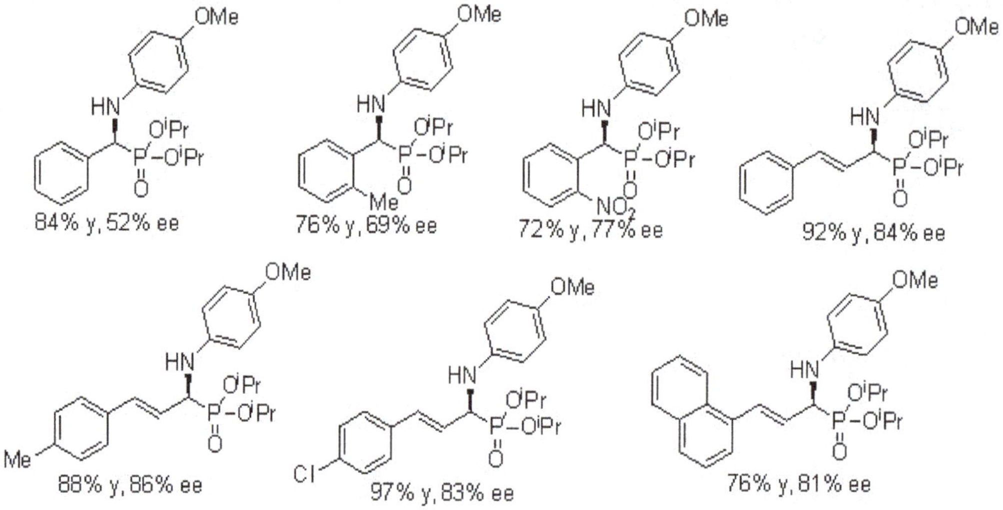

Aza-Friedel-Crafts Reactions

The first organocatalytic aza-Friedel-Crafts reaction of aldimines has been accomplished using PA 5. It is important to note that N-boc-protected aryl imines having electron-donating or –with-drawing groups at either the ortho -, meta -, or para - positions are compatible with the reaction condition.

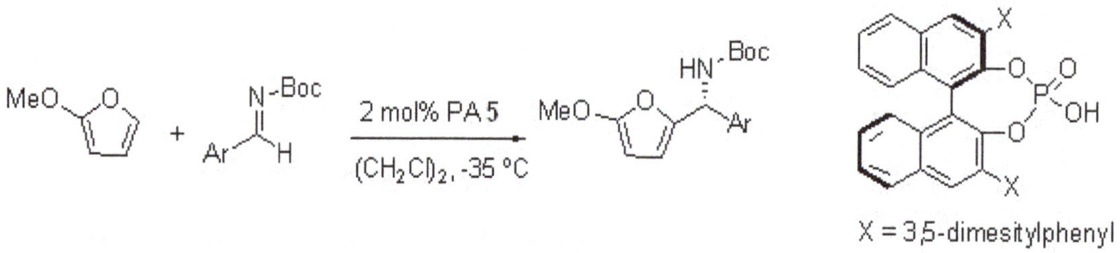

Examples

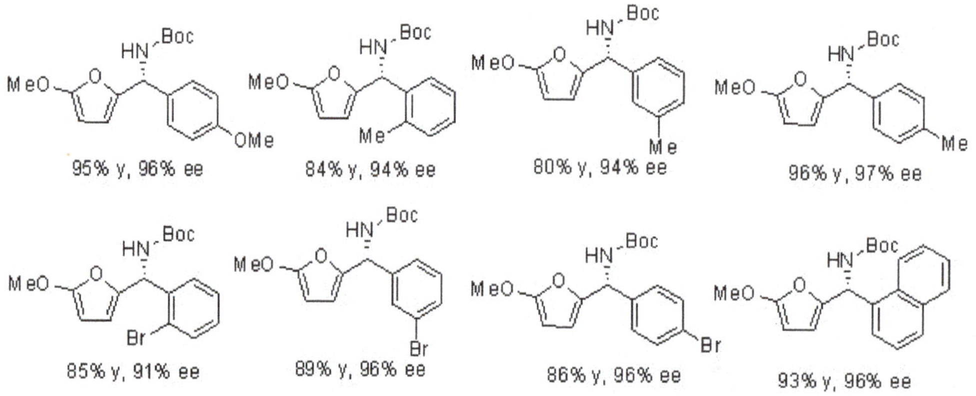

The reaction of indoles with enecarbamates has been successfully accomplished in the presence

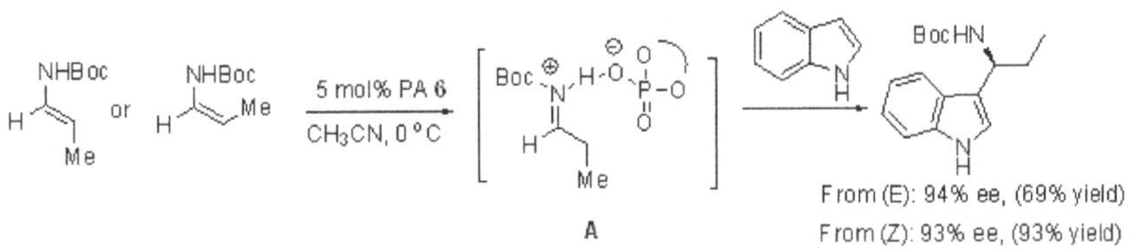

Examples

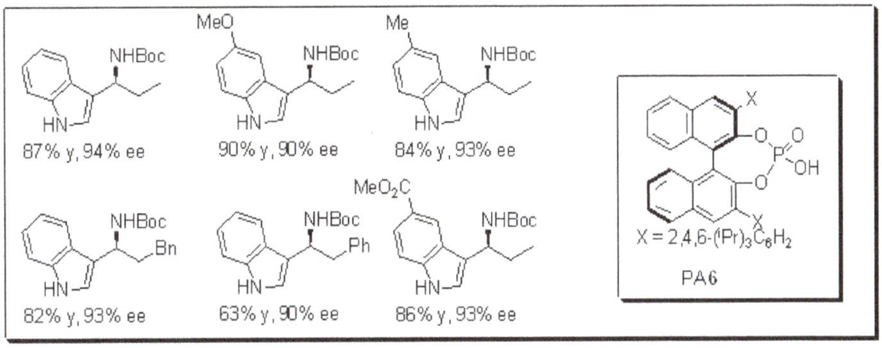

Use of either pure regioisomers (E) or (Z)-enecarbamate gives the same product with similar enantioselectivities. Thus, the reaction is believed to takes place via a common intermediate A that could be generated by the protonation of the enecarbamates.

The reactions of indole with a wide range of imines, derived from aromatic aldehydes, have been demonstrated using PA 7 with excellent enantioselectivities.

The Pictet-Spengler reaction of N-tritylsulfenyl tryptamines with various alphatic and aromatic aldehydes has been accomplished using PA 7. The sulfenyl substituent stabilizes the intermediate iminium ion and favours the Pictet-Spengler cyclization compared to the undesired enamine formation.

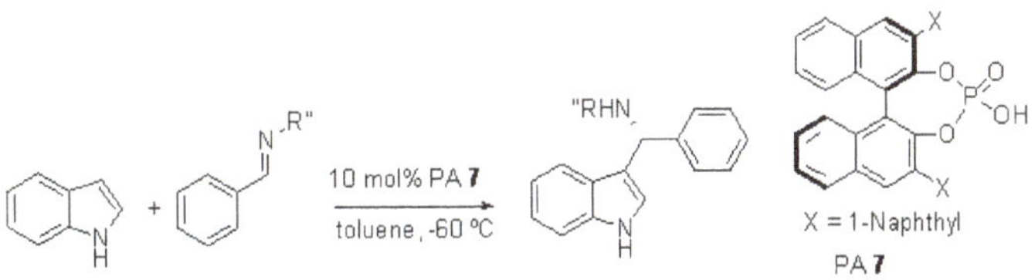

Examples

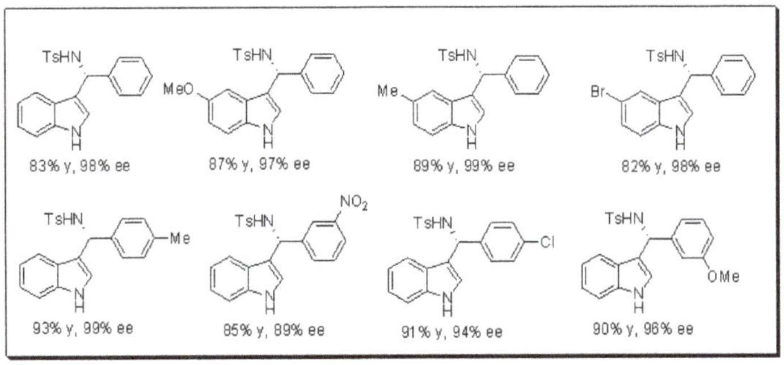

The Pictet-Spengler reaction of N-tritylsulfenyl tryptamines with various alphatic and aromatic aldehydes has been accomplished using PA 7. The sulfenyl substituent stabilizes the intermediate iminium ion and favours the Pictet-Spengler cyclization compared to the undesired enamine formation.

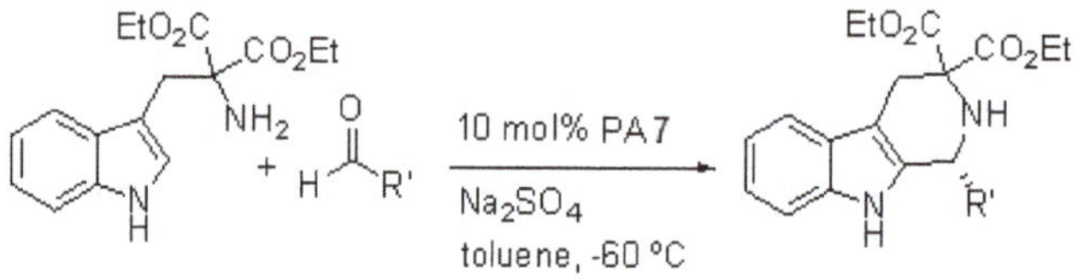

Examples

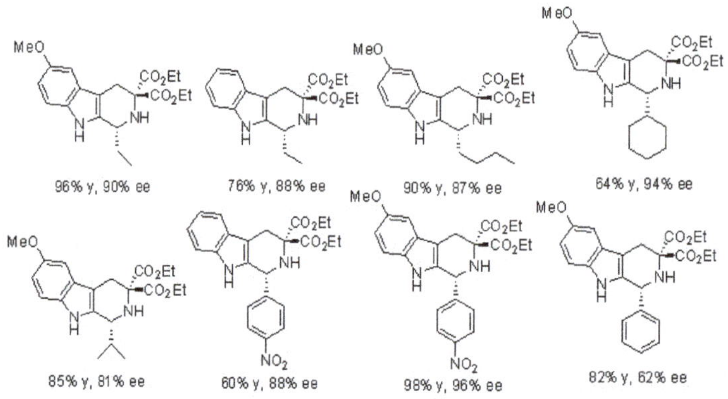

The quite interesting alkylation of α -diazoesters with N -acyl imines has been shown using PA 8 with high enantioselectivities. Diazoacetate is generally used in aziridine formation in the presence of Lewis acidic and Brønsted acidic conditions. Under these conditions, the competing aziridine formation has been eliminated by decreasing nucleophilicity of resulting amine intermediates and thus, the Friedel-Crafts adduct could be formed via C-H bond cleavage by the phosphoryl oxygen of phosphoric acid.

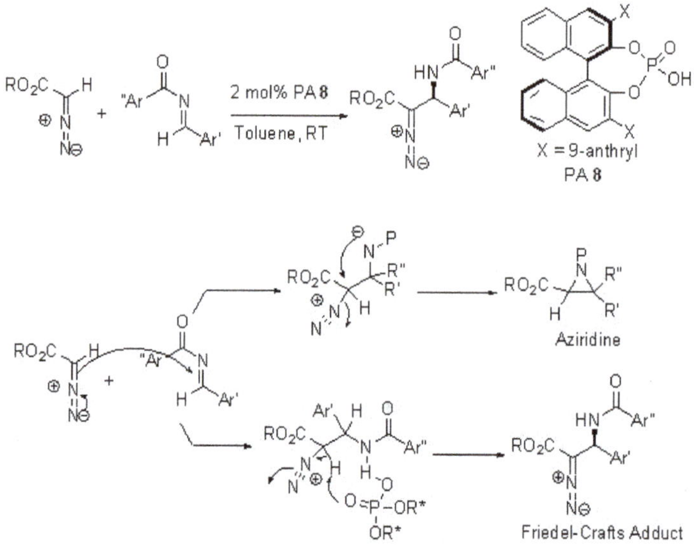

Examples

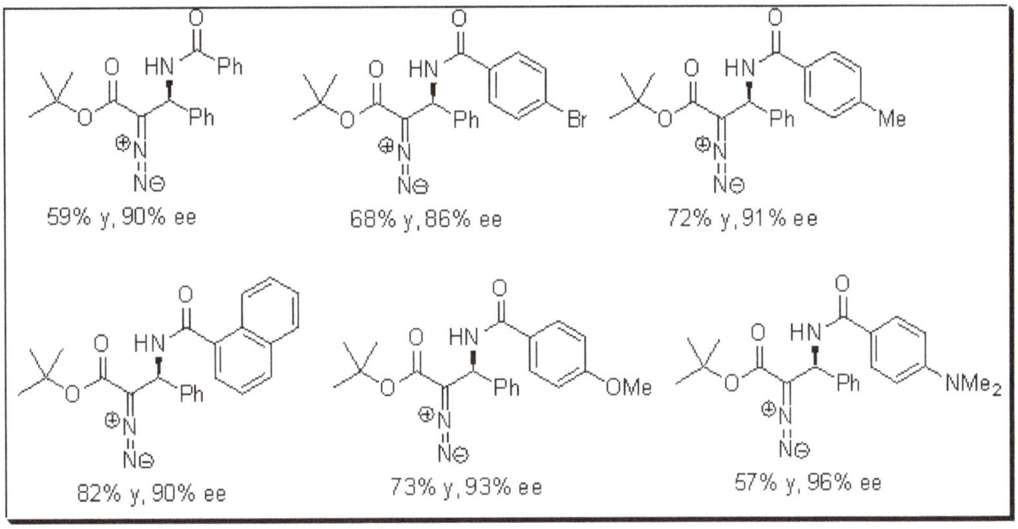

Diels-Alder Reaction

Chiral phosphoric acids (PAs) are excellent catalysts for the Diels-Alder reaction. For examples, the aza- Diels Alder reaction of Danishefsky's diene with aldimines is effective using PA 1 with good enantioselectivities. The addition of acetic acid leads to increase significantly the yield and enantioselectivities.

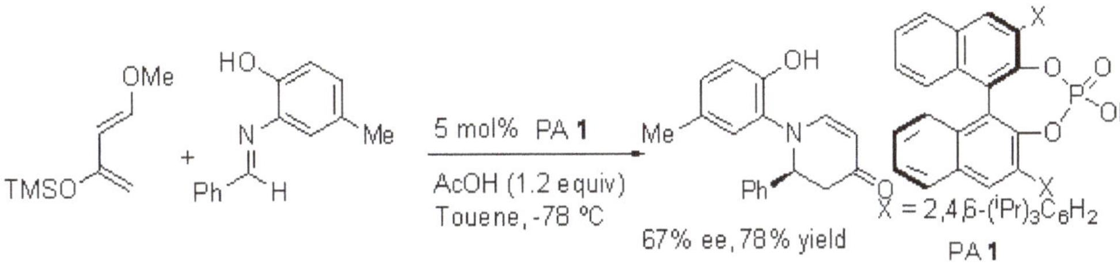

Although the aza- Diels Alder reaction of Brassard's diene using a Brønsted acid is rare due to the labilitiy of the diene in the presence of a strong Brønsted acid, PA 2 has been found to be an effective catalyst for the aza- Diels Alder reaction of Brassard's diene. The yield of the product could be improved using the pyridinium salt of the phosphoric acid as catalyst.

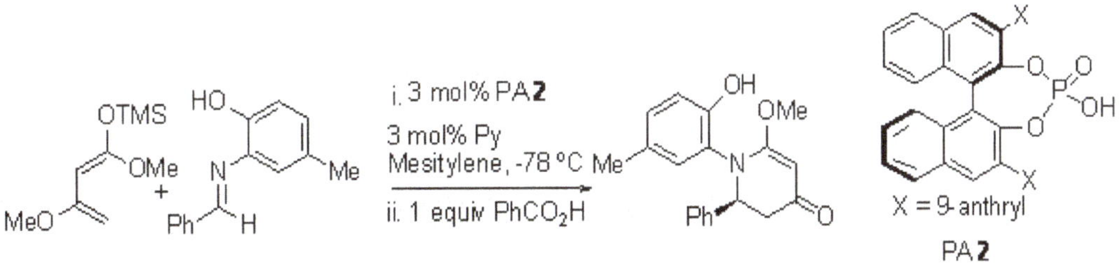

The PA 2 has also been found to effective for the inverse electron-demand aza -Diels Alder reaction of electron-rich alkenes with 2-aza dienes with excellent enantioselectivities. The presence of OH group is crucial for the cis selectivity in the products.

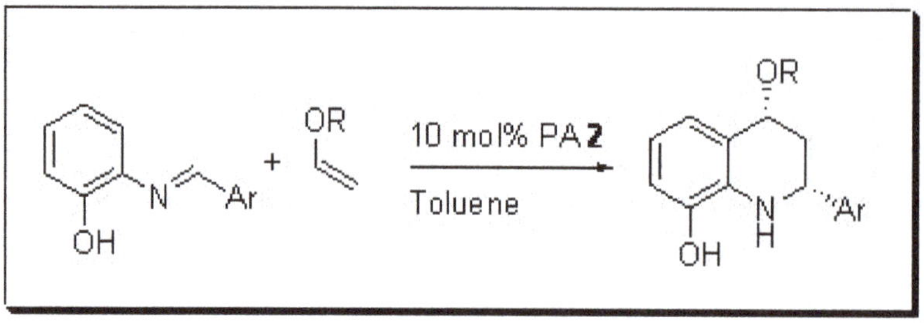

Examples

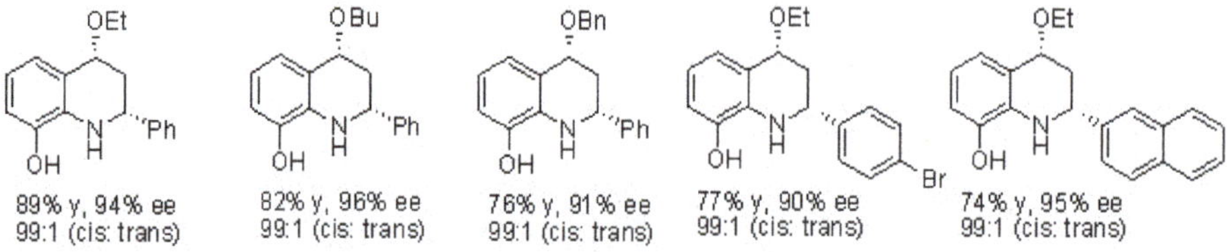

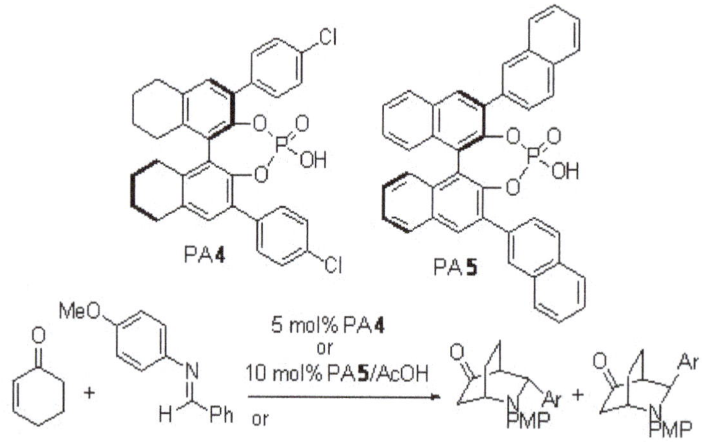

The aza- Diels Alder reaction of aldimines with cyclohexenone has been accomplished using either PA 4 or PA 5 /AcOH. A cooperative catalytic is proposed for the reaction using PA 5 /AcOH, where both the activation of an electrophile and a nucleophile takes place cooperatively.

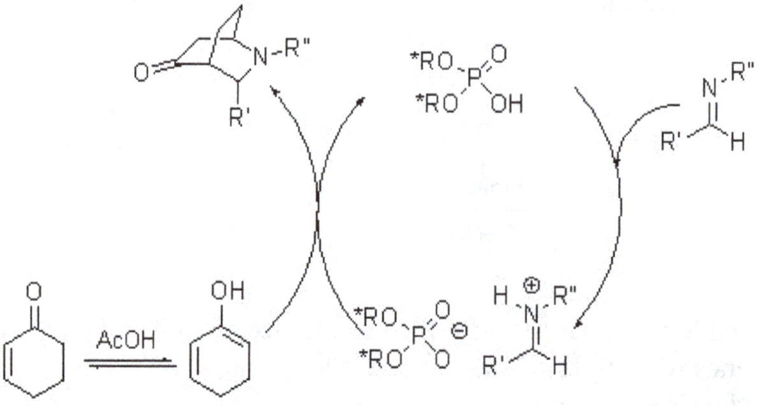

Transfer Hydrogenation

Chiral phosphoric acids (PAs) are effective catalysts for the biomimetic hydrogenation using Hantzsch ester as a hydride source. For examples, the reduction of ketimines using Hantzsch ester can be accomplished using PA 6 with good yield and enantioselectivities. PA 1 bearing bulky 2,4,6-(i-Pr)$_3$ C$_6$ H$_3$ at the 3,3'-positions of BINOL is found to superior to PA 6 for this purpose.

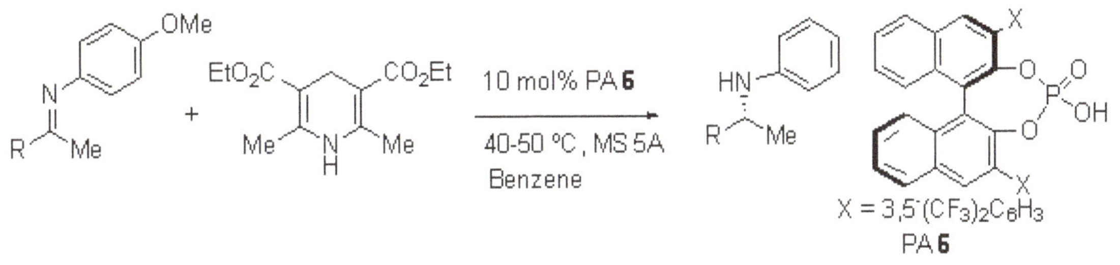

Examples

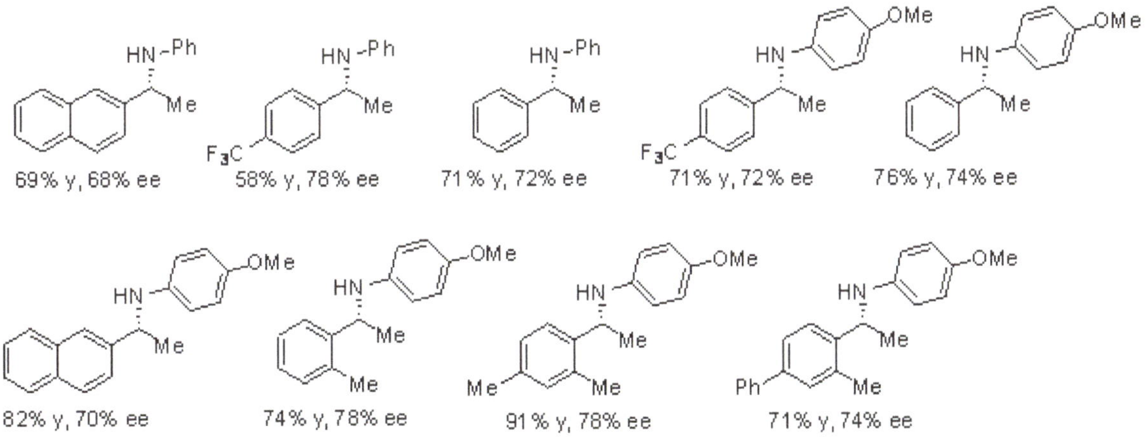

A three-component reductive amination reactions starting from ketones, amines and Hantzsch ester can be accomplished using PA 7 with excellent yield and enantioselectivities. This method is also compatible for the reactions of methyl phenyl ketones as well as methyl alkyl ketones.

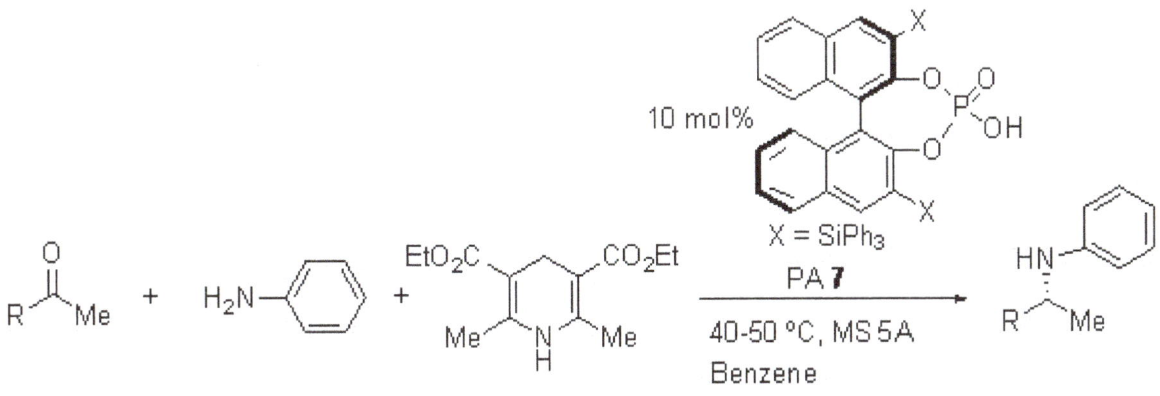

Examples

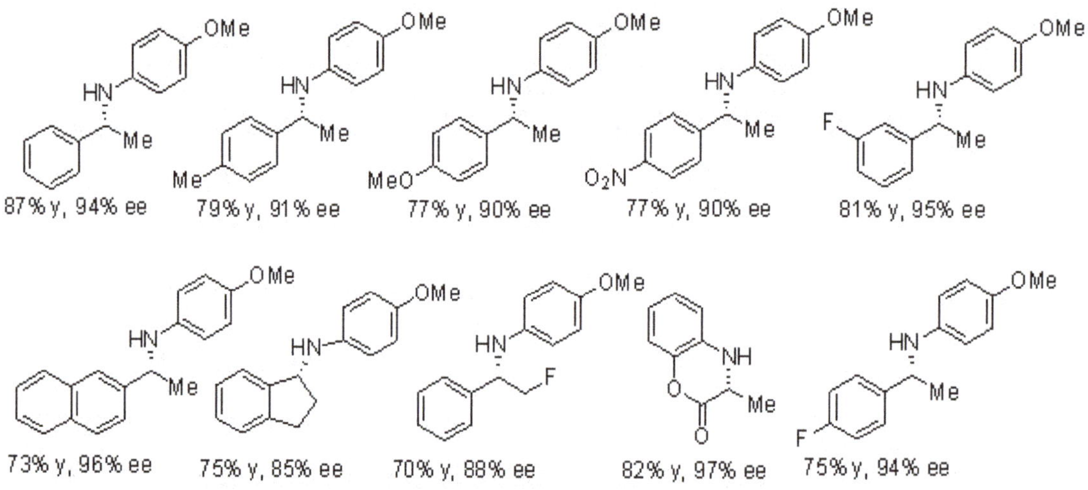

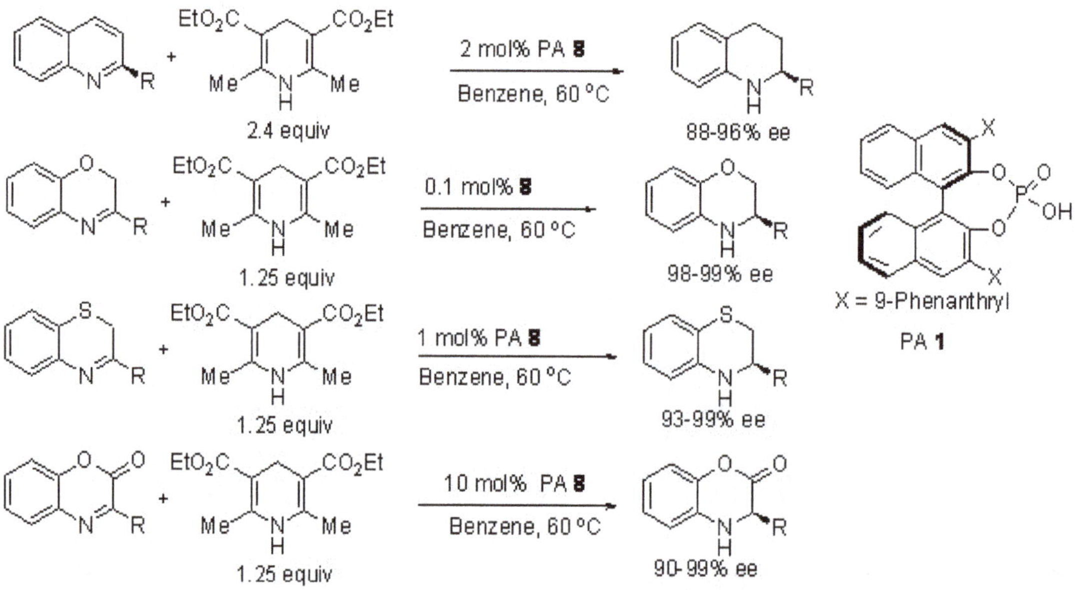

Following these initial studies, the reduction of wide of range of heterocycles has been explored. For examples, the reduction of a series of substituted quinonlines, benzoxazines, benzothiazines and benzoxazinones can be accomplished using PA 8 with excellent enantioselectivities.

Asymmetric reductive amination of α -branched aldehydes and p -anisidine with Hantzch ester can be performed employing PA 1 with high enantioselectivities. The observed results suggest that the reaction proceeds via a dynamic kinetic resolution.

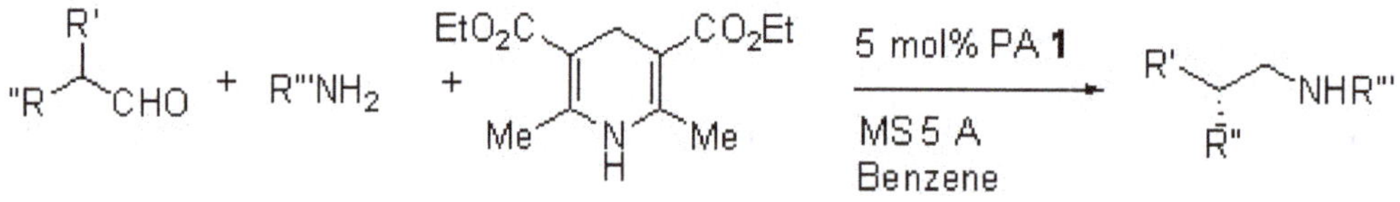

Examples

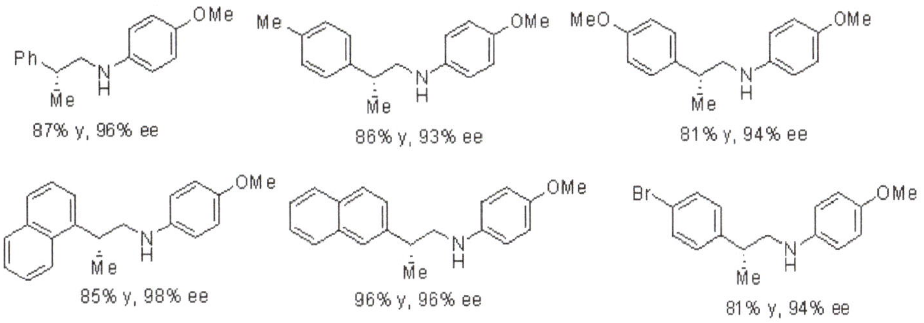

Proposed Mechanism

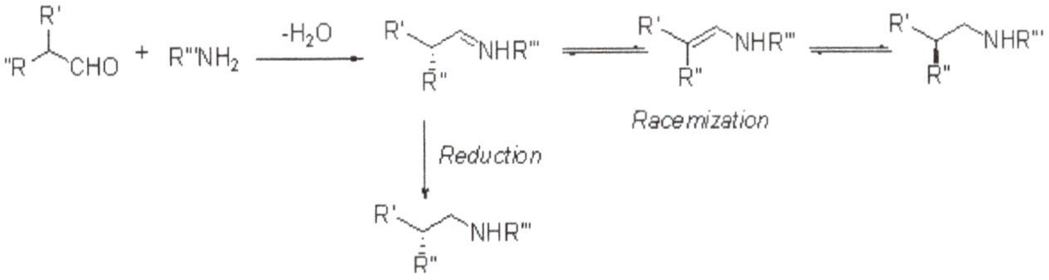

Chiral phosphoric acid PA 9 derived from (S)-VAPOL is found to superior to PAs derived from BINOL for the reduction of α -imino esters using Hantzsh ester to afford α -amino esters with higher enantioselectivities.

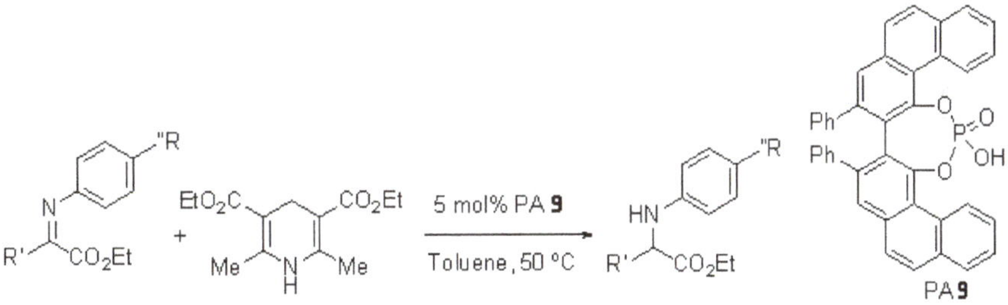

Examples

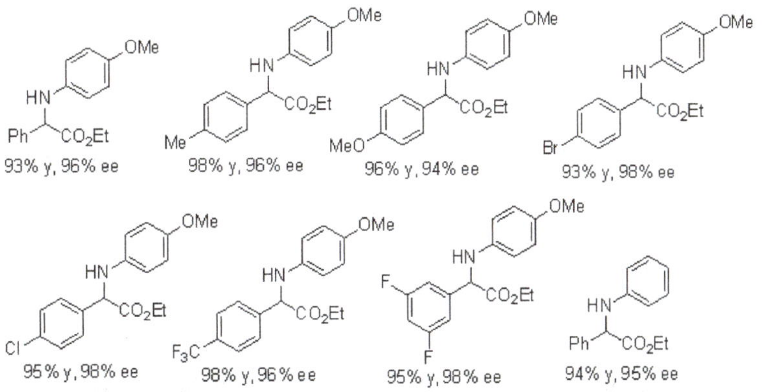

Mannich-type Reaction

The utility of PA 9 has been further extended as excellent catalyst for the addition of nitrogen nucleophiles such as sulfonamides and imides to imines to give protected aminals. The procedure has wide substrate scope to give the target products in 73-99% ee and 80-99% yield.

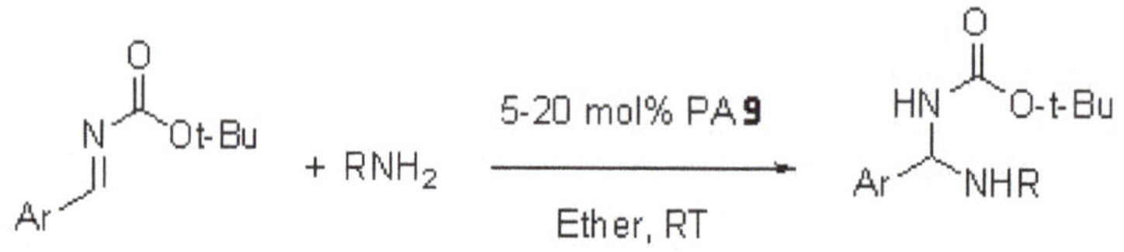

Examples

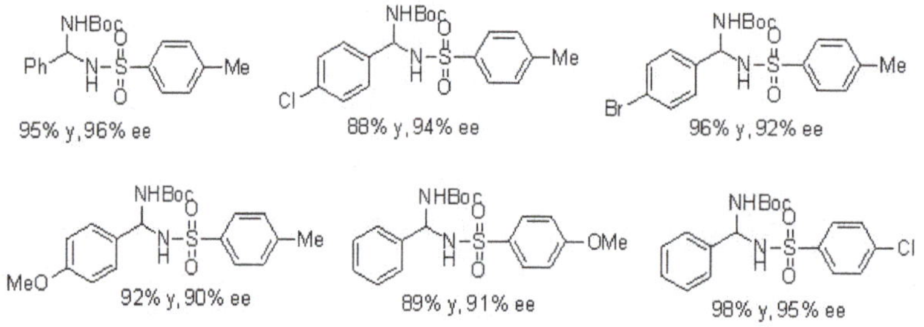

Asymmetric Desymmetrization of Meso-Aziridines

The application of PA 9 has been further extended to ring opening of meso -pyridines. This is the first example of organocatalyic desymmetrization of meso -aziridines. The substrates having electron-withdrawing protecting groups on the nitrogen proceed reaction with enhanced yields and enantioselectivity of the products.

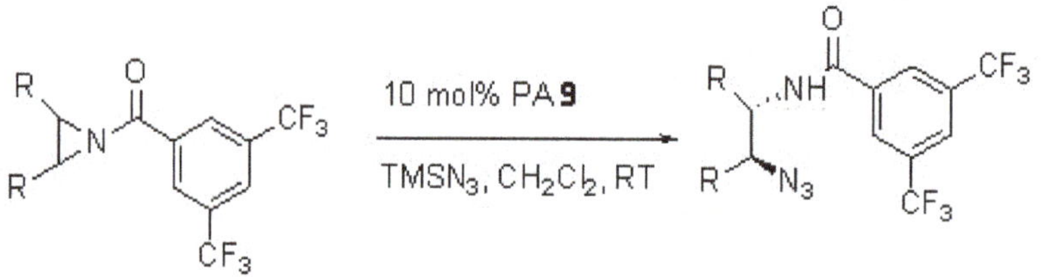

Examples

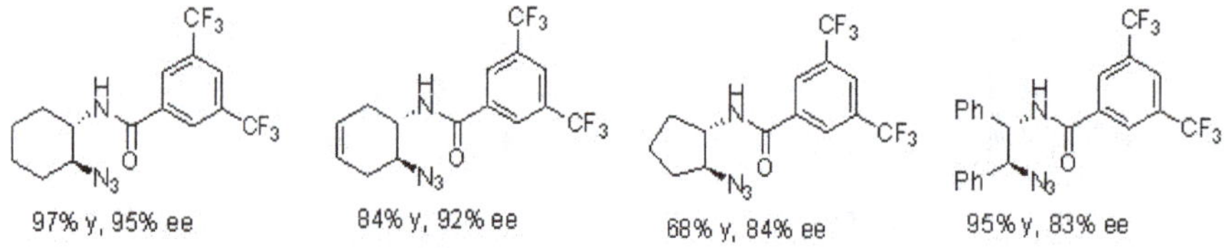

Proposed Mechanism

The phosphoric acid first reacts with TMSN 3 to give silylated phosphoric acid as the active catalyst. The latter activates the aziridine by coordination of its carbonyl group, and subsequent attack of azide affords the precursor of the product and regeneration of the phosphoric acid.

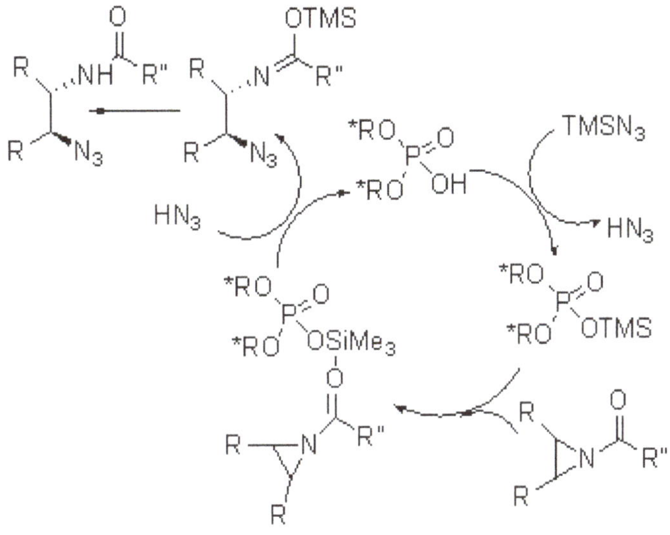

References

- Hyttel, J.; Bøgesø, K. P.; Perregaard, J.; Sánchez, C. (1992). "The pharmacological effect of citalopram resides in the (S)-(+)-enantiomer". Journal of Neural Transmission. 88 (2): 157–160. PMID 1632943. doi:10.1007/BF01244820
- Clayden, Jonathan; Greeves, Nick; Warren, Stuart; Wothers, Peter (2001). Organic Chemistry (1st ed.). Oxford University Press. ISBN 978-0-19-850346-0. Page 1226
- Narasaka, K.; Shimada, S.; Osoda, K.; Iwasawa, N. (1991). "Phenylboronic Acid as a Template in the Diels-Alder Reaction". Synthesis. 1991 (12): 1171–1172. doi:10.1055/s-1991-28413
- Knowles, W. S. (March 1986). "Application of organometallic catalysis to the commercial production of L-DOPA". Journal of Chemical Education. 63 (3): 222. doi:10.1021/ed063p222
- M. Heitbaum; F. Glorius; I. Escher (2006). "Asymmetric Heterogeneous Catalysis". Angewandte Chemie International Edition. 45 (29): 4732–4762. PMID 16802397. doi:10.1002/anie.200504212
- Breslow, R.; Rizzo, C. J. (1991). "Chaotropic salt effects in a hydrophobically accelerated Diels-Alder reaction". Journal of the American Chemical Society. 113 (11): 4340–4341. doi:10.1021/ja00011a052
- Koskinen, Ari M.P. (2013). Asymmetric synthesis of natural products (Second ed.). Hoboken, N.J.: Wiley. pp. 17, 28–29. ISBN 1118347331
- Kozmin, S. A.; Rawal, V. H. (1998). "A General Strategy to Aspidosperma Alkaloids: Efficient, Stereocontrolled Synthesis of Tabersonine". Journal of the American Chemical Society. 120 (51): 13523–13524. doi:10.1021/ja983198k
- Bertelsen, Søren; Jørgensen, Karl Anker (2009). "Organocatalysis—after the gold rush". Chemical Society Reviews. 38 (8): 2178–89. PMID 19623342. doi:10.1039/b903816g
- Wender, P. A.; Schaus, J. M.; White, A. W. (1980). "General methodology for cis-hydroisoquinoline synthesis: Synthesis of reserpine". Journal of the American Chemical Society. 102 (19): 6157–6159. doi:10.1021/ja00539a038
- Holmes, H. L. (1948). "The Diels-Alder Reaction Ethylenic and Acetylenic Dienophiles". Organic Reactions. 4: 60–173. ISBN 0471264180. doi:10.1002/0471264180.or004.02

- Wandrey, Christian; Liese, Andreas; Kihumbu, David (2000). "Industrial Biocatalysis: Past, Present, and Future". Organic Process Research & Development. 4 (4): 286–290. doi:10.1021/op990101l
- Woodward, R. B.; Sondheimer, F.; Taub, D.; Heusler, K.; McLamore, W. M. (1952). "The Total Synthesis of Steroids". Journal of the American Chemical Society. 74 (17): 4223–4251. doi:10.1021/ja01137a001
- Nicolaou, K. C.; Sorensen, E. J. (1996). Classics in Total Synthesis: Targets, Strategies, Methods. Wiley VCH. ISBN 978-3-527-29231-8
- Klundt, I. L. (1970). "Benzocyclobutene and its derivatives". Chemical Reviews. 70 (4): 471–487. doi:10.1021/cr60266a002

2

Organic Reactions of Asymmetric Synthesis

An Alder-ene reaction is a chemical reaction between an allylic hydrogen and an alkene and an enophile, so that σ-bond is formed. These reactions are especially favored when the enophile is electrophilic. This chapter elucidates the crucial theories and principles of asymmetric synthesis.

Ene Reaction

The ene reaction (also known as the Alder-ene reaction) is a chemical reaction between an alkene with an allylic hydrogen (the ene) and a compound containing a multiple bond (the enophile), in order to form a new σ-bond with migration of the ene double bond and 1,5 hydrogen shift. The product is a substituted alkene with the double bond shifted to the allylic position.

Ene: alkene, alkyne, allene, arene, C-heteroatom bond

Enophile: C=C, C=O, C=N, C=S, O=O, N=N, C≡C

The ene reaction

This transformation is a group transfer pericyclic reaction, and therefore, usually requires highly activated substrates and/or high temperatures. Nonetheless, the reaction is compatible with a wide variety of functional groups that can be appended to the ene and enophile moieties. Also,many useful Lewis acid-catalyzed ene reactions have been developed which can afford high yields and selectivities at significantly lower temperatures, making the ene reaction a useful C–C forming tool for the synthesis of complex molecules and natural products.

Ene Component

Enes are π-bonded molecules that contain at least one active hydrogen atom at the allylic, propargylic, or α-position. Possible ene components include olefinic, acetylenic, allenic, aromatic, cyclopropyl, and carbon-hetero bonds. Usually, the allylic hydrogen of allenic components participates in ene reactions, but in the case of allenyl silanes, the allenic hydrogen atom α to the silicon substituent is the one transferred, affording a silylalkyne. Phenol can act as an ene component, for example in the reaction with dihydropyran, but high temperatures are required (150–170 °C).

Nonetheless, strained enes and fused small ring systems undergo ene reactions at much lower temperatures. In addition, ene components containing C=O, C=N and C=S bonds have been reported, but such cases are rare.

Enophile

Enophiles are π-bonded molecules which have electron-withdrawing substituents that lower significantly the LUMO of the π-bond. Possible enophiles contain carbon-carbon multiple bonds (olefins, acetylenes, benzynes), carbon-hetero multiple bonds (C=O in the case of carbonyl-ene reactions, C=N, C=S, C≡P), hetero-hetero multiple bonds (N=N, O=O, Si=Si, N=O, S=O), cumulene systems (N=S=O, N=S=N, C=C=O, C=C=S, SO_2) and charged π systems ($C=N^+$, $C=S^+$, $C≡O^+$, $C≡N^+$).

Mechanism

Concerted Pathway and Transition States

The main frontier-orbital interaction occurring in an ene reaction is between the HOMO of the ene and the LUMO of the enophile. The HOMO of the ene results from the combination of the pi-bonding orbital in the vinyl moiety and the C-H bonding orbital for the allylic H. Concerted, all-carbon-ene reactions have, in general, a high activation barrier, which was approximated at 33 kcal/mol in the case of propene and ethene, as computed at the M06-2X/def2-TZVPP level of theory. However, if the enophile becomes more polar (going from ethane to formaldehyde), its LUMO has a larger amplitude on C, yielding a better C–C overlap and a worse H–O one, determining the reaction to proceed in an asynchronous fashion. This translates into a lowering of the activation barrier until 14.7 kcal/mol (M06-2X/def2-TZVPP), if S replaces O on the enophile. By computationally examining both the activation barriers and the activation strains of several different ene reactions involving propene as the ene component, Fernandez and co-workers have found that the barrier decreases along the enophiles in the order $H_2C=CH_2$ > $H_2C=NH$ > $H_2C=CH(COOCH_3)$ > $H_2C=O$ > $H_2C=PH$ > $H_2C=S$, as the reaction becomes more and more asynchronous and/or the activation strain decreases.

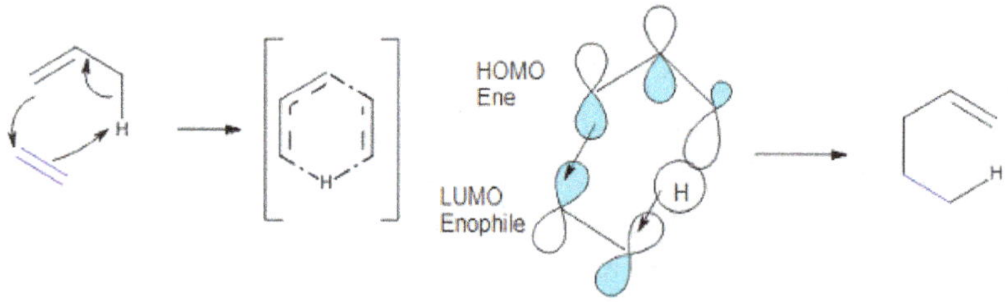

Concerted mechanism for the ene reaction

The concerted nature of the ene process has been supported experimentally, and the reaction can be designated as $[_{\sigma}2_s + _{\pi}2_s + _{\pi}2_s]$ in the Woodward-Hoffmann notation. The early transition state proposed for the thermal ene reaction of propene with formaldehyde has an envelope conformation, with a C–O–H angle of 155°, as calculated at the 3-21G level of theory.

Schnabel and co-workers have studied an uncatalyzed intramolecular carbonyl-ene reaction, which was used to prepare the cyclopentane fragment of natural and non-natural jatropha-5,12-dienes, members of a family of P-glycoprotein modulators. Their DFT calculations, at the B1B95/6-31G* level of theory for the reaction presented in Figure, propose that the reaction can proceed through one of two competing concerted and envelope-like transition states. The development of 1,3-transannular interactions in the disfavored transition state provides a good explanation for the selectivity of this process.

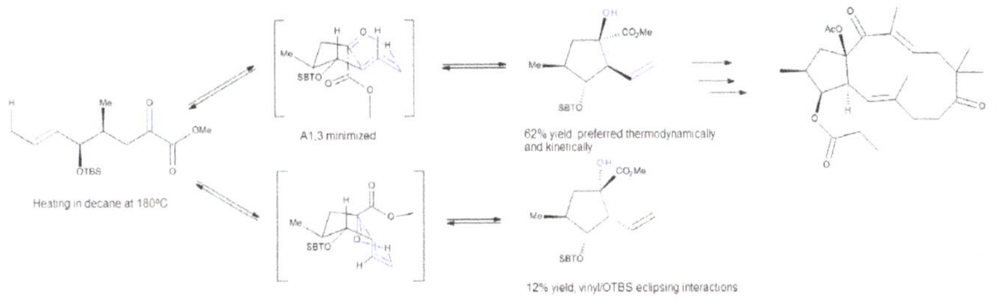

DFT study (B1B95/6-31G*) of a thermal intramolecular carbonyl–ene reaction and its use in the synthesis of jatropha-5,12-dienes

The study of Lewis acid promoted carbonyl-ene reactions, such as aluminum-catalyzed glyoxylate-ene processes, prompted researchers to consider a chair-like conformation for the transition state of ene reactions which proceed with relatively late transition states. The advantage of such a model is the fact that steric parameters such as 1,3-diaxial and 1,2-diequatorial repulsions are easy to visualize, which allows for accurate predictions regarding the diastereoselectivity of many reactions.

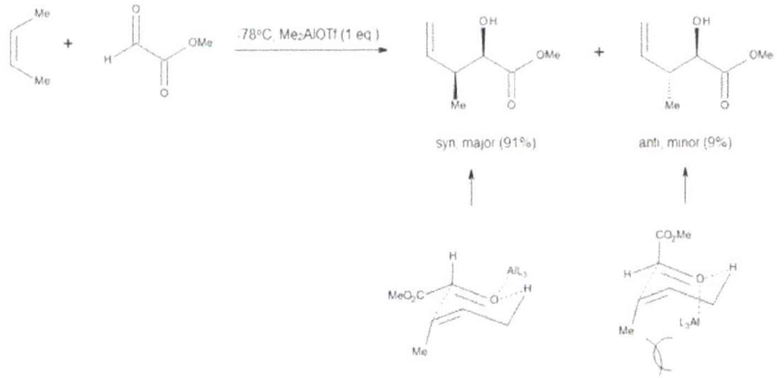

Chair-like transition state proposed for Lewis-acid catalyzed carbonyl-ene additions

Radical Mechanism

When a concerted mechanism is geometrically unfavorable, a thermal ene reaction can occur through a stepwise biradical pathway. Another possibility is a free-radical process, if radical initiators are present in the reaction mixture. For example, the ene reaction of cyclopentene and cyclohexene with diethyl azodicarboxylate can be catalyzed by free-radical initiators. As seen in Figure, the stepwise nature of the process is favored by the stability of the cyclopentenyl or cyclohexenyl radicals, as well as the difficulty of cyclopentene and cyclohexene in achieving the optimum geometry for a concerted process.

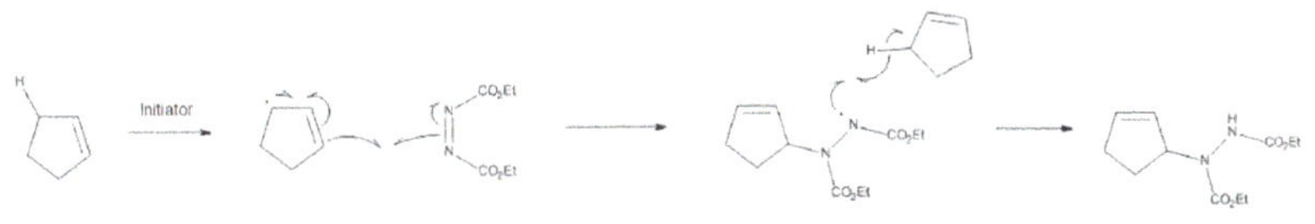

Stepwise, free-radical pathway for the ene reaction

Regioselection

Just as in the case of any cycloaddition, the success of an ene reaction is largely determined by the steric accessibility of the ene allylic hydrogen. In general, methyl and methylene H atoms are abstracted much more easily than methine hydrogens. In thermal ene reactions, the order of reactivity for the abstracted H atom is primary> secondary> tertiary, irrespective of the thermodynamic stability of the internal olefin product. In Lewis-acid promoted reactions, the pair enophile/ Lewis acid employed determines largely the relative ease of abstraction of methyl vs. methylene hydrogens.

The orientation of ene addition can be predicted from the relative stabilization of the developing partial charges in an unsymmetrical transition state with early formation of the σ bond. The major regioisomer will come from the transition state in which transient charges are best stabilized by the orientation of the ene and enophile.

Internal Asymmetric Induction

In terms of the diastereoselection with respect to the newly created chiral centers, an endo preference has been qualitatively observed, but steric effects can easily modify this preference.

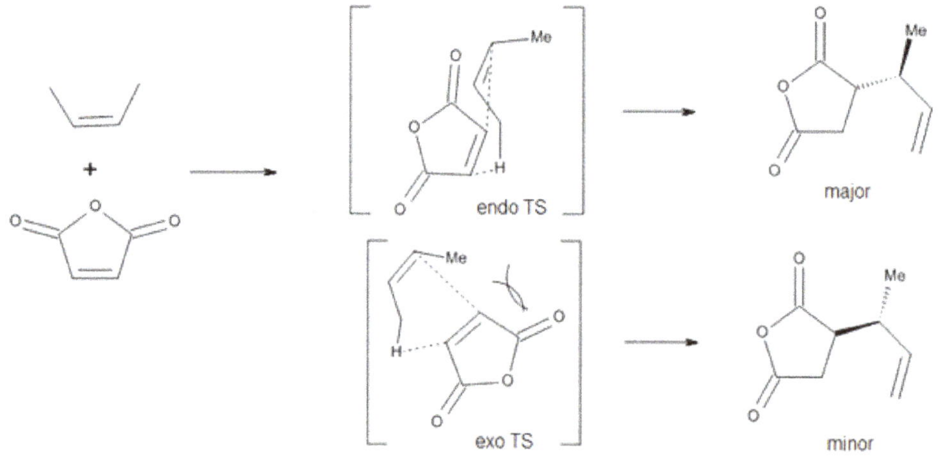

Endo preference for the ene reaction

Intramolecular Ene Reactions

Intramolecular ene reactions benefit from less negative entropies of activation than their intrermolecular counterparts, so are usually more facile, occurring even in the case of simple enophiles, such as unactivated alkenes and alkynes. The high regio- and stereoselectivities that can be obtained in these reactions can offer considerable control in the synthesis of intricate ring systems.

Considering the position of attachment of the tether connecting the ene and enophile, Oppolzer has classified both thermal and Lewis acid-catalyzed intramolecular ene reactions as types I, II and III, and Snider has added a type IV reaction. In these reactions, the orbital overlap between the ene and enophile is largely controlled by the geometry of the approach of components.

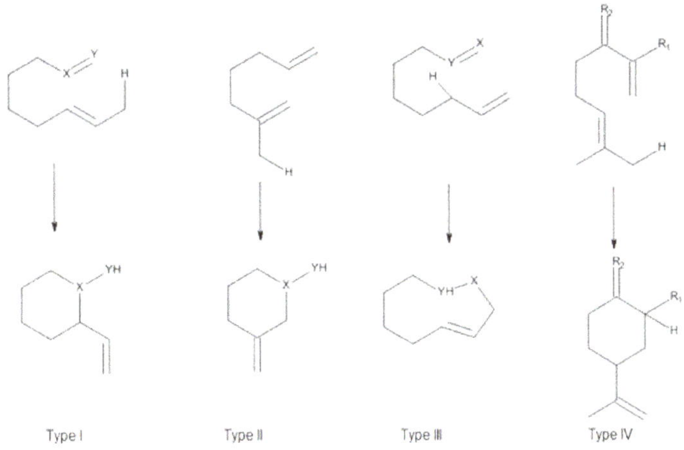

Types I, II, III: X=Y: RC=CR'R'', HC=O, HC=NR
Type IV: R^1: H, CO_2Et; R^2: CH_2, O

Types of intramolecular ene reactions.

Lewis Acid – catalyzed Ene Reactions

Advantages and Rationale

Thermal ene reactions have several drawbacks, such as the need for very high temperatures and the possibility of side reactions, like proton-catalyzed olefin polymerization or isomerization reactions. Since enophiles are electron-deficient, it was reasoned that their complexation with Lewis acids should accelerate the ene reaction, as it occurred for the reaction shown in Figure.

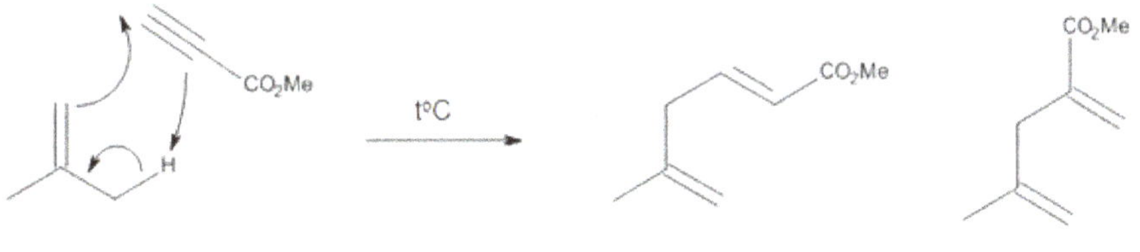

t°C=220°C	47%	3%
t°C=25°C, $AlCl_3$	61%	-

Improvements brought to the ene reaction by Lewis acid catalysis.

Alkylaluminum halides are well known as proton scavengers, and their use as Lewis acid catalysts in ene reactions has greatly expanded the scope of these reactions and has allowed their study and development under significantly milder conditions.

Since a Lewis acid can directly complex to a carbonyl oxygen, numerous trialkylaluminum catalysts have been developed for enophiles that contain a C=O bond. In particular, it was found that Me_2AlCl is a very useful catalyst for the ene reactions of α,β-unsaturated aldehydes and ketones, as well as of other aliphatic and aromatic aldehydes. The reason behind the success of this catalyst is the fact that the ene-adduct- Me_2AlCl complex can further react to afford methane and aluminum alkoxide, which can prevent proton-catalyzed rearrangements and solvolysis.

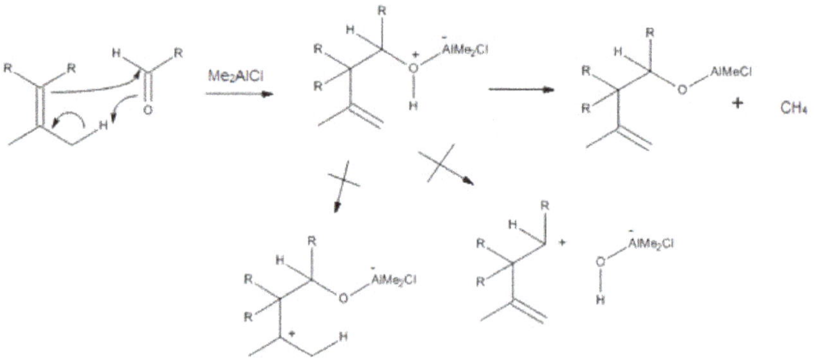

Me_2AlCl-catalyzed carbonyl-ene reactions

In the case of directed carbonyl-ene reactions, high levels of regio- and stereo-selectivity have been observed upon addition of a Lewis Acid, which can be explained through chair-like transition states. Notably, some of these reactions can run at very low temperatures and still afford very good yields of a single regioisomer.

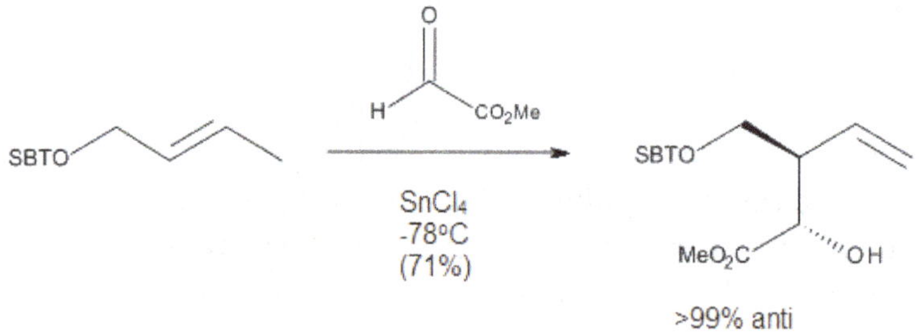

Lewis acid catalyzed, directed carbonyl-ene reaction.

Reaction Conditions

As long as the nucleophilicity of the alkyl group does not lead to side reactions, catalytic amounts of Lewis acid are sufficient for many ene reactions with reactive enophiles. Nonetheless, the amount of Lewis acid can widely vary, as it largely depends on the relative basicity of the enophile and the ene adduct. In terms of solvent choice for the reactions, the highest rates are usually achieved using halocarbons as solvents; polar solvents such as ethers are not suitable, as they would complex to the Lewis acid, rendering the catalyst inactive.

Reactivity of Enes

While steric effects are still important in determining the outcome of a Lewis acid catalyzed ene reaction, electronic effects are also significant, since in such a reaction, there will be a considerable positive charge developed at the central carbon of the ene. As a result, alkenes with at least one disubstituted vinylic carbon are much more reactive than mono or 1,2 disubstituted ones.

Mechanism

As seen in Figure, Lewis acid-catalyzed ene reactions can proceed either through a concerted mechanism that has a polar transition state, or through a stepwise mechanism with a zwitterionic intermediate. The ene, enophile and choice of catalyst can all influence which pathway is the lower energy process. In general, the more reactive the ene or enophile-Lewis acid complex is, the more likely the reaction is to be stepwise.

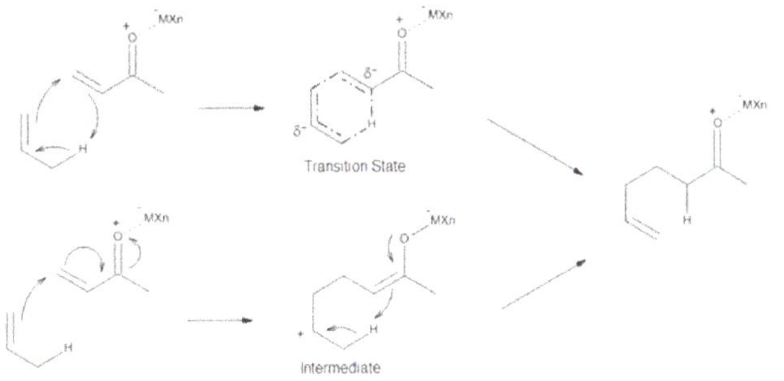

Mechanisms of Lewis acid-catalyzed ene reactions

Chiral Lewis Acids for the Asymmetric Catalysis of Carbonyl-ene Reactions

Chiral Dialkoxytitanium Complexes and the Synthesis of Laulimalide

A current direction in the study of Lewis acid-catalyzed ene reactions is the development of asymmetric catalysts for C–C bond formation. Mikami has reported the use of a chiral titanium complex in asymmetric ene reactions involving prochiral glyoxylat esters. The catalyst is prepared in situ from $(i\text{-PrO})_2TiX_2$ and optically pure binaphthol, the alkoxy-ligand exchange being facilitated by the use of molecular sieves. The method affords α-hydroxy esters of high enantiomeric purities, compounds that represent a class of biological and synthetic importance.

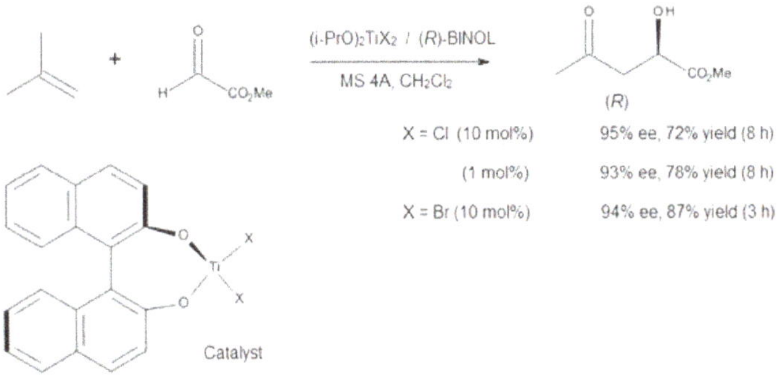

Asymmetric glyoxylate-ene reaction catalyzed by a chiral titanium complex.

Since both (R)- and (S)-BINOL are commercially available in optically pure form, this asymmetric process allows the synthesis of both enantiomers of α-hydroxy esters and their derivatives. However, this method is only applicable to 1,1-disubstituted olefins, due to the modest Lewis acidity of the titanium-BINOL complex.

As shown in Figure, Corey and co-workers propose an early transition state for this reaction, with the goal of explaining the high enantioselectivity observed (assuming that the reaction is exothermic as calculated from standard bond energies). Even if the structure of the active catalyst is not known, Corey's model proposes the following: the aldehyde is activated by complexation with the chiral catalyst (R)-BINOL-TiX_2 via the formyl lone electron pair syn to the formyl hydrogen to form a pentacoordinate Ti structure. CH—O hydrogen bonding occurs to the stereoelectronically most favorable oxygen lone pair of the BINOL ligand. In such a structure, the top (re) face of the formyl group is much more accessible to a nucleophile attack, as the bottom (si) face is shielded by the neighboring naphtol moiety, thus affording the observed configuration of the product.

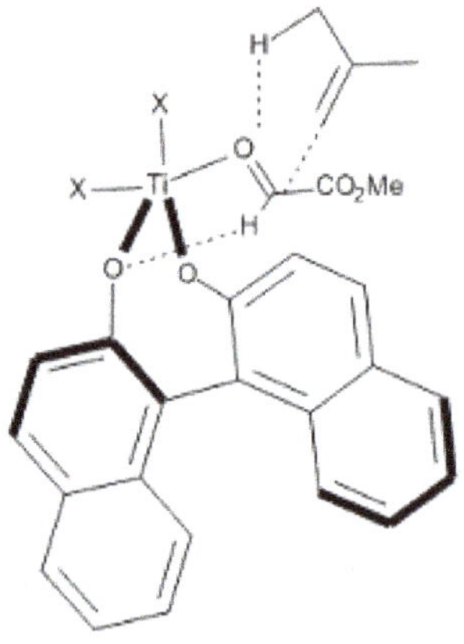

Transition state proposed for the reaction.

The formal total synthesis of laulimalide illustrates the robustness of the reaction developed by Mikami. Laulimalide is a marine natural product, a metabolite of various sponges that could find a potential use as an anti-tumor agent, due to its ability to stabilize microtubuli. One of the key steps in the strategy used for the synthesis of the C3-C16 fragment was a chirally catalyzed ene reaction that installed the C15 stereocenter. Treatment of the terminal allyl group of compound 1 with ethyl glyoxylate in the presence of catalytic (S)-BINOL-$TiBr_2$ provided the required alcohol in 74% yield and >95% ds. This method eliminated the need for a protecting group or any other functionality at the end of the molecule. In addition, by carrying out this reaction, Pitts et al. managed to avoid the harsh conditions and low yields associated with installing exo-methylene units late in the synthesis.

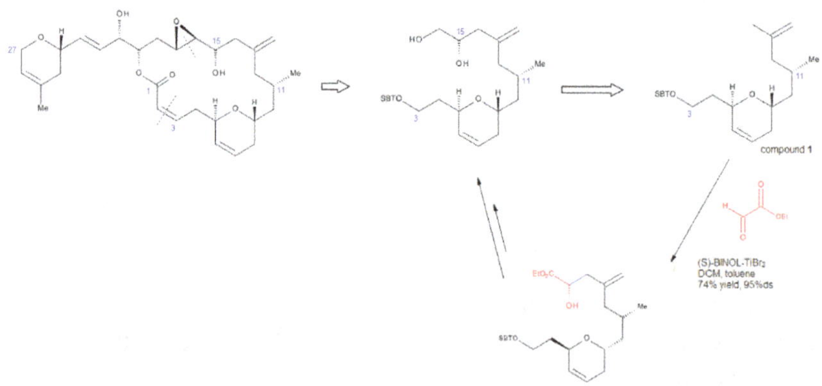

Retrosynthetic analysis of the C3-C16 fragment of laulimalide and use of the ene reaction in its synthesis.

Chiral C2-symmetric Cu(II) Complexes and the Synthesis of (+)-azaspiracid-1

Evans and co-workers have devised a new type of enantioselective C2-symmetric Cu(II) catalysts to which substrates can chelate through two carbonyl groups. The catalysts were found to afford high levels of asymmetric induction in several processes, including the ene reaction of ethyl glyoxylate with different unactivated olefins. Figure reveals the three catalysts they found to be the most effective in affording gamma-delta-unsaturated alpha-hydroxy esters in high yields and excellent enantio-selectivities. What is special about compound 2 is that it is bench-stable and can be stored indefinitely, making it convenient to use. The reaction has a wide scope, as shown in Figure, owing to the high Lewis acidity of the catalysts, which can activate even weakly nucleophilic olefins, such as 1-hexene and cyclohexene.

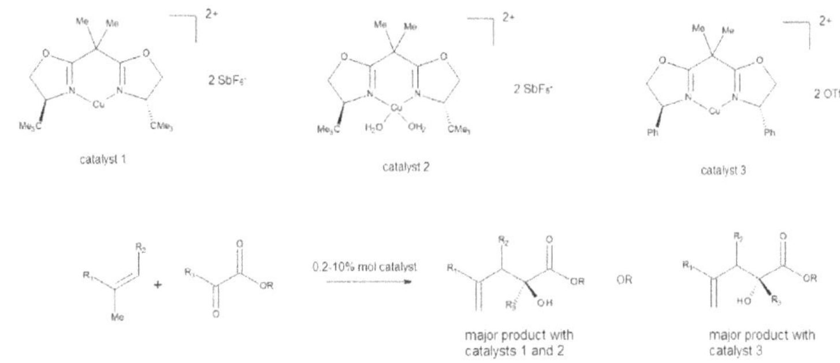

C2-symmetric Cu(II) catalysts developed for the enantioselective carbonyl-ene reactions of olefins and ethyl glyoxylate

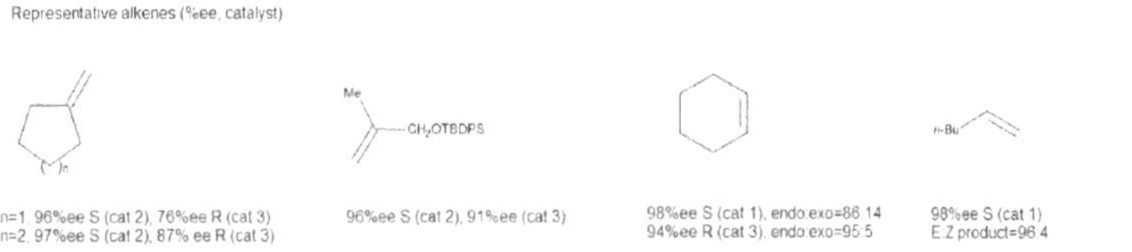

Scope of the reaction catalyzed by C2-symmetric Cu(II) chiral Lewis acids

In the case of catalysts 1 and 2, it has been proposed that asymmetric induction by the catalysts results from the formation of a square-planar catalyst-glyoxylate complex, in which the Re face of the aldehyde is blocked by the tert-butyl substituents, thus allowing incoming olefins to attack only the Si face. This model does not account however for the induction observed when catalyst 3 was employed. The current view is that the geometry of the metal center becomes tetrahedral, such that the sterically shielded face of the aldehyde moiety is the Re face.

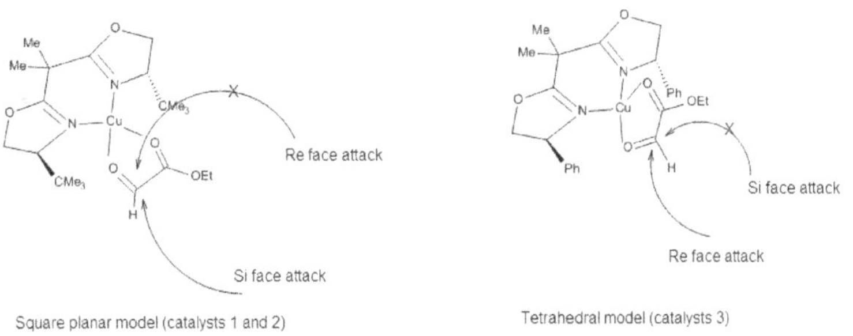

Square planar and tetrahedral Cu (II) stereochemical models.

Initially, the value of the method developed by Evans and coworkers was proved by successfully converting the resulting alpha-hydroxy ester into the corresponding methyl ester, free acid, Weinreb amide and alpha-azido ester, without any racemization, as shown in Figure. It is worth noting that the azide displacement of the alcohol that results from the carbonyl ene reaction provides a facile route towards the synthesis of orthogonally protected amino acids.

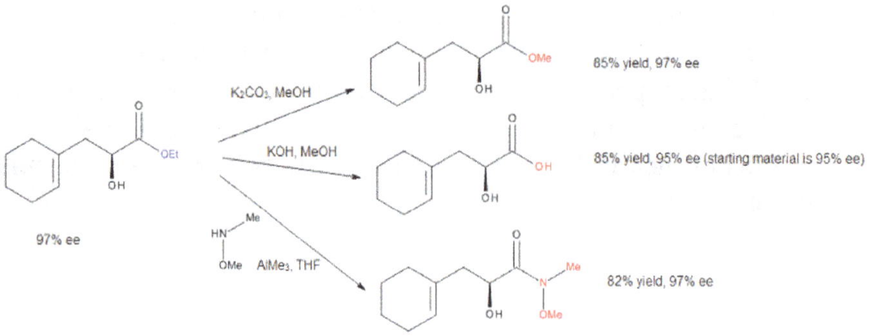

Derivatization of the alcohols afforded by C2-symmetric Cu(II) chiral Lewis acids.

The synthetic utility of the chiral C2-Symmetric Cu(II) catalysts was truly revealed in the formation of the C17 stereocenter of the CD ring fragment of (+)-azaspiracid-1, a very potent toxin (cytotoxic to mammalian cells) produced in minute quantities by multiple shellfish species including mussels, oysters, scallops, clams, and cockles. As shown in Figure, the reaction that establishes the C17 stereocenter is catalyzed by 1 mol % Cu(II) complex 2, and the authors note that it can be conducted on a 20 g scale and still give very good yields and excellent enantioselectivities. Furthermore, the product can be easily converted into the corresponding Weinreb amide, without any loss of selectivity, allowing for the facile introduction of the C14 methyl group. Thus, this novel catalytic enantioselective process developed by Evans and coworkers can be easily integrated into complex synthesis projects, particularly early on in the synthesis, when high yields and enantioselectivites are of utmost importance.

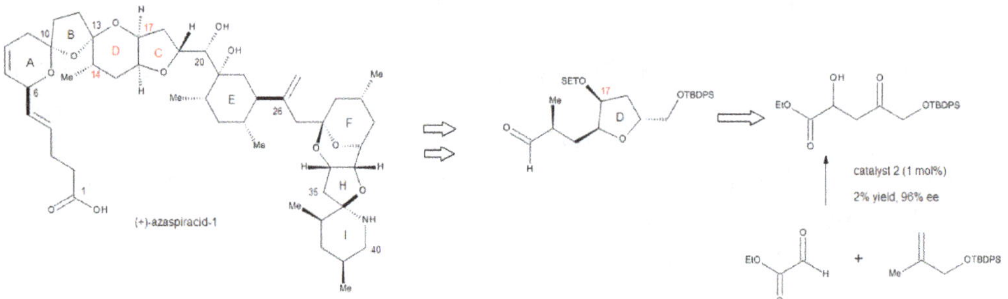

Structure of (+)-azaspiracid-1 and the ene reaction used to introduce the C17 stereocenter.

Cycloaddition

A cycloaddition is a pericyclic chemical reaction, in which "two or more unsaturated molecules (or parts of the same molecule) combine with the formation of a cyclic adduct in which there is a net reduction of the bond multiplicity." The resulting reaction is a cyclization reaction. Many but not all cycloadditions are concerted. As a class of addition reaction, cycloadditions permit carbon–carbon bond formation without the use of a nucleophile or electrophile.

Cycloadditions can be described using two systems of notation. An older, but still common, notation is based on the size of linear arrangements of atoms in the reactants. It uses parentheses: $(i + j + ...)$ where the variables are the numbers of linear atoms in each reactant. The product is a cycle of size ($i + j + ...$). In this system, the standard Diels-Alder reaction a (4 + 2)cycloaddition, the 1,3-dipolar cycloaddition is a (3 + 2)cycloaddition and cyclopropanation of a carbene with an alkene a (2 + 1)cycloaddition.

A more recent, IUPAC-preferred notation uses square brackets to indicate the number of *electrons*, rather than carbon atoms, involved in the formation of the product. In the [$i + j + ...$] notation, the standard Diels-Alder reaction is a [4 + 2]cycloaddition, the 1,3-dipolar cycloaddition is [4 + 2].

Thermal Cycloadditions and their Stereochemistry

Thermal cycloadditions are those cycloadditions where the reactants are in the ground electronic state. They usually have ($4n + 2$) π electrons participating in the starting material, for some integer n. These reactions occur, for reasons of orbital symmetry, in a suprafacial-suprafacial or antarafacial-antarafacial manner (rare). There are a few examples of thermal cycloadditions which have $4n$ π electrons (for example the [2 + 2] cycloaddition); these proceed in a suprafacial-antarafacial sense, such as the dimerisation of ketene, in which the orthogonal set of p orbitals allows the reaction to proceed via a crossed transition state.

Photochemical Cycloadditions and their Stereochemistry

Cycloadditions in which 4n π electrons participate can also occur via photochemical activation. Here, one component has an electron promoted from the HOMO (π bonding) to the LUMO (π* antibonding). Orbital symmetry is then such that the reaction can proceed in a suprafacial-suprafacial manner. An example is the DeMayo reaction. Another example is shown below, the photochemical dimerization of cinnamic acid. The two *trans* alkenes react head-to-tail, and the isolated isomers are called *truxillic acids*.

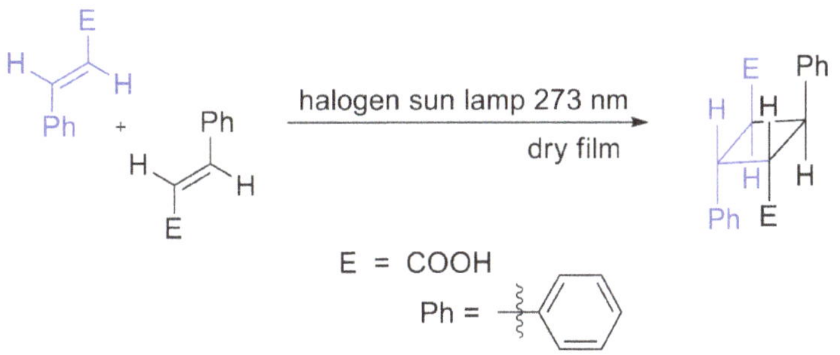

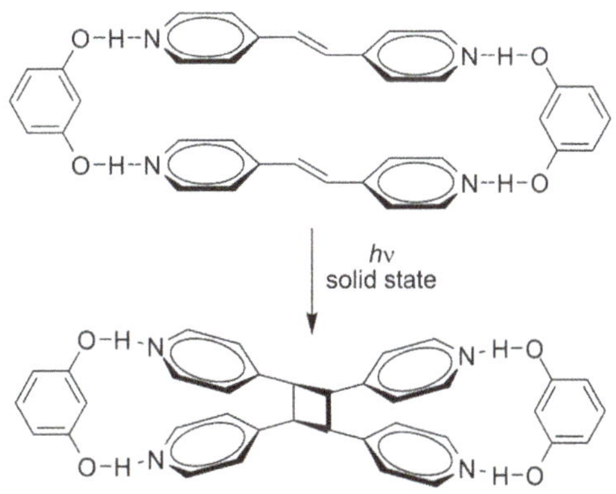

Cycloaddition of *trans*-1,2-bis(4-pyridyl)ethene

Supramolecular effects can influence these cycloadditions. The cycloaddition of *trans*-1,2-bis(4-pyridyl)ethene is directed by resorcinol in the solid-state in 100% yield.

Some cycloadditions instead of π bonds operate through strained cyclopropane rings; as these have significant π character. For example, an analog for the Diels-Alder reaction is the quadricyclane-DMAD reaction:

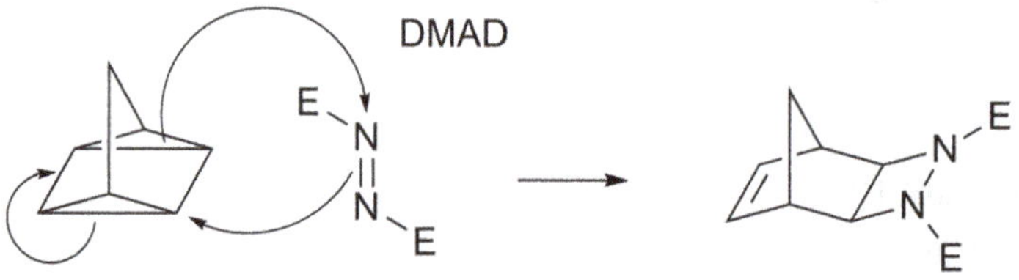

In the (i+j+...) cycloaddition notation i and j refer to the number of atoms involved in the cycloaddition. In this notation a Diels-Alder reaction is a (4+2)cycloaddition and a 1,3-dipolar addition such as the first step in ozonolysis is a (3+2)cycloaddition. The IUPAC preferred notation however, with [i+j+...] takes electrons into account and not atoms. In this notation the DA reaction and the dipolar reaction both become a [4+2]cycloaddition. The reaction between norbornadiene and an activated alkyne is a [2+2+2]cycloaddition.

Types of Cycloaddition

Diels-Alder Reactions

The Diels-Alder reaction is perhaps the most important and commonly taught cycloaddition reaction. Formally it is a [4+2] cycloaddition reaction and exists in a huge range of forms, including the inverse electron-demand Diels–Alder reaction, Hexadehydro Diels-Alder reaction and the related alkyne trimerisation. The reaction can also be run in reverse in the retro-Diels–Alder reaction.

diene + dienophile

Reactions involving heteroatoms are known; including the aza-Diels–Alder and Imine Diels–Alder reaction.

Huisgen Cycloadditions

The Huisgen cycloaddition reaction is a (2+3)cycloaddition.

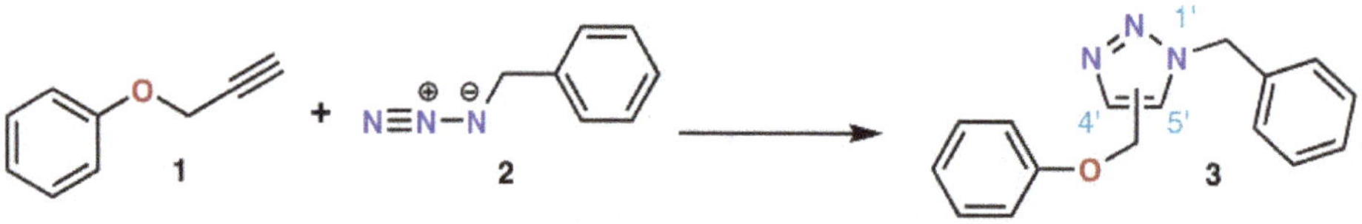

Nitrone-olefin Cycloaddition

The Nitrone-olefin cycloaddition is a (3+2)cycloaddition.

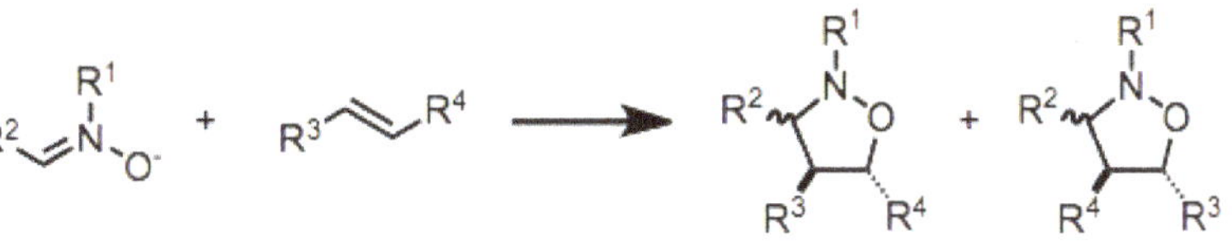

Iron-catalyzed 2+2 Olefin Cycloaddition

Iron[pyridine(diimine)] catalysts contain a redox active ligand in which the central iron atom can coordinate with two simple, unfunctionalized olefin double bonds. The catalyst can be written as a resonance between a structure containing unpaired electrons with the central iron atom in the II oxidation state, and one in which the iron is in the 0 oxidation state. This gives it the flexibility to engage in binding the double bonds as they undergo a cyclization reaction, generating a cyclobutane structure via C-C reductive elimination; alternatively a cyclobutene structure may be produced by beta-hydrogen elimination. Efficiency of the reaction varies substantially depending on the alkenes used, but rational ligand design may permit expansion of the range of reactions that can be catalyzed.

Cheletropic Reactions

Cheletropic reactions are a subclass of cycloadditions. The key distinguishing feature of cheletropic reactions is that on one of the reagents, both new bonds are being made to the same atom. The classic example is the reaction of sulfur dioxide with a diene.

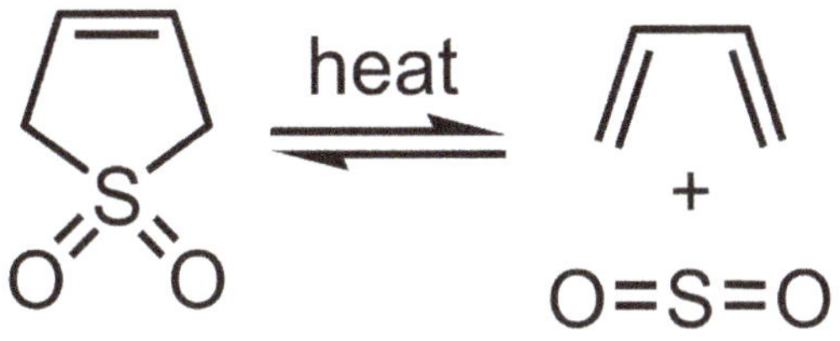

Other

Other cycloaddition reactions exist: [4+3] cycloadditions, [6+4] cycloadditions, [2+2]photocycloadditions and [4+4] photocycloadditions.

Formal Cycloadditions

Cycloadditions often have metal-catalyzed and stepwise radical analogs, however these are not strictly speaking pericyclic reactions. When in a cycloaddition charged or radical intermediates are involved or when the cycloaddition result is obtained in a series of reaction steps they are sometimes called formal cycloadditions to make the distinction with true pericyclic cycloadditions.

One example of a formal [3+3]cycloaddition between a cyclic enone and an enamine catalyzed by *n*-butyllithium is a Stork enamine / 1,2-addition cascade reaction.

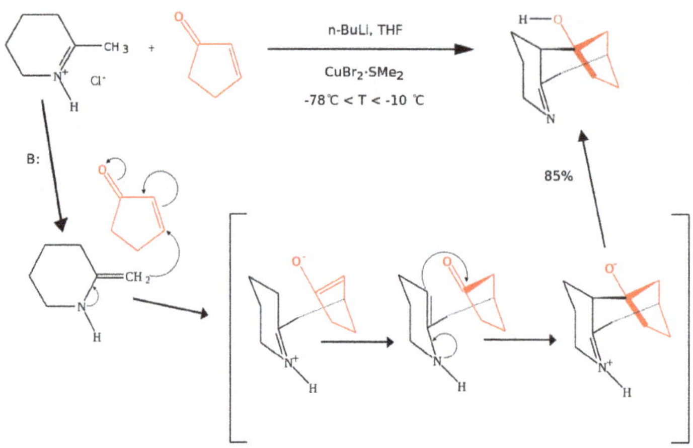

Intra- and Intermolecular Diels-Alder Type Reactions

Enantioselective Ene and Cycloaddition Reactions

Alder-ene and Diels-Alder reactions are six electron pericyclic processes between a "diene" or an alkene bearing an allylic hydrogen and an electron-deficient multiple bond to form two bonds σ with migration of the π bond. The section covers the examples of recent developments in enantioselective intermolecular Alder-ene glyoxylates with alkenes. Few studies on intra- and intermolecular Diels-Alder type reactions are also covered in the latter part of the section.

Carbonyl-Ene Reaction

Chiral Lewis acid catalyzed enantioselective ene reaction is one of the efficient methods for atom economical carbon-carbon bond formation. For example, Ti-BINOL prepared *in situ* catalyzes efficiently the carbonyl-ene reaction of glyoxylate with α -methylstyrene in the presence of molecular sieves with high enantioselectivity.

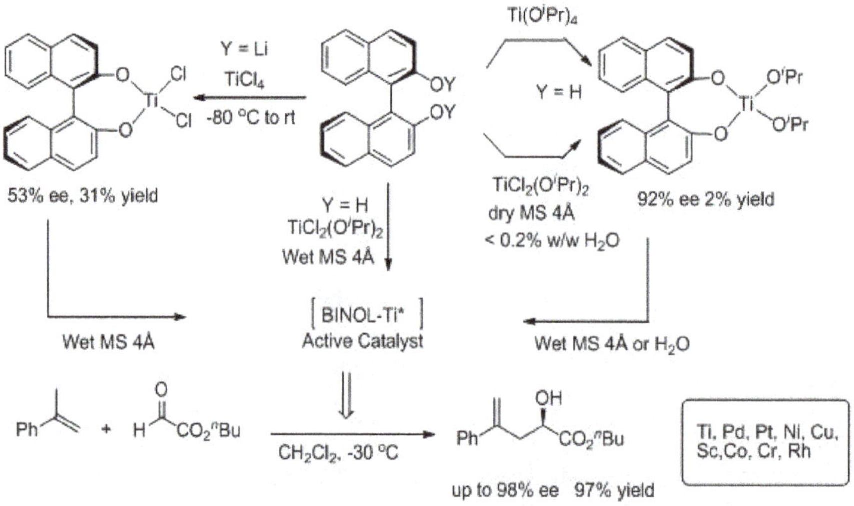

Besides the early transition metal based Lewis acid catalysts, square planar dicationic late transition metal complexes bearing C_2-symmetric diphosphine ligands have also been considerably studied as chiral Lewis acids for carbonyl-ene reactions. For example, the isolated MeO-BIPHEP-Pd complex 1a bearing electron withdrawing benzonitrile as the labile, stabilizing ligands has been used for the ene reaction of ethyl glyoxylate with up to 81% ee. The isolated 1a exhibits more catalytic activity compared to that 1b which is in situ generated although both offer similar enantioselectivity.

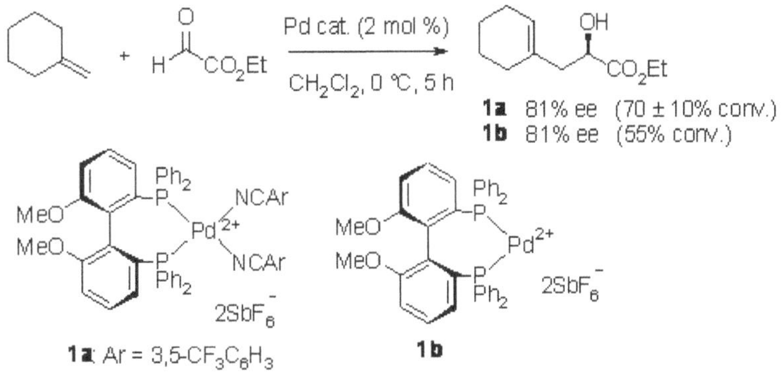

MeO-BIPHEPs-Pt complexes 3 with OTf - as counter anion also exhibit similar catalytic activity and selectivity in the asymmetric glyoxylate ene reaction. The addition of phenol facilitates the reaction by trapping the OTf anion and traces of water.

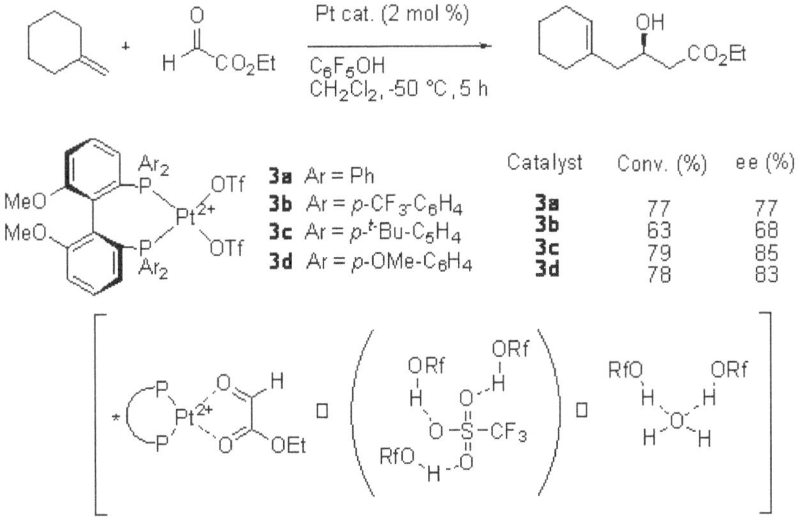

The glyoxylate ene reaction is also effective using tropox dicationic DPPF-Ni complex 4 with enantioselectivity up to 90% ee.

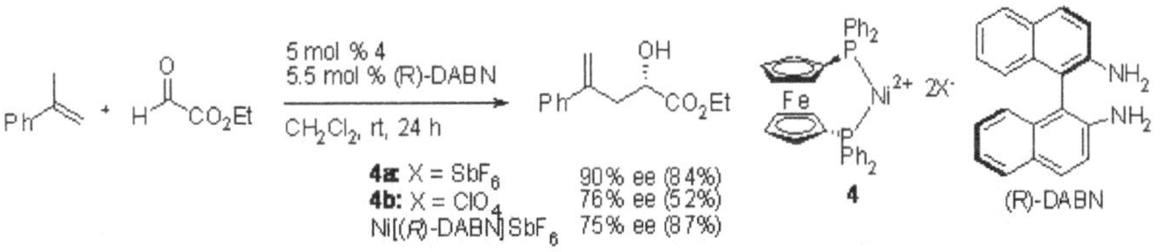

The glyoxylate-ene reaction can also be carried out using chiral C_2-symmetric bisoxazolinyl copper(II) complexes 5 and 6 as Lewis acid catalysts. The aqua complex is air and water stable and exhibits only slight decrease in the reaction rate compared to the anhydrous complex 6. The sense of asymmetric induction depends on the oxazoline ring substituents, which can be rationalized by the tetrahedral and square-planer intermediates to account for the absolute configuration of the products.

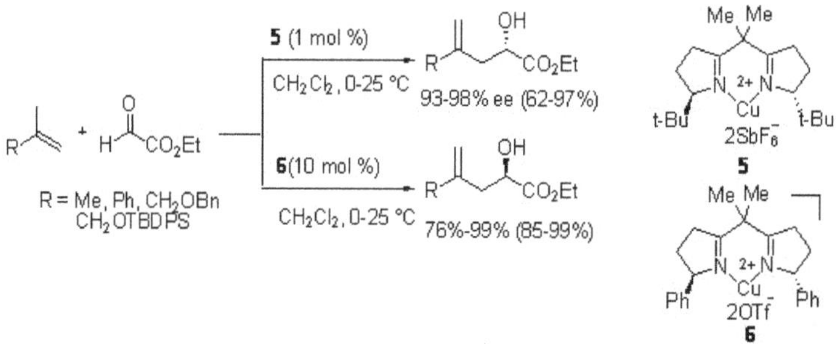

In addition, chiral C_2-symmetric trivalent pybox-Sc complex 7 is studied for the carbonyl-ene reactions with N -phenyl glyoxamides. The ene products are obtained with excellent diastereo- and enantioselectivity. Presumably, the products are formed via proton transfer from the β - cis substituent through an exo -transition state.

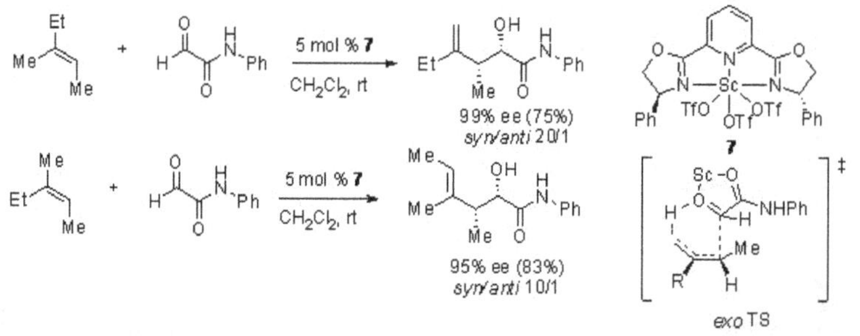

Co and Cr-based chiral complexes have also been explored for the carbonyl-ene reaction with glyoxylates. For example, chiral β-ketoiminato complex 8 catalyzes efficiently the reaction of 1,1-disubstiuted alkene and glyoxyl derivative in high enantioselectivity. Similar to the earlier described Pd, Pt and Ni-based catalysts, hexafluoroantimonate as a counter anion is found to be the most effective.

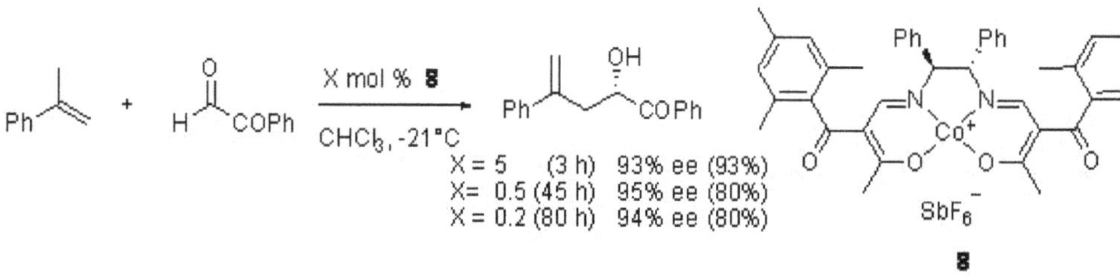

Chiral Cr(III)-salen complex 9 bearing adamantyl group in the salen ligand has been used for the reaction of ethyl glyoxylate with 1,2-disubstituted alkenes. The catalyst can be prepared in multigram scale and the ene products are obtained with up to 92% ee. The presence of adamantyl substituent essential for the enhancement in the enantioselectivity.

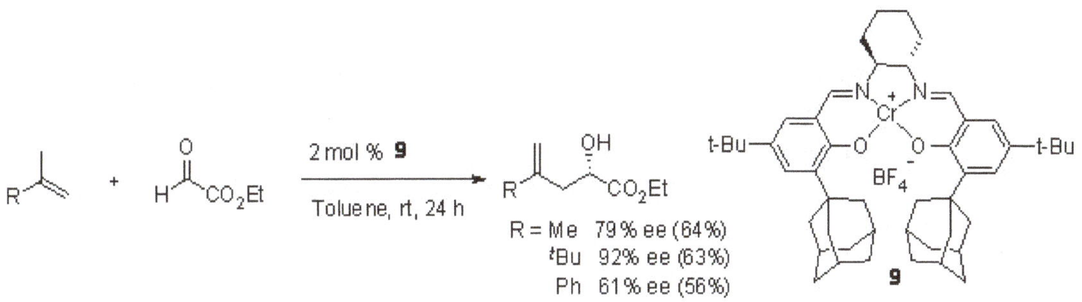

Besides the metal based catalysts, chiral organocatalysts have also been considerably explored during the recent years for the carbonyl-ene reactions. For example, the chiral phosphoric acid 10 as a chiral Bronsted acid catalyzes readily the enantioselective aza-ene reaction of enamides to imines with excellent enantioselectivity even on a gram scale.

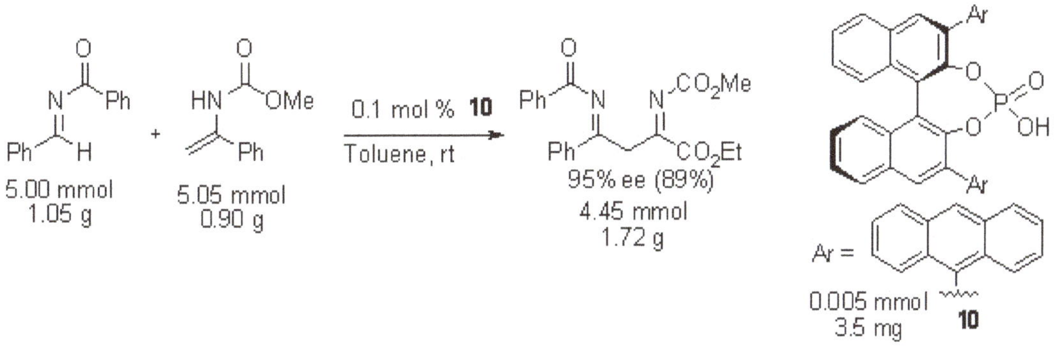

Besides the intermolecular reactions, intramolecular version of this reaction has also been well explored using chiral metal as well as chiral phosphoric acids as catalysts. For example, the palladium-phosphine complex catalyzed cyclization of 1,7-enyenes bearing benzene ring takes place efficiently to afford six membered quinoline derivatives with quaternary stereogenic centers as single enantiomer.

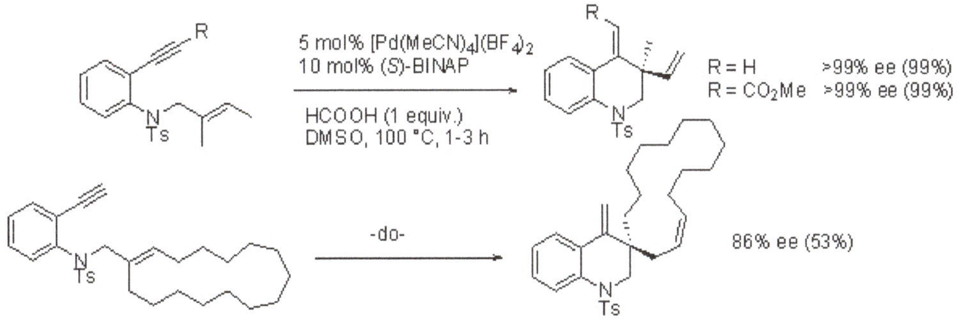

Diels-Alder Type Reactions

Asymmetric intra- and intermolecular Diels-Alder reactions have made remarkable progress using chiral metal complexes as catalysts. Subsequently, several studies are focused on the use of chiral organocatalysis for this reaction. Since the organocatalysis based reactions are covered, this topic covers recent examples of the metal catalyzed reactions.

Intramolecular [4+2]-Cycloaddition

Intramolecular Diels-Alder reactions of unactivated dieneynes provide powerful tool to construct 5,6- or 6,6-fuzed rings. These fuzed rings can be inducted in the synthesis of many natural products. Therefore, a number of methods using transition metal catalysis have been developed over the past two decades. The chiral Rh complex bearing chiral diene and chiral phosphine has been shown to give better enantioselectivity compared to that bear achiral diene and chiral phosphine complex.

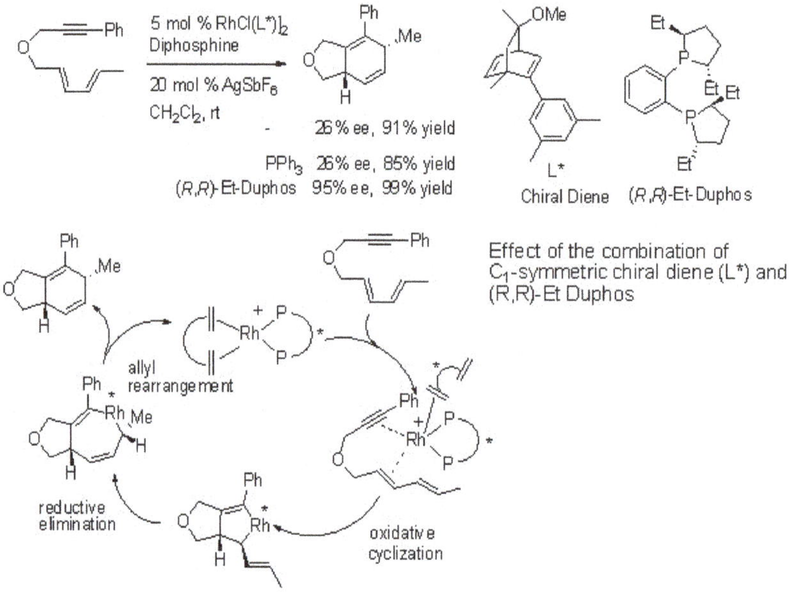

Intermolecular Diels-Alder Reactions

Intermolecular hetero Diels-Alder reactions have also been extensively explored using both chiral metal complexes as well as chiral organocompounds as catalysts. Since the use of chiral organocatalysis has been covered, this topic focuses on few examples using chiral metal complexes as the catalysts. The reaction of benzaldehyde with Danishefsky's diene proceeds in the presence of BINOL/diimine/Zn complex with excellent enantioselectivity and yield.

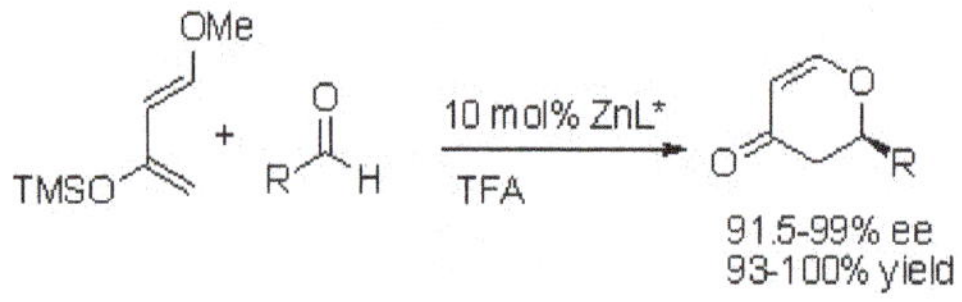

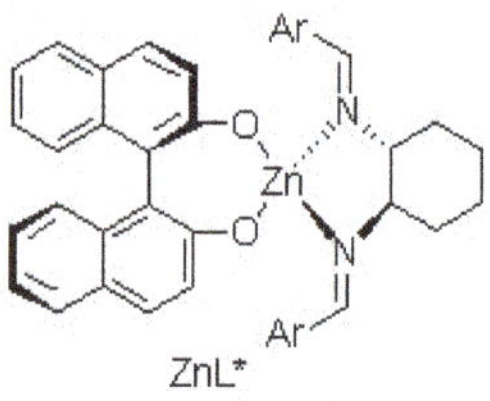

R	Yield [%]	ee [%]
Ph	100	99.0
3-MeOC_6H_4	100	96.4
4-ClC_6H_4	99	98.5
4-MeOC_6H_4	100	98.5
α-napthyl	93	91.5
β-napthyl	100	93.8

Chiral box-Cu(II) complexes are found to be excellent catalysts for a variety of hetero Diels-Alder reactions.

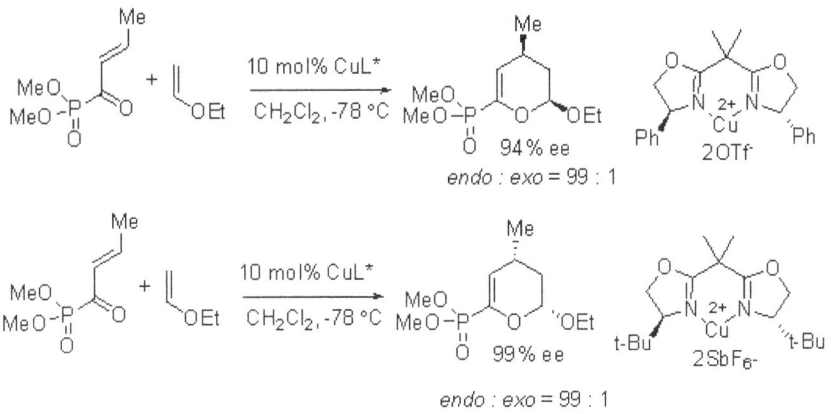

The readily accessible oxazaborolidine-aluminum bromide catalyst catalyzes the reaction of furan with diethyl fumarate with excellent enantioselectivity.

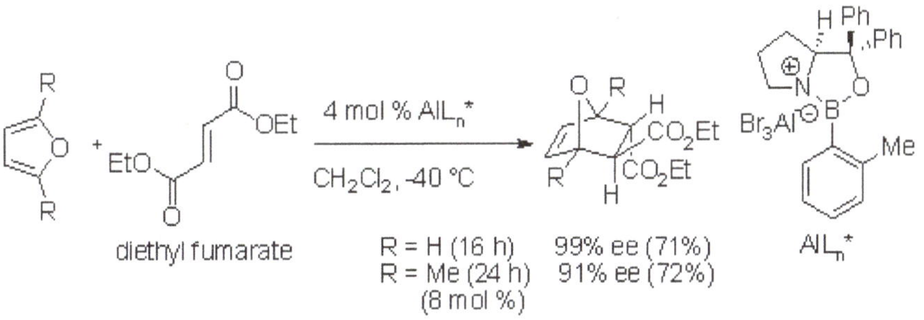

Alkene Metathesis Catalysts

Enantioselective Alkene Metathesis

Among the alkene metathesis catalysts, Mo and Ru-based complexes have emerged as powerful exhibiting complementary reactivity and functional group tolerance. The asymmetric alkene metathesis provides access to enantiomerically enriched molecules that can not be generally prepared through the commonly practiced strategy. Unlike most of the other enantioselective processes, alkene metathesis, which entails the formation and cleavage of carbon-carbon double bonds, does not involve the direct construction of *sp* 3 -hybridized stereogenic center. Instead, the stereochemistry is created indirectly, often by desymmetrization of an achiral substrate, wherein the chiral catalyst has to discriminate between enaniotopic groups or sites of the molecule.

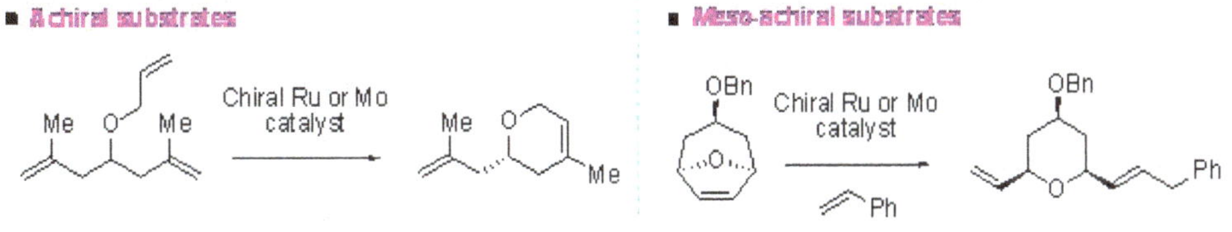

Desymmetrization in catalytic enantioselective alkene metathesis .

Ring-Closing Metathesis (RCM) Reactions

Ru-Catalyzed Reactions

RCM is most commonly used in organic synthesis to construct cyclic systems, which are sometimes difficult to prepare by most of the other methods. During the past decade, several Ru and Mo-based chiral catalysts have been developed for the enantioselective RCM process and made remarkable progress. Figure summarizes examples for enantioselective RCM employing monodentate chiral NHC-Ru and chiral Mo-diolate complexes. The Ru-based catalysts are selective compared to Mo-based one, which catalyzes a wide range of substrates.

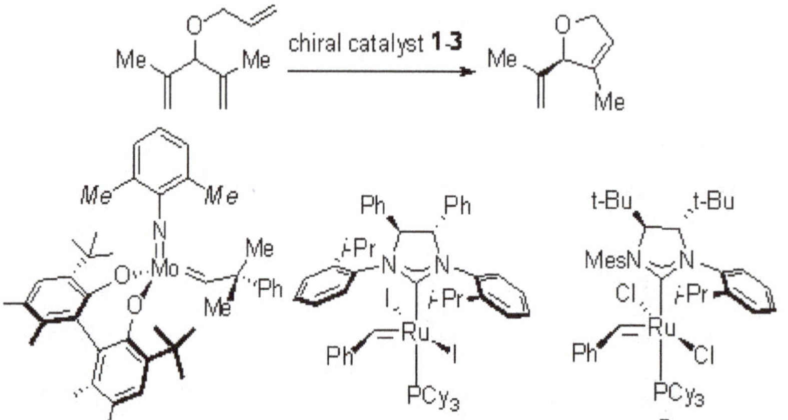

1 2 3

Product	Catalyst (mol %)	Time (h)	Temp (°C)	Yield (%)	ee (%)
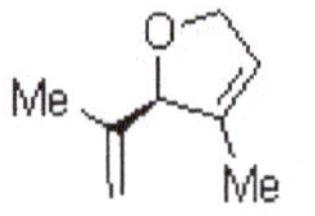	**1** (1)	5	22	85	93
	2 (5)	2	38	20	39
	3 (2.5)	2	40	80	68
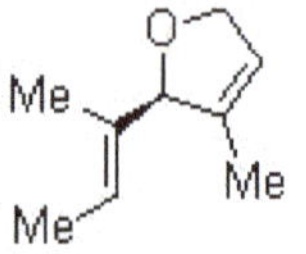	**1**(2)	5	22	93	98
	2 (4)	2	38	64	90
	3(2.5)	2	40	81	82

Comparison of Chiral Mo and Ru Catalysts in Enantioselective RCM

The mechanism of the Ru-catalyzed RCM is outlined in Figure. Initiation of the reaction may take place *via* the dissociation of either the phosphine ligand or chelated etherate moiety. Subsequently, the less substituted alkene may make coordination to the Ru center, which could proceed [2+2]-cycloaddition, followed by cycloreversion and ruthenacyclobutane formation that could lead to the target product. The formation and cleavage of the cyclobutanes are crucial for the enantioselectivity of the products.

The Synthesis of Cyclic Enol Ethers using Mo-Catalyzed RCM

Mo-based RCM is found to be successful for the synthesis of furan and pyran products with up to 98% ee. Although high catalyst loading is required, the products can be constructed with tertiary and quaternary stereogenic centers. In contrast, the Ru-based catalysts are not successful for this transformation.

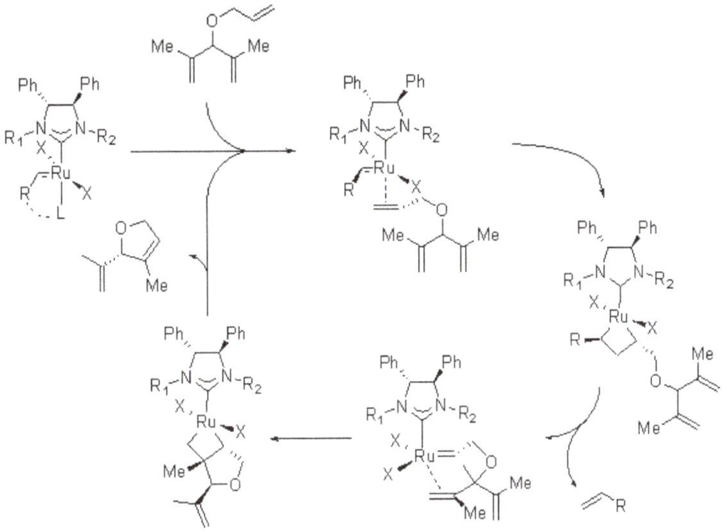

Mechanism for Ru-catalyzed enantioselective RCM

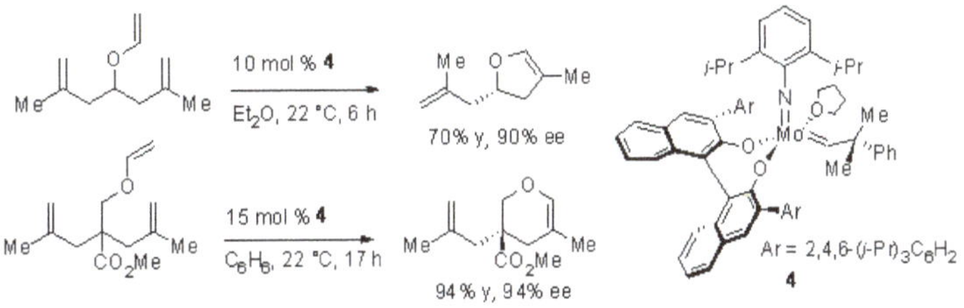

Synthesis of Cyclic Enol Ethers with Tertiary and Quaternary Stereogenic Centers

Ring-Opening/Ring-Closing Metathesis (RORCM) and Ring-Opening/Cross Metathesis (ROCM)

Following the ring opening, the resulting carbene intermediate can be traped intramolecularly by a pendant alkene (RORCM, Path A) or intermolecularly using a cross-partner (ROCM, Path B). These reaction pathways can be controlled by selection of the appropriate catalyst and cross partner, which can lead to a wide range of enantiomerically enriched products from common starting material. In the absence of intramolecular trap (ROCM process), a number of complex mixture of products can be generated.

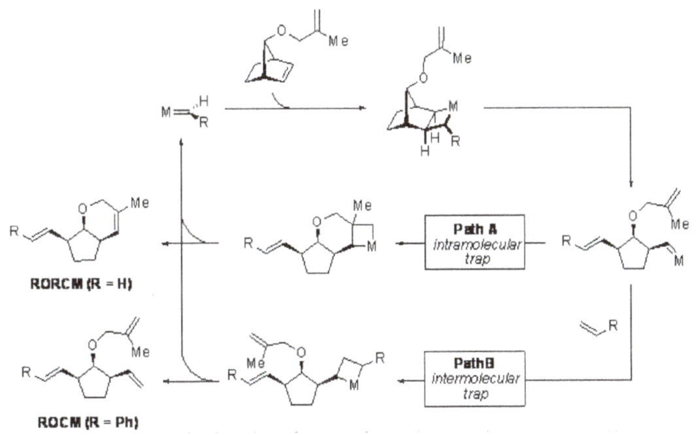

Pathways for Enantioselective RORCM versus ROCM Products

Figure presents examples for the Mo and Ru-catalyzed enantioselective ROCM processes. Norbornenes react with styrene via ROCM with high enantioselectivities. In both cases, E -alkenes are generated. In the absence of styrene, in the case of Mo-based system, RORCM product is formed with 92% ee. The substrate used for the Ru-catalyzed ROCM process, proceed polymerization in the presence of Mo-catalyst instead of ROCM process.

Figure shows the comparison of the Ru-catalyzed ROCM of norbornenes. The catalysts 7 and 8 bearing monodendate NHC ligands exhibit greater reactivity (i.e., lower catalyst loading) compared to the complex bearing bidendate NHC ligand 6 . But the systems using 7 and 8 produce poor E / Z selectivity, whereas the reaction using 6 gives exclusively E -isomer.

The synthesis of isoindole has been recently shown using chiral Ru-catalyzed RORCM with moderate enantioselectivity. In this reaction the use of ethylene is to facilitate the release of the catalyst. The direct alkene metathesis product is unstable and thus it was isolated after hydrogenation.

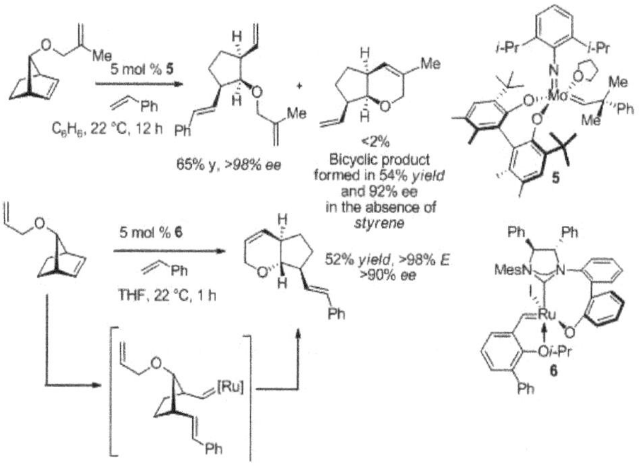

Mo-and Ru-Catalyzed Enantioselective ROCM of Norbornenes.

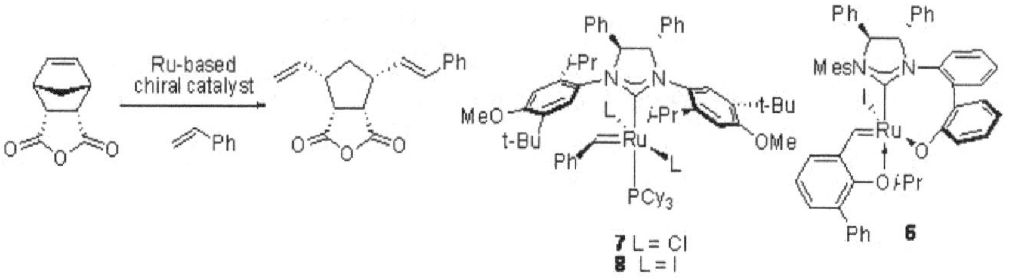

Catalyst (mol %)	Styrene (equiv)	Time	Temp. (°C)	Yield (%)	E/Z	ee(%)
7 (10)	1	10	22	95	1:1	76
8 (10)	3	10	22	96	1:1	80
6 (5)	5	5	22	50	98:2	90

Comparison of the Activity of Chiral Ru Catalysts in ROCM of Norbornene

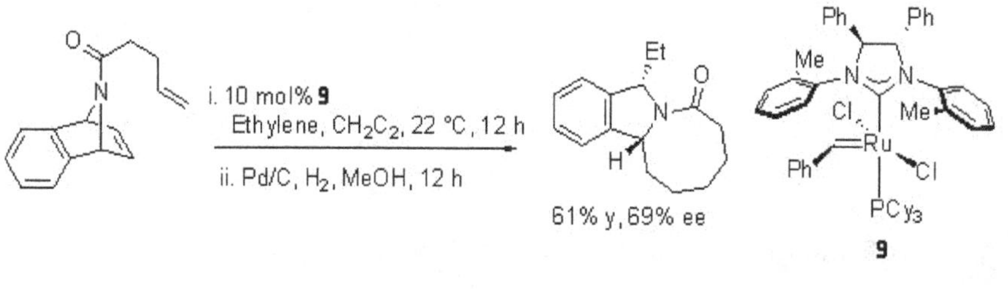

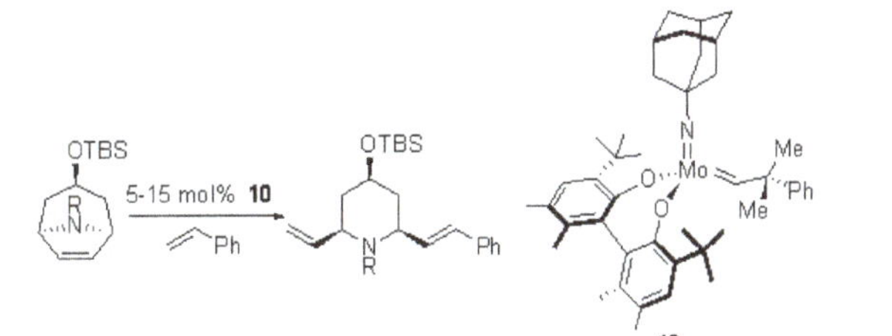

10

Product	10 (mol%)	Time (h)	Yield (%)	ee (%)
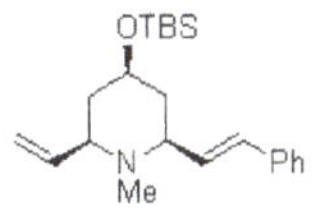	5	1	95	94
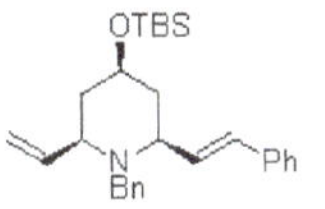	5	1	85	21
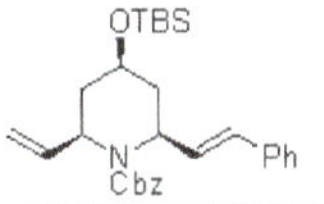	10	24	80	82

Enantioselective ROCM Reaction of meso -Azabicyles

2,6-Disubstituted piperidines are important structural unit present in medicinally significant compounds. Using the Mo-based enantioselective ROCM reactions, the synthesis of the *N* -protected 2,6-substituted piperidines can be accomplished from of *meso* -azabicycles with moderate to high enantioselectivities.

Cross-Metathesis (CM)

Catalytic enantioselective CM is least developed in enantioselective alkene metathesis reactions. Unlike the ring-closing and ring-opening metatheses that are thermodynamically driven, there is minimal driving force for the CM. In addition, selectivity between two different cross partners leads to complex. Figure presents some examples of CM using chiral Ru complexes with moderate enantioselectivity. These substrates don't proceed RCM due to ring strain of the products.

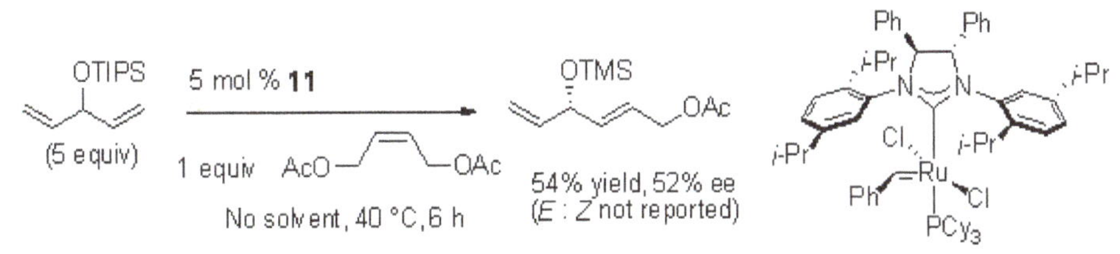

11

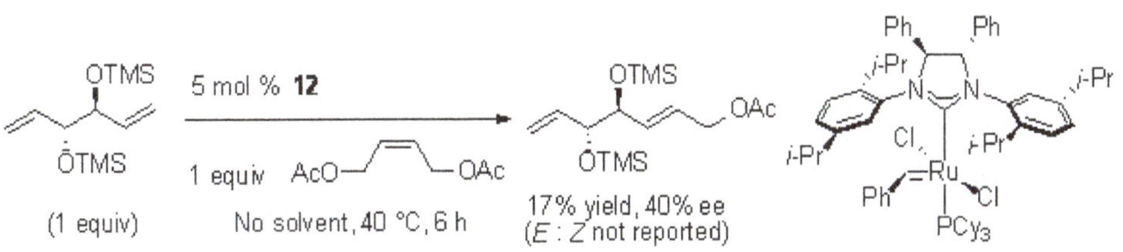

12

Catalytic enantioselective cross-metathesis reactions

Carbometallation Reactions

Carbometallation and Carbocyclization Reactions

Organometallic compounds add to carbon-carbon multiple bonds to give a new organometallic species, which could be further modified to yield new carbon-carbon bonds. These processes are called as "carbometallation reactions". It primarily refers to the relationship between the reactants and products. This topic covers some examples of the asymmetric carbometallation reactions using Rh, Cu and Pd-based systems.

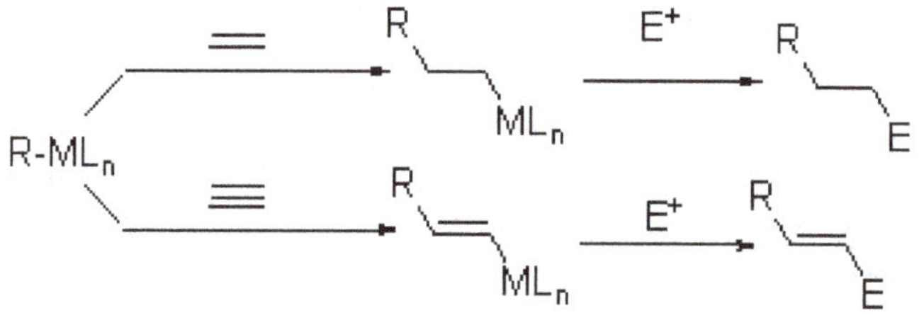

Rhodium-Catalyzed Reactions

Hydrogen-mediated carbon-carbon bond formation has emerged as powerful industrial process in chemical industries. For example, the hydroformylation and Fischer-Tropsch reactions are well known for the hydrogen-mediated carbon-carbon bond formation reactions. These processes require heterolytic activation of molecular hydrogen to give monohydride species, where the C-H reductive elimination pathway is disabled. Addition of a metal hydride to carbon-carbon multiples bonds (i.e., alkene and alkyne) give organometallic species that could be rapidly captured by an electrophile (i.e., aldehydes and imine) prior to its reaction with molecular hydrogen *via* oxidative addition or σ-bond metathesis of carbon-metal bond. Figure illustrates metal-dihydride route (leading to hydrogenation) and metal-monohydride route (leading to C-C bond formation) with an alkyne.

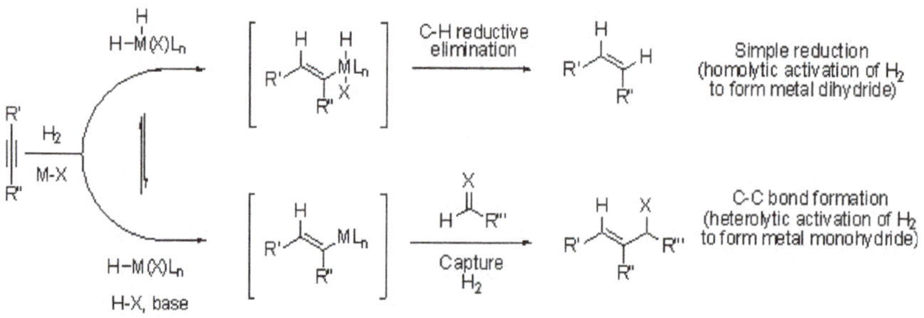

Formation of the monohydride organometallic species depends on the choice of the catalytic system. For example, the heterolytic activation of molecular hydrogen is observed with cationic rhodium complexes in the presence of base. The reaction takes place via the oxidation addition of the molecular hydrogen with metal species followed by a base induced reductive elimination of HX.

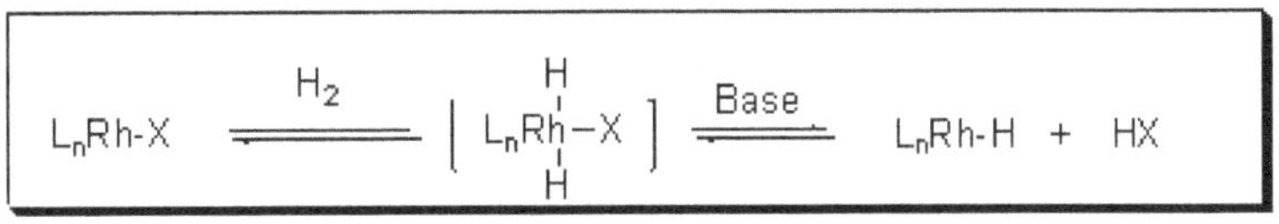

For example, Figure presents enantioselective reductive cyclization of 1,6-enynes using $Rh(COD)_2OTf$ and (*R*)-BINAP in the presence of molecular hydrogen. This carbocyclization reaction is compatible with various functional groups, however, the yield and enantioselectivity of the product depends on the structure of 1,6-enynes and the ligands.

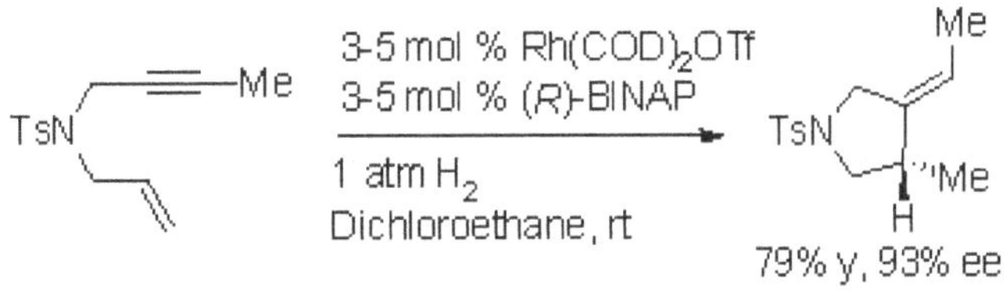

A possible mechanism has been proposed for this reaction based on deuterium labeling control experiments. The catalytic cycle starts with cycloaddition of RhL_n and 1,6-enyne forming rhodacyclopentene. Homolytic hydrogen activation via oxidative addition of molecular hydrogen or σ-bond metathesis may lead to the formation of vinyl-rhodium vinyl species that could afford cyclization product by reductive elimination to complete the catalytic cycle.

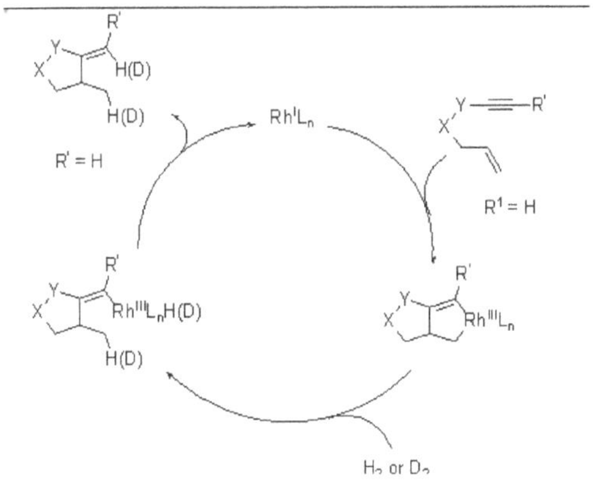

1,4-Conjugate addition of organometallic reagents to α,β -unsaturated carbonyl compounds afford effective method for carbon-carbon bond formation. Much effort has been on the development of asymmetric version of the reaction using a series of catalytic systems. The first reductive aldol cyclization of keto-enone with phenylboronic acid has been shown utilizing $Rh[(COD)Cl]_2$ and (R)-BINAP with yield and enantioselectivity.

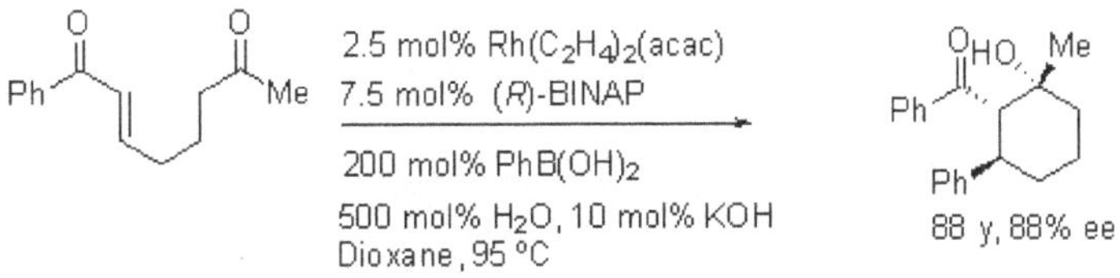

The mechanism of the reaction is presented in Figure. The observed stereochemistry has been rationalized by assuming Z -enolate formation.

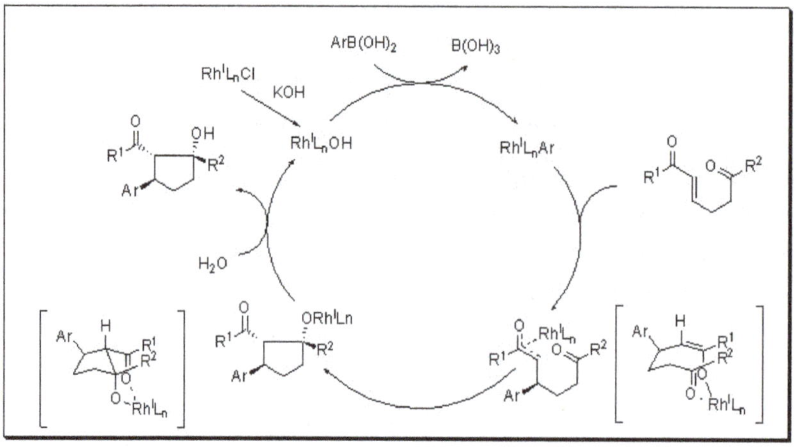

Copper-Catalyzed Reactions

CuH is found to be highly efficient catalyst for the asymmetric reductive aldol cyclization of keto-enones to give the target product as a singly diastereoisomer with high enantiopuritiy. These reactions use ferrocenylphosphines, (S, R)-PPE-P(t -Bu)2, as effective chiral ligands in the presence of silane as a hydride source. These reactions can also be carried out under heterogeneous as well as aqueous conditions with surfactant.

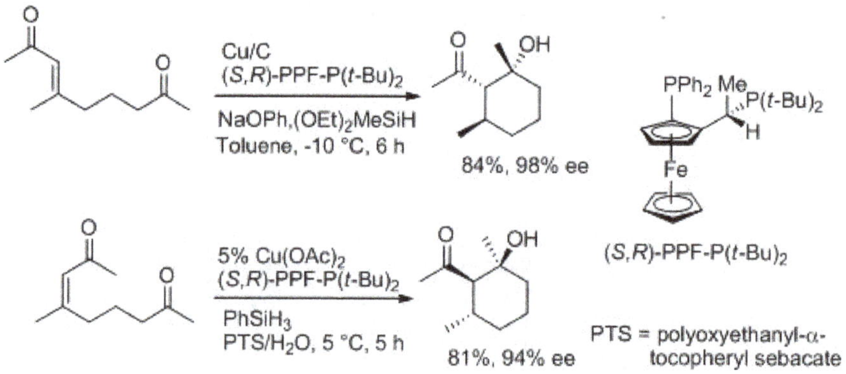

Palladium-Catalyzed Reactions

The palladium-catalyzed cross-coupling reactions of aryl or alkenyl halides with alkenes in the presence of base are among the powerful reactions in organic synthesis to construct carbon-carbon bonds. The asymmetric version of the reaction is also well explored. Figures illustrates some examples for the intramolecular and intermolecular Heck reactions. In 1970, the Heck reaction was discovered and, in 1989, the first example of asymmetric intramolecular Heck reactions appeared using $Pd(OAc)_2$ with (R)-BINAP with moderate enantioselectivity.

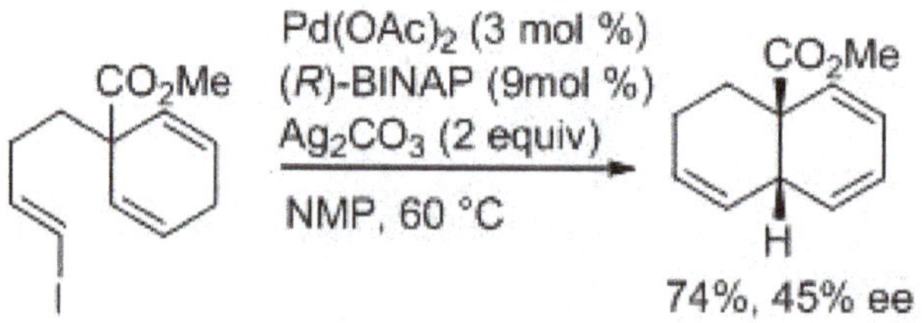

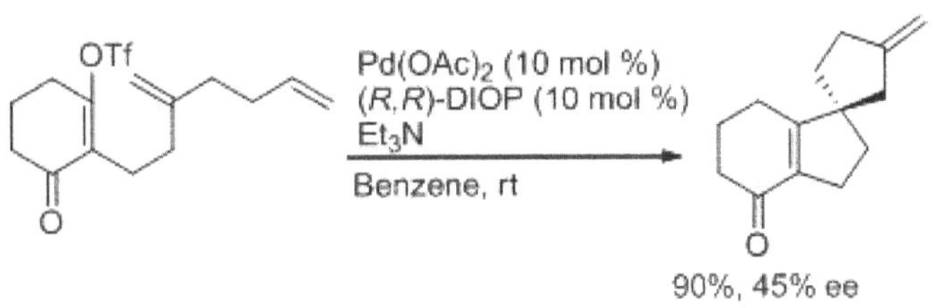

The intramolecular Heck reaction finds wide applications in organic synthesis. Among those applications the synthesis of optically active oxindoles having a quaternary asymmetric center has been considerably explored. Because the oxindole moiety serves as useful synthetic intermediate in the synthesis of numerous natural products. For example, (E)- α,β -unsaturated-2-iodoanilide undergoes cyclization in the presence of Pd_2 $(dba)_3$-$CHCl_3$ and (R)-BINAP to give oxindoles with (S) or (R) configuration under cationic and neutral conditions, respectively. It is noteworthy that a dramatic switching in the direction of asymmetric induction has been observed between the two conditions even though the same chiral ligand (R)-BIANP is employed. In these reactions, Ag_3PO_4 and PMP act as HI scavenger.

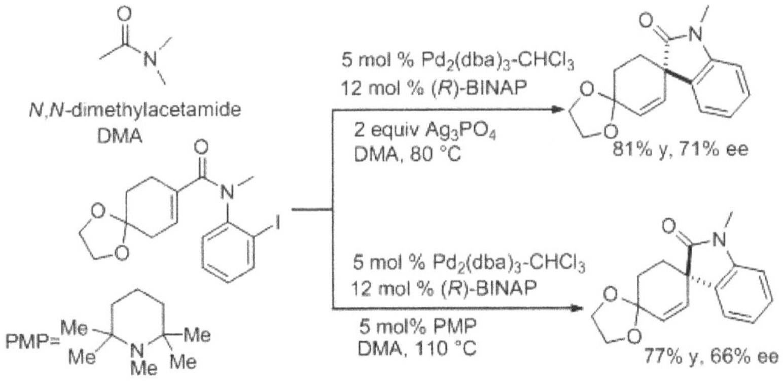

The use of TADDOL-based monophosphoramide has been demonstrated instead of BINAP in the reaction of intramolecular cyclization of cyclohexadienone derivatives. This reaction can be performed in the absence of silver salt.

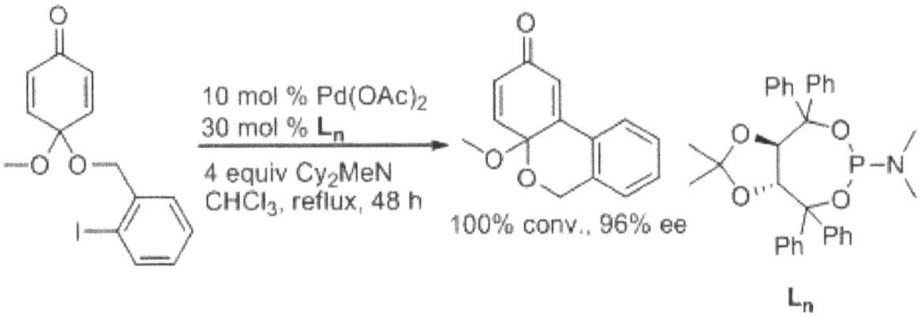

Intermolecular Heck reaction is also well studied. For example, dihydrofuran reacts with phenyl triflate to give 2-phenyl-2,3-dihydrofuran along with small amount 2-phenyl-2,5-dihydrofuran in the presence of Pd-BINAP with excellent enantioselectivity. A mechanism has been proposed to explain the high enantioselectivity of the major product and inversion configuration of the minor product. It involves a kinetic resolution process that enhances the enantioselectivity of the major product.

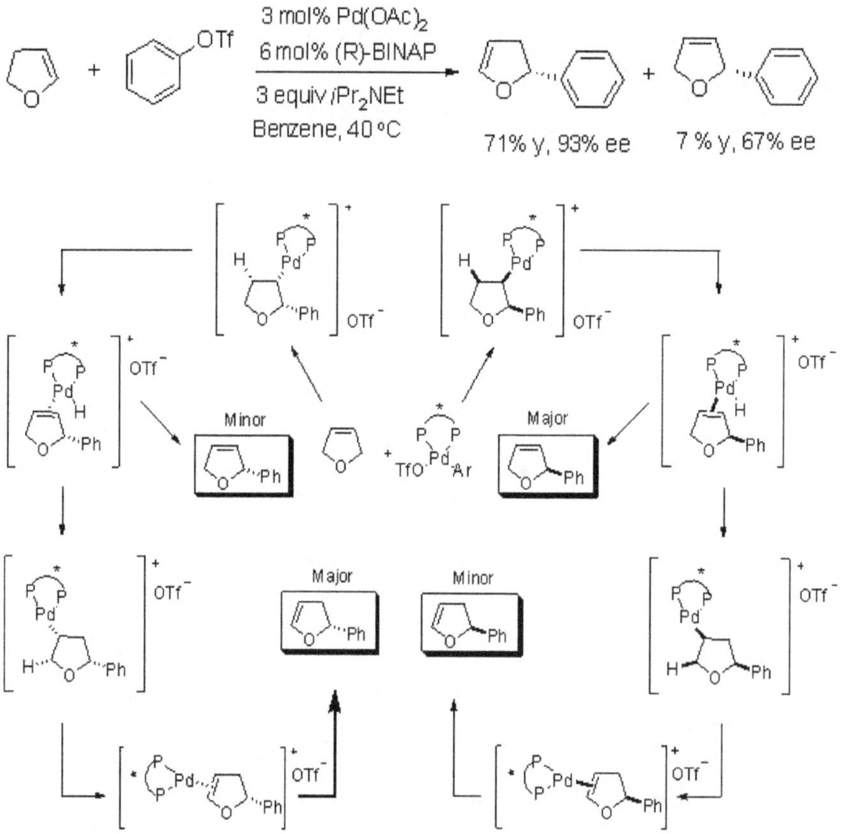

Figure exemplifies the reactions of 2,3-dihydrofuran and 2,2-dimethyl-2,3-dihydrofuran with phenyl triflate, 2-carbethoxy cyclohexenyl triflate and cyclohexenyl trilfate using palladium complexes with oxazoline based aminophosphine and (D-glucosamine)phosphiteoxazoline as the ligands. The reactions are effective affording the products with excellent enantioselectivity.

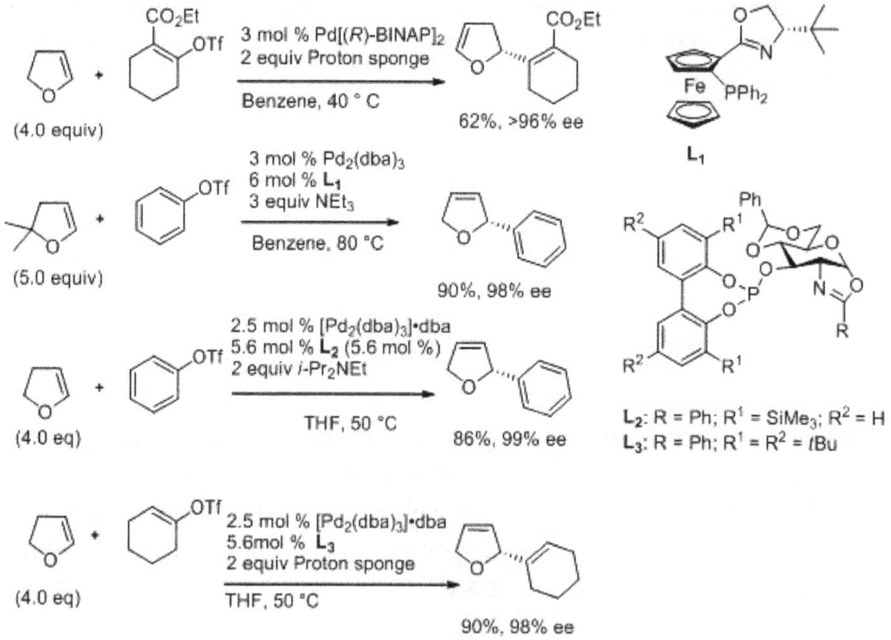

Conjugate Addition Reactions

Metal-Catalyzed Asymmetric Conjugate Addition Reactions

Asymmetric conjugate addition is one of the powerful tools for the construction carbon-carbon and carbon-heteroatom bonds in organic synthesis. This reaction finds extensive applications for the construction enantioenriched carbon skeletons for the total synthesis of numerous biologically active compounds. Sometimes possible to construct multiple stereocentres in single synthetic operation. This topic covers some examples for the recent developments in the conjugate addition of Grignard, organozinc, organolithium, organocopper and organoborane reagents with activated alkenes in the presence of chiral ligand or chiral catalysts.

Grignard Reaction

A solution of a carbonyl compound is added to a Grignard reagent.

The Grignard reaction is an organometallic chemical reaction in which alkyl, vinyl, or aryl-magnesium halides (Grignard reagents) add to a carbonyl group in an aldehyde or ketone. This reaction is an important tool for the formation of carbon–carbon bonds. The reaction of an organic halide with magnesium is *not* a Grignard reaction, but provides a Grignard reagent.

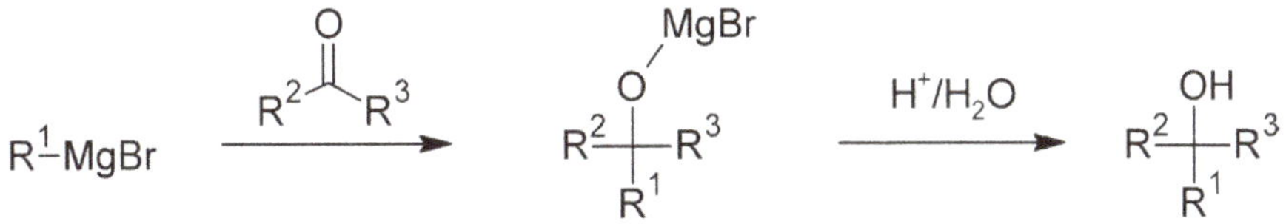

Grignard reactions and reagents were discovered by and are named after the French chemist François Auguste Victor Grignard (University of Nancy, France), who published it in 1900 and was awarded the 1912 Nobel Prize in Chemistry for this work. Grignard reagents are similar to organolithium reagents because both are strong nucleophiles that can form new carbon–carbon bonds. The nucleophilicity increases if the alkyl substituent is replaced by an amido group. These amido magnesium halides are called Hauser bases.

Reaction Mechanism

The Grignard reagent functions as a nucleophile, attacking the electrophilic carbon atom that is present within the polar bond of a carbonyl group. The addition of the Grignard reagent to the carbonyl typically proceeds through a six-membered ring transition state.

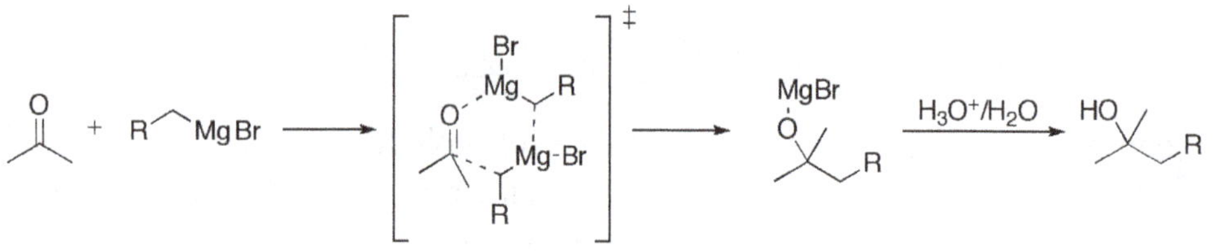

However, with hindered Grignard reagents, the reaction may proceed by single-electron transfer. Similar pathways are assumed for other reactions of Grignard reagents, e.g., in the formation of carbon–phosphorus, carbon–tin, carbon–silicon, carbon–boron and other carbon–heteroatom bonds.

Preparation of Grignard Reagent

Grignard reagents form via the reaction of an alkyl or alkyl halide with magnesium metal. The reaction is conducted by adding the organic halide to a suspension of magnesium in an etherial solvent, which provides ligands required to stabilize the organomagnesium compound. Empirical evidence suggests that the reaction takes place on the surface of the metal. The reaction proceeds through single electron transfer: In the Grignard formation reaction, radicals may be converted into carbanions through a second electron transfer.

$$R - X + Mg \rightarrow R - X^{\bullet -} + Mg^{\bullet +}$$

$$R - X^{\bullet -} \rightarrow R^{\bullet} + X^{-}$$

$$R^{\bullet} + Mg^{\bullet +} \rightarrow RMg^{+}$$

$$RMg^{+} + X^{-} \rightarrow RMgX$$

A limitation of Grignard reagents is that they do not readily react with alkyl halides via an S_N2 mechanism. On the other hand, they readily participate in transmetalation reactions:

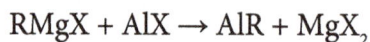

$$\text{RMgX} + \text{AlX} \rightarrow \text{AlR} + \text{MgX}_2$$

For this purpose, commercially available Grignard reagents are especially useful because this route avoids the problem with initiation.

Reaction Conditions

In reactions involving Grignard reagents, it is important to exclude water and air, which rapidly destroy the reagent by protonolysis or oxidation. Since most Grignard reactions are conducted in anhydrous diethyl ether or tetrahydrofuran, side-reactions with air are limited by the protective blanket provided by solvent vapors. Small-scale or quantitative preparations should be conducted under nitrogen or argon atmospheres, using air-free techniques. Although the reagents still need to be dry, ultrasound can allow Grignard reagents to form in wet solvents by activating the magnesium such that it consumes the water.

The Organic Halide

Grignard reactions often start slowly. As is common for reactions involving solids and solution, initiation follows an induction period during which reactive magnesium becomes exposed to the organic reagents. After this induction period, the reactions can be highly exothermic. Alkyl and aryl bromides and iodides are common, with chlorides being seen as well. However, fluorides are generally unreactive, except with specially activated magnesium (through Rieke metals).

Magnesium

Typical Grignard reactions involve the use of magnesium ribbon. All magnesium is coated with a passivating layer of magnesium oxide, which inhibits reactions with the organic halide. Specially activated magnesium, such as Rieke magnesium, circumvents this problem. The oxide layer can also be broken up using ultrasound, or by adding a few drops of iodine or 1,2-Diiodoethane.

Solvent

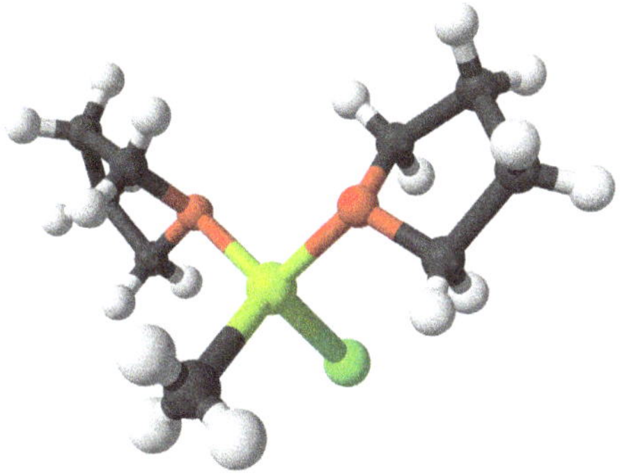

Usually Grignard reagents are written as RMgX, but in fact the magnesium(II) centre is tetrahedral when dissolved in Lewis basic solvents, as shown here for the bis-adduct of methylmagnesium chloride and THF.

Most Grignard reactions are conducted in ethereal solvents, especially diethyl ether and THF. With the chelating diether dioxane, some Grignard reagents undergo a redistribution reaction to give diorganomagnesium compounds (R = organic group, X = halide):

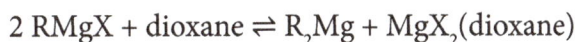

$$2\ RMgX + \text{dioxane} \rightleftharpoons R_2Mg + MgX_2(\text{dioxane})$$

This reaction is known as the Schlenk equilibrium.

Testing Grignard Reagents

Because Grignard reagents are so sensitive to moisture and oxygen, many methods have been developed to test the quality of a batch. Typical tests involve titrations with weighable, anhydrous protic reagents, e.g. menthol in the presence of a color-indicator. The interaction of the Grignard reagent with phenanthroline or 2,2'-bipyridine causes a color change.

Initiation

Many methods have been developed to initiate sluggish Grignard reactions. These methods weaken the passivating layer of MgO, thereby exposing highly reactive magnesium to the organic halide. Mechanical methods include crushing of the Mg pieces in situ, rapid stirring, and sonication of the suspension. Iodine, methyl iodide, and 1,2-Dibromoethane are common activating agents. The use of 1,2-dibromoethane is particularly advantageous as its action can be monitored by the observation of bubbles of ethylene. Furthermore, the side-products are innocuous:

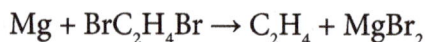

$$Mg + BrC_2H_4Br \rightarrow C_2H_4 + MgBr_2$$

The amount of Mg consumed by these activating agents is usually insignificant. A small amount of mercuric chloride will amalgamate the surface of the metal, allowing it to react.

Industrial Production

Grignard reagents are produced in industry for use *in situ*, or for sale. As with a bench-scale, the main problem is that of initiation; a portion of a previous batch of Grignard reagent is often used as the initiator. Grignard reactions are exothermic, and this exothermicity must be considered when a reaction is scaled-up from laboratory to production plant.

Many Grignard reagents such as methylmagnesium bromide, methylmagnesium chloride, phenylmagnesium bromide, and allylmagnesium bromide are available commercially as tetrahydrofuran or diethyl ether solutions.

Mg Transfer Reaction (Halogen–Mg Exchange)

An alternative preparation of Grignard reagents involves transfer of Mg from a preformed Grignard reagent to an organic halide. This method offers the advantage that the Mg transfer tolerates many functional groups. A typical reaction involves isopropylmagnesium chloride and aryl bromide or iodides.

Reactions of Grignard Reagents

With Carbonyl Compounds

Grignard reagents react with a variety of carbonyl derivatives.

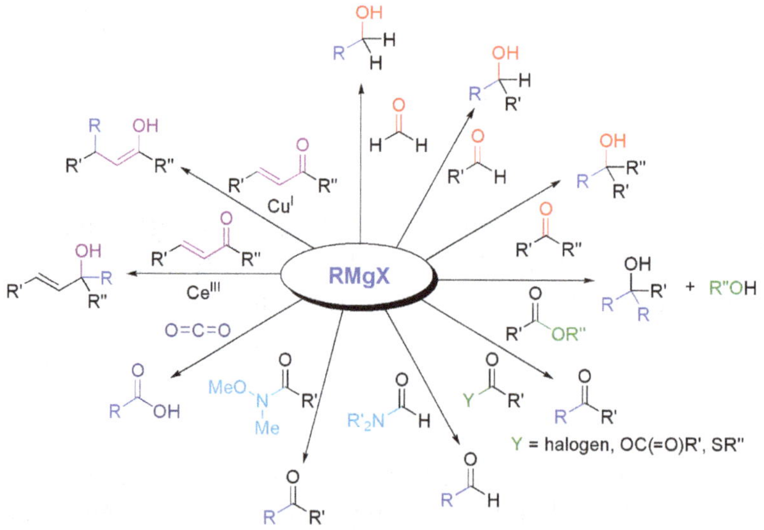

The most common application of Grignard reagents is the alkylation of aldehydes and ketones, i.e. the Grignard reaction:

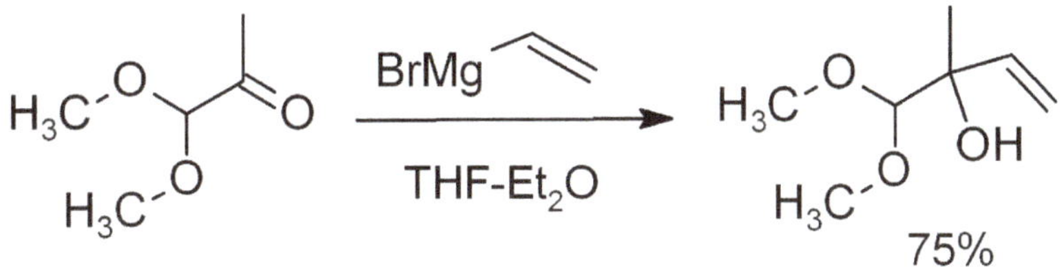

Note that the acetal function (a protected carbonyl) does not react.

Such reactions usually involve an aqueous acidic workup, though this step is rarely shown in reaction Figure. In cases where the Grignard reagent is adding to an aldehyde or a prochiral ketone, the Felkin-Anh model or Cram's Rule can usually predict which stereoisomer will be formed. With easily deprotonated 1,3-diketones and related acidic substrates, the Grignard reagent RMgX functions merely as a base, giving the enolate anion and liberating the alkane RH.

Grignard reagents are nucleophiles in nucleophilic aliphatic substitutions for instance with alkyl halides in a key step in industrial Naproxen production:

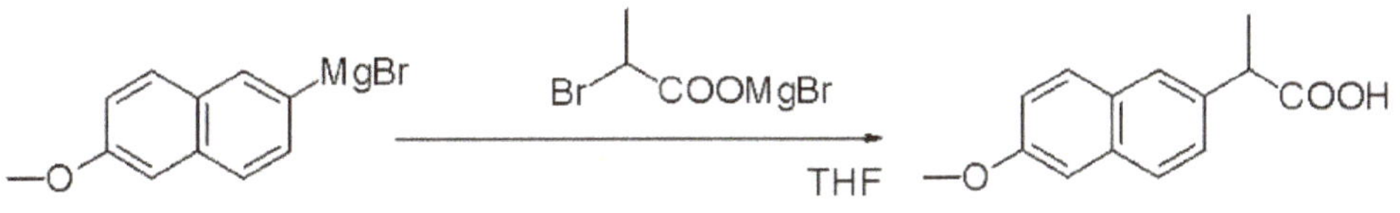

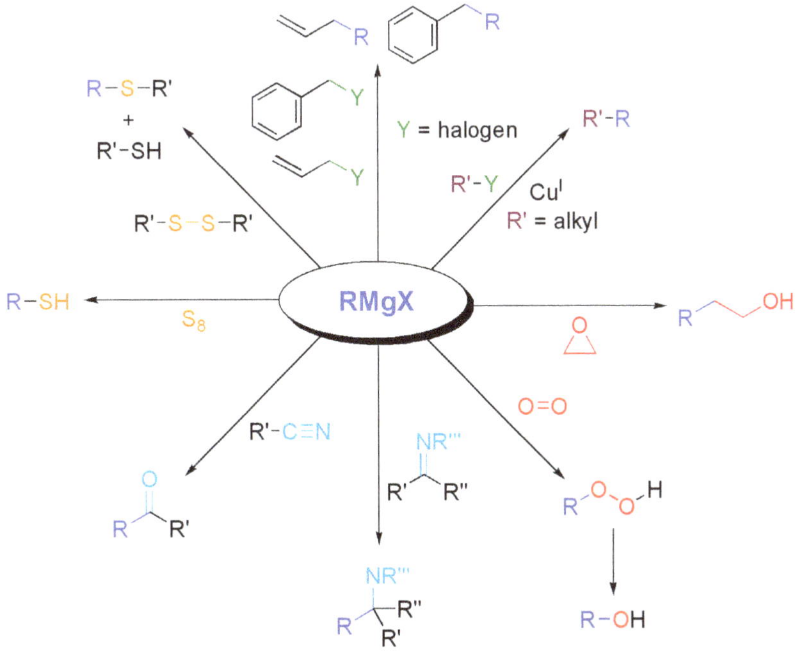

Reactions as a Base

Grignard reagents serve as a base for protic substrates. Grignard reagents are basic and react with alcohols, phenols, etc. to give alkoxides (ROMgBr). The phenoxide derivative is susceptible to formylation paraformaldehyde to give salicylaldehyde.

Formation of Bonds to B, Si, P, Sn

Like organolithium compounds, Grignard reagents are useful for forming carbon–heteroatom bonds.

$$R_4B^-$$

$$\uparrow Et_2O.BF_3 \text{ or } NaBF_4$$

$$Ph_2PR \xleftarrow{\ Ph_2PCl\ } RMgX \xrightarrow{\ Bu_3SnCl\ } Bu_3SnR$$

$$\downarrow B(OMe)_3$$

$$RB(OMe)_2$$

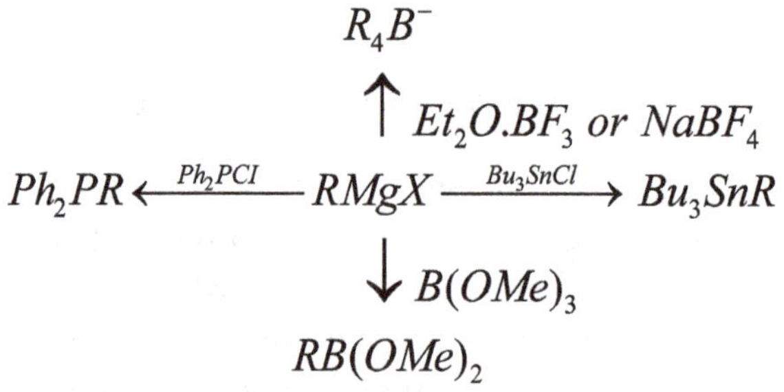

Reaction with Transition Metal Halides

Grignard reagents react with many metal-based electrophiles. For example, they undergo trans-metallation with cadmium chloride ($CdCl_2$) to give dialkylcadmium:

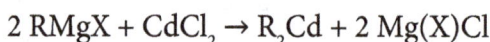

$$2\ RMgX + CdCl_2 \rightarrow R_2Cd + 2\ Mg(X)Cl$$

Dialkylcadmium reagents are used for preparation of ketones from acyl halides:

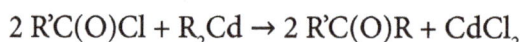

$$2\ R'C(O)Cl + R_2Cd \rightarrow 2\ R'C(O)R + CdCl_2$$

With Organic Halides

Grignard reagents do *not* typically react with organic halides, in contrast with their high reactivity with other main group halides. In the presence of metal catalysts, however, Grignard reagents participate in C-C coupling reactions. For example, nonylmagnesium bromide reacts with methyl *p*-chlorobenzoate to give *p*-nonylbenzoic acid, in the presence of Tris(acetylacetonato)iron(III) (Fe(acac)$_3$), after workup with NaOH to hydrolyze the ester, shown as follows. Without the Fe(a-cac)$_3$, the Grignard reagent would attack the ester group over the aryl halide.

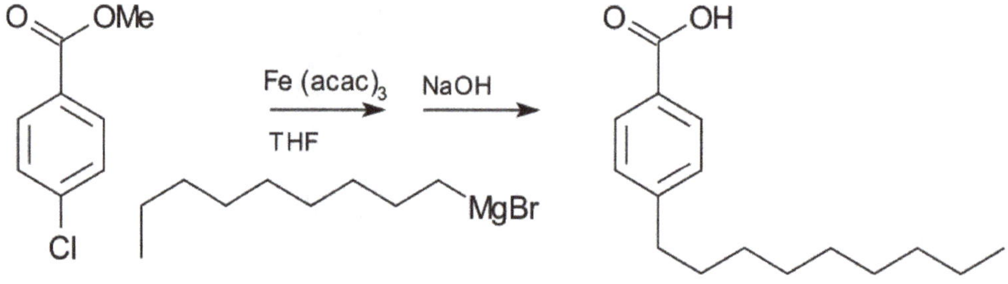

For the coupling of aryl halides with aryl Grignard reagents, nickel chloride in tetrahydrofuran (THF) is also a good catalyst. Additionally, an effective catalyst for the couplings of alkyl halides is dilithium tetrachlorocuprate (Li_2CuCl_4), prepared by mixing lithium chloride (LiCl) and copper(II) chloride ($CuCl_2$) in THF. The Kumada-Corriu coupling gives access to [substituted] styrenes.

Oxidation

Treatment of a Grignard reagent with oxygen gives the magnesium organoperoxide. Hydrolysis of this material yields hydroperoxides or alcohol. These reactions involve radical intermediates.

$$R-MgX+O_2 \rightarrow R.+O_2^{\cdot-}+MgX^+ \rightarrow R-O-O-MgX \quad +H_3O^+ \quad \rightarrow R-O-O-H \quad +HO-MgX+H^+$$

$$\downarrow R-MgX$$

$$R-O-MgX \quad +H_3O^+ \quad \rightarrow R-O-H \quad +HO-MgX+H^+$$

The simple oxidation of Grignard reagents to give alcohols is of little practical import as yields are generally poor. In contrast, two-step sequence via a borane (*vide supra*) that is subsequently oxidized to the alcohol with hydrogen peroxide is of synthetic utility.

The synthetic utility of Grignard oxidations can be increased by a reaction of Grignard reagents with oxygen in presence of an alkene to an ethylene extended alcohol. This modification requires aryl or vinyl Grignards. Adding just the Grignard and the alkene does not result in a reaction demonstrating that the presence of oxygen is essential. The only drawback is the requirement of at least two equivalents of Grignard although this can partly be circumvented by the use of a dual Grignard system with a cheap reducing Grignard such as n-butylmagnesium bromide.

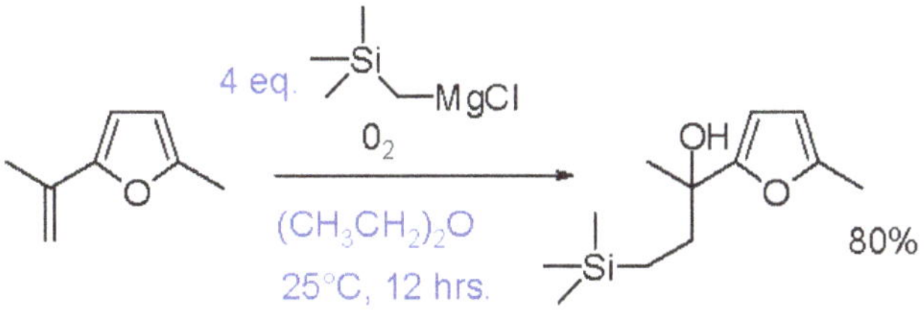

Elimination

In the Boord olefin synthesis, the addition of magnesium to certain β-haloethers results in an elimination reaction to the alkene. This reaction can limit the utility of Grignard reactions.

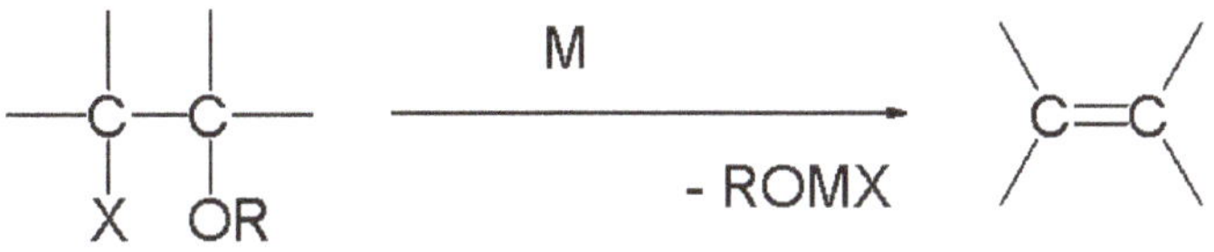

Degradation of Grignard Reagents

At one time, the formation and hydrolysis of Grignard reagents was used in the determination of the number of halogen atoms in an organic compound. In modern usage Grignard degradation is used in the chemical analysis of certain triacylglycerols.

Industrial use

An example of the Grignard reaction is a key step in the (non-stereospecific) industrial production of Tamoxifen (currently used for the treatment of estrogen receptor positive breast cancer in women)

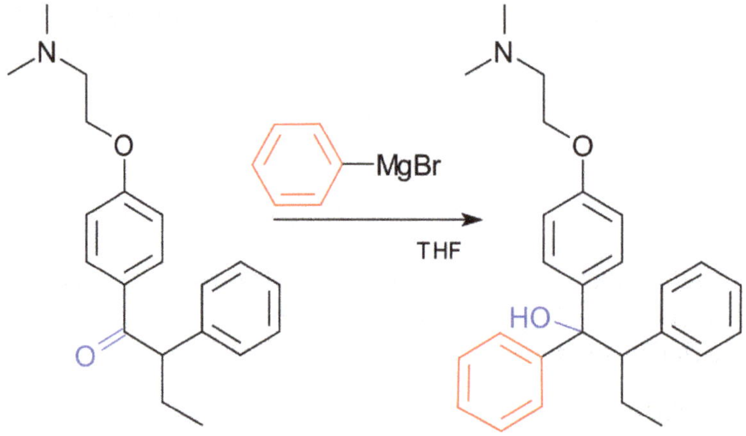

Reactions of Organozinc Reagents

The asymmetric conjugate addition of dialkylzinc to prochiral α,β -unsaturated compounds is one of the powerful methods for carbon-carbon bond formation in organic synthesis. Much attention has been made on the development of new ligands for this reaction. Phosphoramidite ligand from BINOL L-4 has been found to be effective for the conjugate addition to cyclic substrates with up to 98% ee.

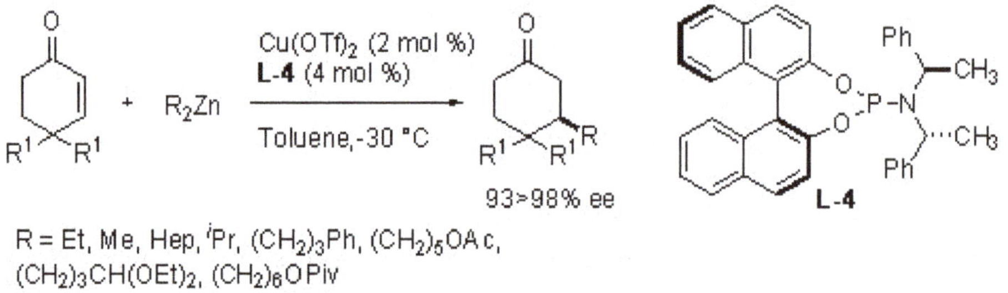

Subsequently, copper(I)-catalyzed enantioselective addition of dialkylzinc to 3-nitroacrolein derivatives has been demonstrated using phosphoramidite ligands L-5 and L-6 with up to 98% ee.

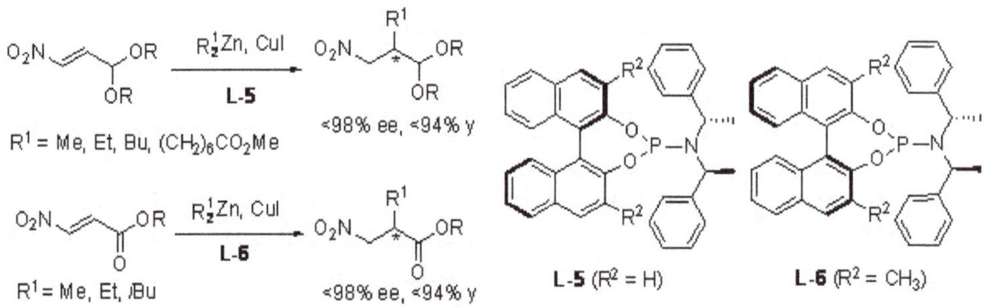

Figure summarizes some of the peptide based ligands for the dialkylzinc addition to α,β -unsaturated compounds. For example, the copper-catalyzed conjugate addition of dialkylzinc reagents to acyclic aliphatic α,β -unsaturated ketones proceed in the presence of L-9 with up to 94% ee, while the reaction using L-10 gives up to 98% ee.

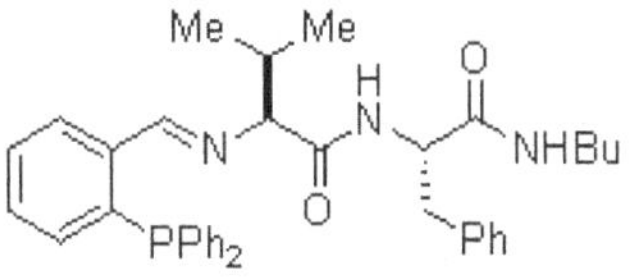

L-7

Ligand for conjugate addition to cyclic disubstituted enones

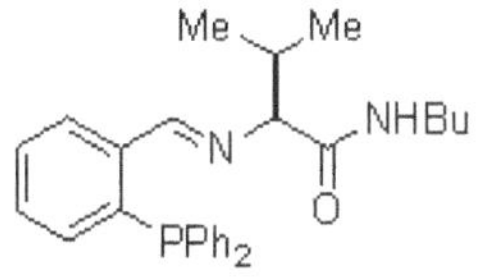

L-8

Ligand for conjugate addition to cyclic trisubstituted enones

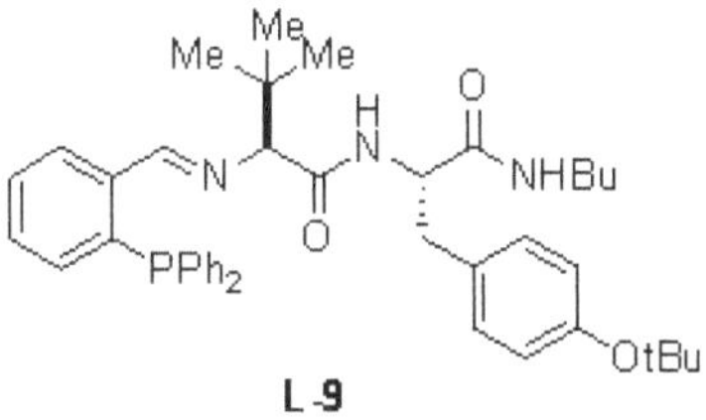

L-9

Ligand for conjugate addition to acyclic disubstituted enones

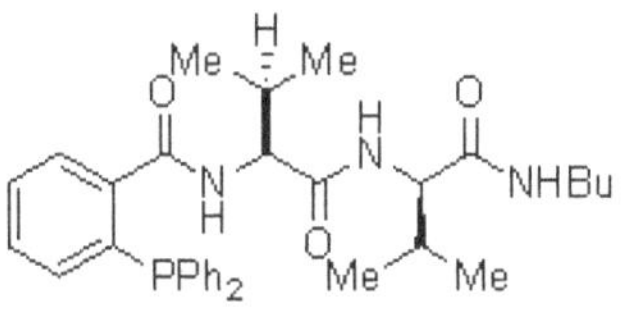

L-10

Ligand for conjugate addition to unsaturated N-acryloxazolidinones

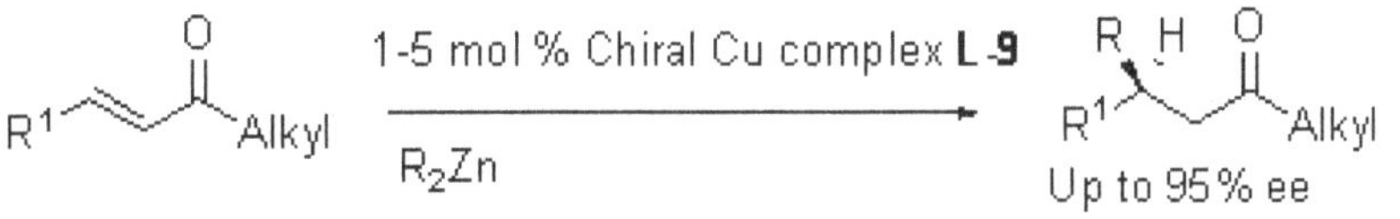

R^1 = Ph, *p*-OMePh, *p*-NO_2Ph, *p*-CF_3Ph, Me, *n*-pent, *i*-Pr, $(CH_2)_3$OAC
Alkyl = Me, *n*-hex, *i*-Pr, *t*-Bu
R = Me, Et

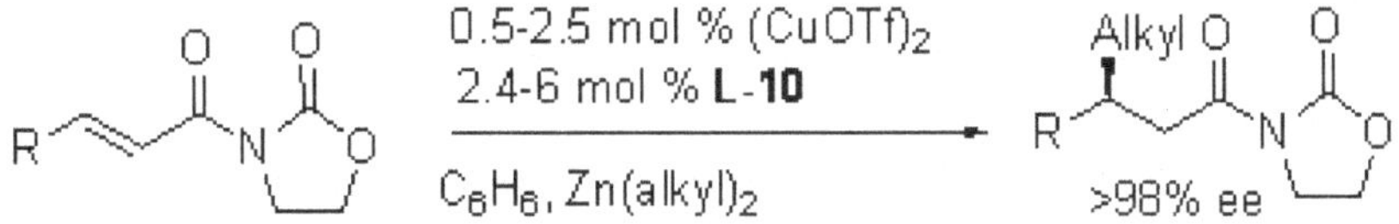

R = Me, *n*-Pr, $(CH_2)_3$OTBS, *i*-Pr
Alkyl = Et, Me, *i*-Pr, *i*-Pr$(CH_2)_3$

Later, the chiral ligands L-10 to L-12 have been studied for the reactions of dialkylzinic reagents to heterocyclic enones such as furanones, pyranones and their derivatives.

Chiral phosphanes:

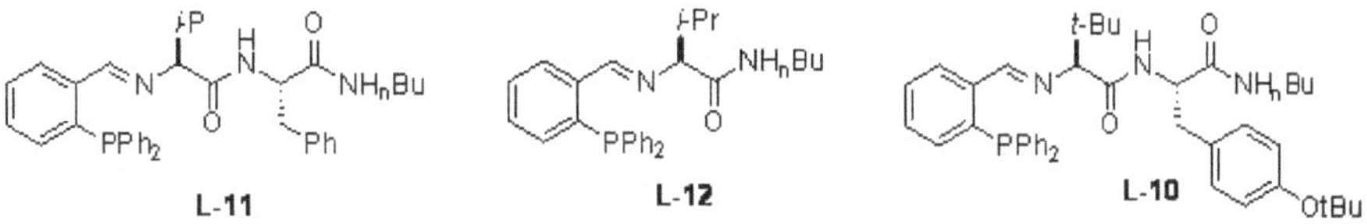

L-11 L-12 L-10

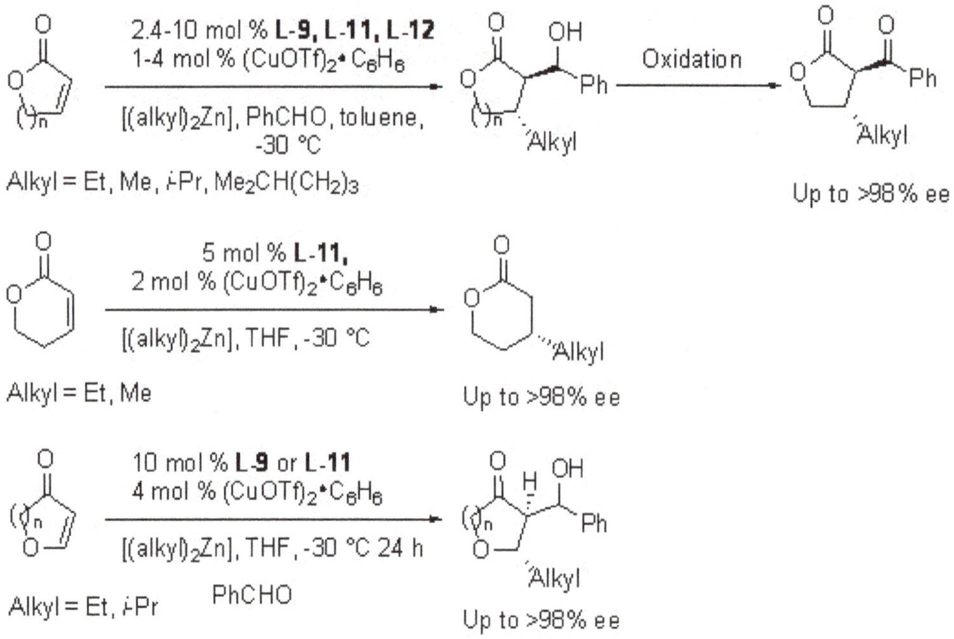

Organolithium Reagent

Organolithium reagents are organometallic compounds that contain carbon – lithium bonds. They are important reagents in organic synthesis, and are frequently used to transfer the organic group or the lithium atom to the substrates in synthetic steps, through nucleophilic addition or simple deprotonation. Organolithium reagents are used in industry as an initiator for anionic polymerization, which leads to the production of various elastomers. They have also been applied in asymmetric synthesis in the pharmaceutical industry.

Due to the large difference in electronegativity between the carbon atom and the lithium atom, the C-Li bond is highly ionic. This extremely polar nature of the C-Li bond makes organolithium reagents good nucleophiles and strong bases. For laboratory organic synthesis, many organolithium reagents are commercially available in solution form. These reagents are highly reactive, and are sometimes pyrophoric.

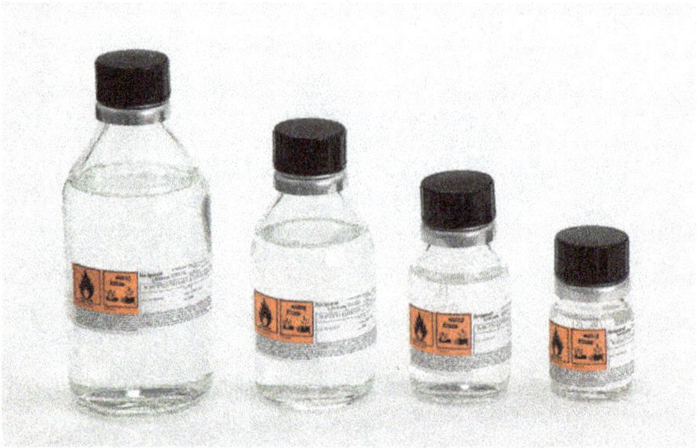

Glass bottles containing butyllithium

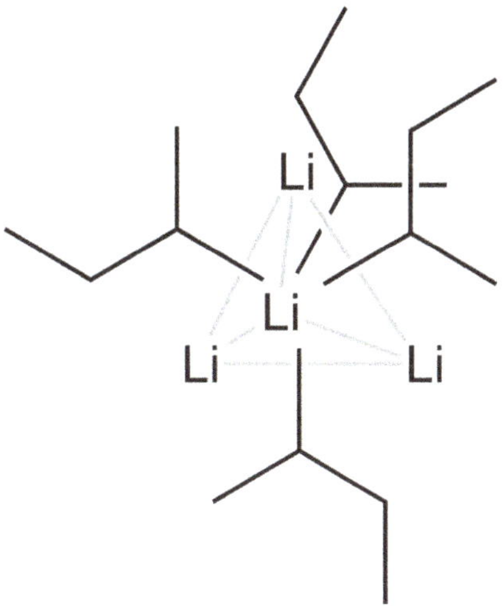

sec-Butyllithium aggregate

History and Development

Studies of organolithium reagents began in the 1930 and were pioneered by Karl Ziegler, Georg Wittig, and Henry Gilman. These chemists found that in comparison with Grignard reagents, organolithium reagents can often perform the same reactions with increased rates and higher yields, such as in the case of metalation. Since then, organolithium reagents have surpassed Grignard reagents in usage. Ongoing research focuses on the nature of carbon-lithium bonding, structural studies of organolithium aggregates, chiral organolithium reagents and asymmetric synthesis, and the role of organolithium reagents in the preparation of new organometallic species.

Structure

Although simple alkyllithium species are often represented as monomer RLi, they exist as aggregates (oligomers) or polymers. Their structures depend on the nature of organic substituent and the presence of other ligands. These structures have been elucidated by a variety of methods, notably ^{6}Li, ^{7}Li, and ^{13}C NMR spectroscopy and X-ray diffraction analysis. Computational chemistry supports these assignments.

Nature of Carbon-lithium Bond

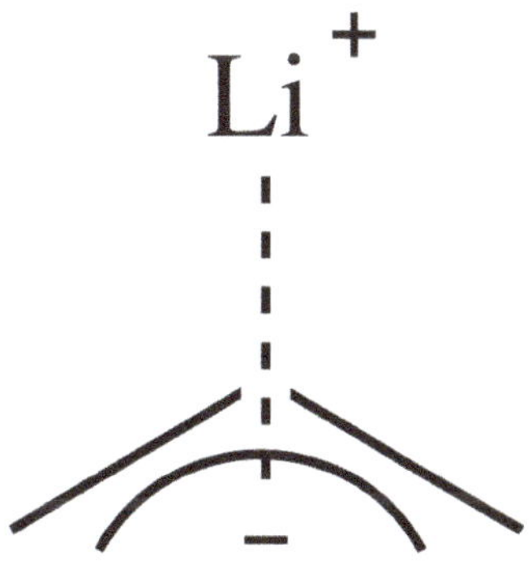

Delocalized electron density in allyllithium reagents

The relative electronegativities of carbon and lithium suggests that the C-Li bond will be highly polar. However, certain organolithium compounds possess properties such as solubility in nonpolar solvents that complicate the issue. While most data suggest the C-Li bond to be essentially ionic, there has been debate as to whether a small covalent character exists in the C-Li bond.

In allyl lithium compounds, the lithium cation coordinates to the face of the carbon π bond in an η-$_3$ fashion instead of a localized, carbanionic center, thus, allyllithiums are often less aggregated than alkyllithiums. In aryllithium complexes, the lithium cation coordinates to a single carbanion center through a Li-C σ type bond.

Solid state structures of methyllithium tetramers, *n*-butyllithium hexamers and polymeric ladder of phenyllithium

Solid State Structure

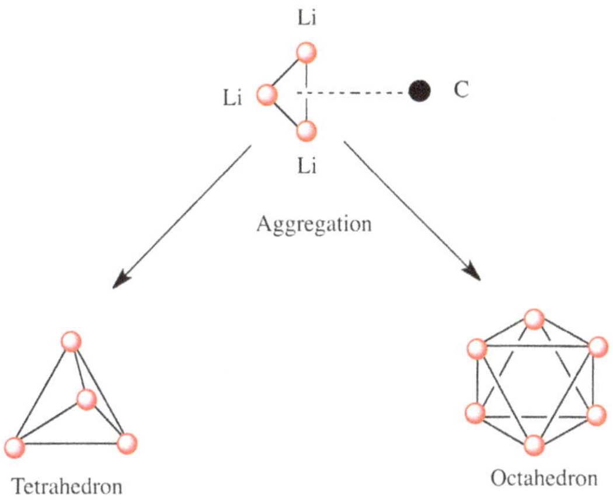

Tetrahedron and octahedron metal cores formed by aggregation of the Li3 triangle - carbanion coordinate complex

Like other species consisting of polar subunits, organolithium species aggregate. Formation of aggregates is influenced by electrostatic interactions, the coordination between lithium and surrounding solvent molecules or polar additives, and steric effects.

A basic building block toward constructing more complex structures is a carbanionic center interacting with a Li_3 triangle in an η- $_3$ fashion. In simple alkyllithium reagents, these triangles aggregate to form tetrahedron or octahedron structures. For example, methyllithium, ethyllithium and *tert*-butyllithium all exist in the tetramer $[RLi]_4$. Methyllithium exists as tetramers in a cubane-type cluster in the solid state, with four lithium centers forming a tetrahedron. Each methanide in the tetramer in methyllithium can have agostic interaction with lithium cations in adjacent tetramers. Ethyllithium and *tert*-butyllithium, on the other hand, do not exhibit this interaction, and are thus soluble in non-polar hydrocarbon solvents. Another class of alkyllithium adopts hexameric structures, such as *n*-butyllithium, isopropyllithium, and cyclohexanyllithium.

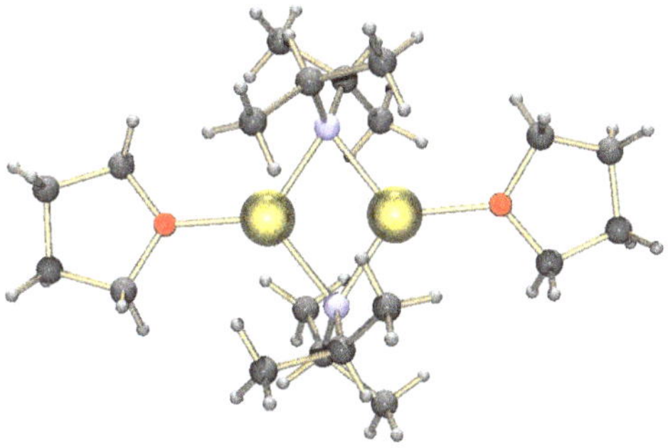

LDA dimer with THF coordinated to Li cations

Common lithium amides, e.g. lithium bis(trimethylsilyl)amide, and lithium diisopropylamide are also subject to aggregation. Lithium amides adopt polymeric-ladder type structures in non-coordinating solvent in the solid state, and they generally exist as dimers in ethereal solvents. In the presence of strongly donating ligands, tri- or tetramerc lithium centers are formed. For example, LDA exists primarily as dimers in THF. The structures of common lithium amides, such as lithium diisopropylamide (LDA) and lithium hexamethyldisilazide (LiHMDS) have been extensively studied by Collum and coworkers using NMR spectroscopy. Another important class of reagents is silyllithiums, extensively used in the synthesis of organometallic complexes and polysilane dendrimers. In the solid state, in contrast with alkyllithium reagents, most silyllithiums tend to form monomeric structures coordinated with solvent molecules such as THF, and only a few silyllithiums have been characterized as higher aggregates. This difference can arise from the method of preparation of silyllithiums, the steric hindrance caused by the bulky alkyl substituents on silicon, and the less polarized nature of Si-Li bonds.The addition of strongly donating ligands, such as TMEDA and (-)-sparteine, can displace coordinating solvent molecules in silyllithiums.

Solution Structure

Relying solely on the structural information of organolithium aggregates obtained in the solid state from crystal structures has certain limits, as it is possible for organolithium reagents to adopt different structures in reaction solution environment. Also, in some cases the crystal structure of an organolithium species can be difficult to isolate. Therefore, studying the structures of organolithium reagents, and the lithium-containing intermediates in solution form is extremely useful in understanding the reactivity of these reagents. NMR spectroscopy has emerged as a powerful tool for the studies of organolithium aggregates in solution. For alkyllithium species, C-Li J coupling can often used to determine the number of lithium interacting with a carbanion center, and whether these interactions are static or dynamic. Separate NMR signals can also differentiate the presence of multiple aggregates from a common monomeric unit.

The structures of organolithium compounds are affected by the presence of Lewis bases such as e.g., tetrahydrofuran (THF), diethyl ether (Et_2O), tetramethylethylene diamine (TMEDA) or hexamethylphosphoramide (HMPA). Methyllithium is a special case, in which solvation with ether, or polar additive HMPA does not deaggregate the tetrameric structure in the solid state. On the other hand, THF deaggregates hexameric butyl lithium: the tetramer is the major species, and ΔG for interconversion between tetramer and dimer is around 11 kcal/mol. TMEDA can also chelate to the lithium cations in n-butyllithium and form solvated dimers such as $[(TMEDA)\ LiBu\text{-}n)]_2$. Phenyllithium has been shown to exist as a distorted tetramer in the crystallized ether solvate, and as a mixture of dimer and tetramer in ether solution.

Aggregates of some alkyllithiums in solvents		
	Solvent	Structure
methyllithium	THF	tetramer
methyllithium	ether/HMPA	tetramer
n-butyllithium	pentane	hexamer
n-butyllithium	ether	tetramer
n-butyllithium	THF	tetramer-dimer
sec-butyllithium	pentane	hexamer-tetramer
isopropyllithium	pentane	hexamer-tetramer
tert-butyllithium	pentane	tetramer
tert-butyllithium	THF	monomer
phenyllithium	ether	tetramer-dimer
phenyllithium	ether/HMPA	dimer

Structure and Reactivity

As the structures of organolithium reagents change according to their chemical environment, so do their reactivity and selectivity. One question surrounding the structure-reactivity relationship is whether there exists a correlation between the degree of aggregation and the reactivity of organolithium reagents. It was originally proposed that lower aggregates such as monomers are more reactive in alkyllithiums. However, reaction pathways in which dimer or other oligomers are the reactive species have also been discovered, and for lithium amides such as LDA, dimer-based reactions are common. A series of solution kinetics studies of LDA - mediated reactions suggest that lower aggregates of enolates do not necessarily lead to higher reactivity.

Also, some Lewis bases increase reactivity of organolithium compounds. However, whether these additives function as strong chelating ligands, and how the observed increase in reactivity relates to structural changes in aggregates caused by these additives are not always clear. For example, TMEDA increases rates and efficiencies in many reactions involving organolithium reagents. Toward alkyllithium reagents, TMEDA functions as a donor ligand, reduces the degree of aggregation, and increases the nucleophilicity of these species. However, TMEDA does not always function as a donor ligand to lithium cation, especially in the presence of anionic oxygen and nitrogen centers. For example, it only weakly interacts with LDA and LiHMDS even in hydrocarbon solvents with no competing donor ligands. In imine lithiation, while THF acts as a strong donating ligand to LiHMDS, the weakly coordinating TMEDA readily dissociates from LiHMDS, leading to the formation of LiHMDS dimers that is the more reactive species. Thus, in the case of LiHMDS, TMEDA does not increase reactivity by reducing aggregation state. Also, as opposed to simple alkyllithium compounds, TMEDA does not deaggregate lithio-acetophenolate in THF solution. The addition of HMPA to lithium amides such as LiHMDS and LDA often results in a mixture of dimer/monomer aggregates in THF. However, the ratio of dimer/monomer species does not change with increased concentration of HMPA, thus, the observed increase in reactivity is not the result of deaggregation. The mechanism of how these additives increase reactivity is still being researched.

Reactivity and Applications

The C-Li bond in organolithium reagents is highly polarized. As a result, the carbon attracts most of the electron density in the bond and resembles a carbanion. Thus, organolithium reagents are strongly basic and nucleophilic. Some of the most common applications of organolithium reagents in synthesis include their use as nucleophiles, strong bases for deprotonation, initiator for polymerization, and starting material for the preparation of other organometallic compounds.

Organolithium Reagent as Nucleophile

Carbolithiation Reactions

As nucleophiles, organolithium reagents undergo carbolithiation reactions, whereby the carbon-lithium bond adds across a carbon-carbon double or triple bond, forming new organolithium species. This reaction is the most widely employed reaction of organolithium compounds. Carbolithiation is key in anionic polymerization processes, and *n*-butyllithium is used as a catalyst to initiate the polymerization of styrene, butadiene, or isoprene or mixtures thereof.

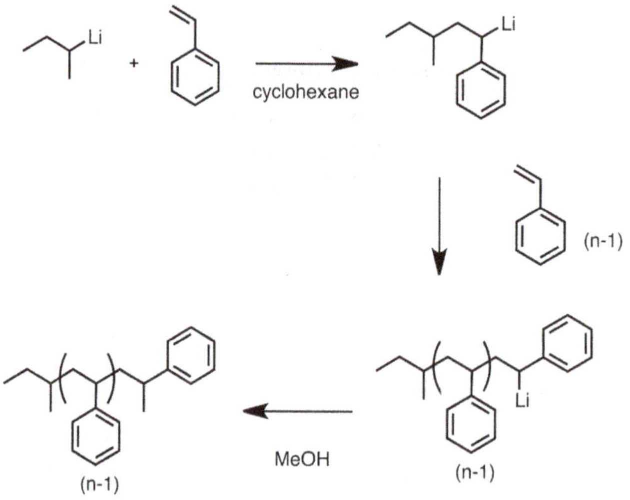

Another application that takes advantage of this reactivity is the formation of carbocyclic and heterocyclic coumpounds via intramolecular carbolithiation. As a form of anionic cyclization, intramolecular carbolithiation reactions offer several advantages over radical cyclization. First, it is possible for the product cyclic organolithium species to react with electrophiles, whereas it is often difficult to trap a radical intermediate of the corresponding structure. Secondly, anionic cyclizations are often more regio- and stereospecific than radical cyclization, particularly in the case of 5-hexenyllithiums. Intramolecular carbolithiation allows addition of the alkyl-, vinyllithium to triple bonds and mono-alkyl substituted double bonds. Aryllithiums can also undergo addition if a 5-membered ring is formed. The limitations of intramolecular carbolithiation include difficulty of forming 3 or 4-membered rings, as the intermediate cyclic organolithium species often tend to undergo ring-openings. Below is an example of intramolecular carbolithiation reaction. The lithium species derived from the lithium-halogen exchange cyclized to form the vinyllithium through 5-exo-trig ring closure. The vinyllithium species further reacts with electrophiles and produce functionalized cyclopentylidene compounds.

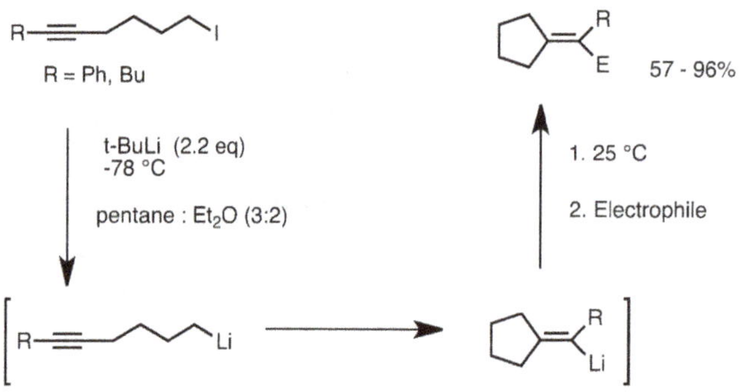

Addition to Carbonyl Compounds

Nucleophilic organolithium reagents can add to electrophilic carbonyl double bonds to form carbon-carbon bonds. They can react with aldehydes and ketones to produce alcohols. The addition proceeds mainly via polar addition, in which the nucleophilic organolithium species attacks from the equatorial direction, and produces the axial alcohol. Addition of lithium salts such as $LiClO_4$ can improve the stereoselectivity of the reaction.

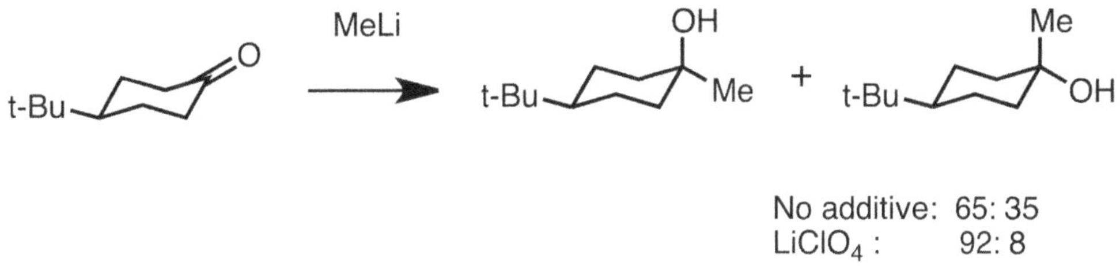

When the ketone is sterically hindered, using Grignard reagents often leads to reduction of the carbonyl group instead of addition. However, alkyllithium reagents are less likely to reduce the ketone, and may be used to synthesize substituted alcohols. Below is an example of ethyllithium addition to adamantone to produce tertiary alcohol.

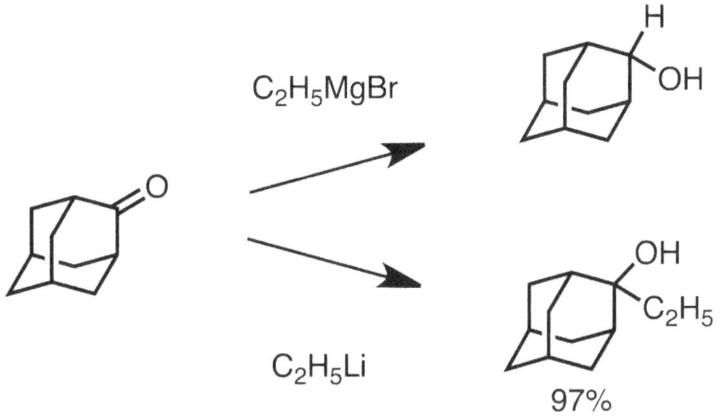

Organolithium reagents are also superior to Grignard reagents in their ability to react with carboxylic acids to form ketones. This reaction can be optimized by carefully controlling the amount of organolithium reagent addition, or using trimethylsilyl chloride to quench excess lithium reagent. A more common way to synthesize ketones is through the addition of organolithium reagents to Weinreb amides (N-methoxy-N-methyl amides). This reaction provides ketones when the organolithium reagents is used in excess, due to chelation of the lithium ion between the N-methoxy oxygen and the carbonyl oxygen, which forms a tetrahedral intermediate that collapses upon acidic work up.

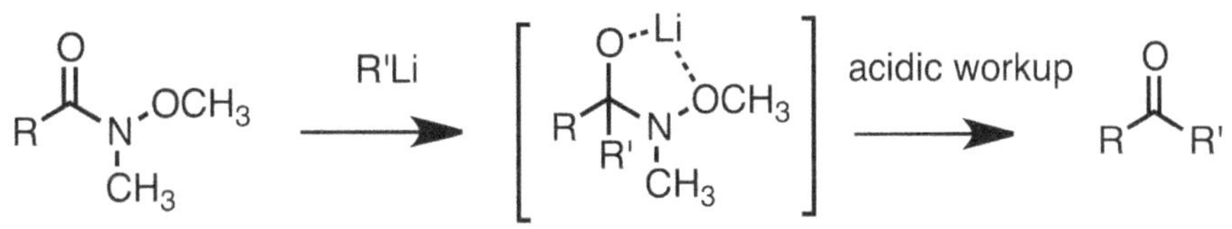

Organolithium reagents can also react with carbon dioxide to form carboxylic acids.

In the case of enone substrates, where two sites of nucleophilic addition are possible (1,2 addition to the carbonyl carbon or 1,4 conjugate addition to the β carbon), most highly reactive organolithium species favor the 1,2 addition, however, there are several ways to propel organolithium reagents to undergo conjugate addition. First, since the 1,4 adduct is the likely to be the more thermodynamically favorable species, conjugate addition can be achieved through equilibration (isomerization of the two product), especially when the lithium nucleophile is weak and 1,2 addition is reversible.

Secondly, adding donor ligands to the reaction forms heteroatom-stabilized lithium species which favors 1,4 conjugate addition. In one example, addition of low-level of HMPA to the solvent favors the 1,4 addition. In the absence of donor ligand, lithium cation is closely coordinated to the oxygen atom, however, when the lithium cation is solvated by HMPA, the coordination between carbonyl oxygen and lithium ion is weakened. This method generally cannot be used to affect the regioselectivity of alkyl- and aryllithium reagents.

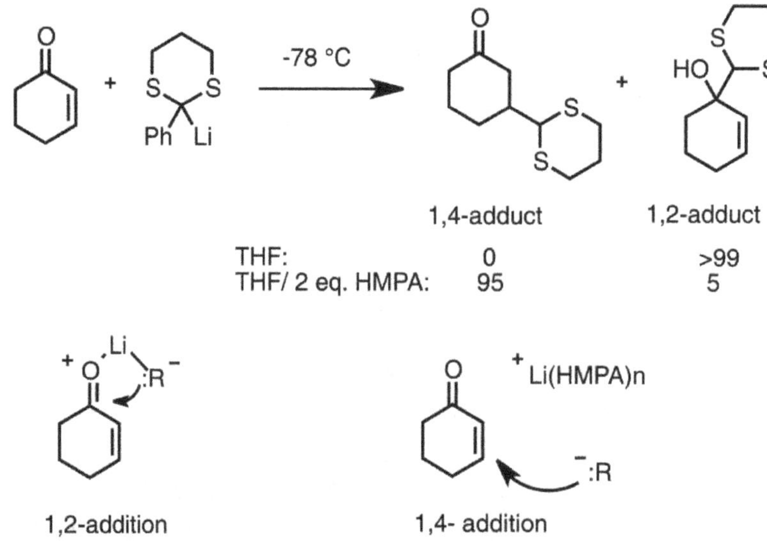

Proposed mechanism for 1,2-and 1,4-addition in the presence of HMPA

Organolithium reagents can also perform enantioselective nucleophilic addition to carbonyl and its derivatives, often in the presence of chiral ligands. This reactivity is widely applied in the industrial syntheses of pharmaceutical compounds. An example is the Merck and Dupont synthesis of Efavirenz, a potent HIV reverse transcriptase inhibitor. Lithium acetylide is added to a prochiral ketone to yield a chiral alcohol product. The structure of the active reaction intermediate was determined by NMR spectroscopy studies in the solution state and X-ray crystallography of the solid state to be a cubic 2:2 tetramer.

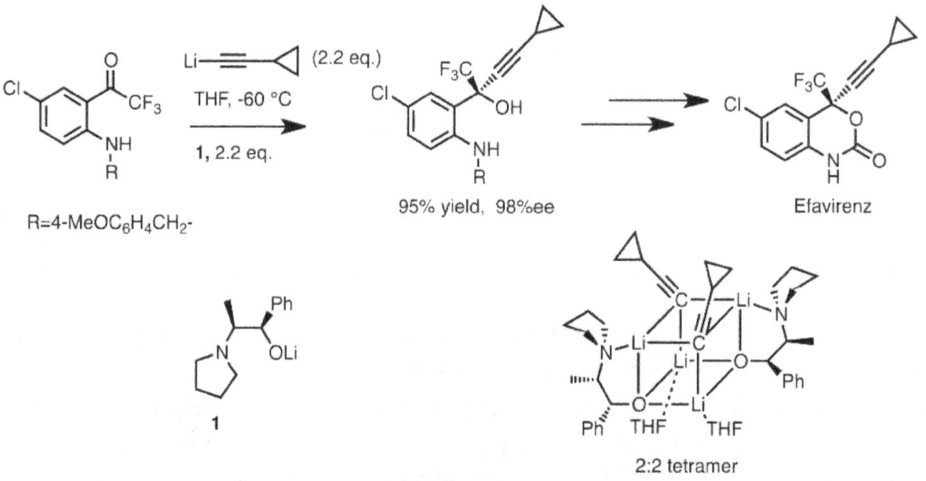

Enantioselective additor of lithium acetylide in the synthesis of HIV drug Efavirenz

S_N2 Type Reactions

Organolithium reagents can serve as nucleophiles and carry out S_N2 type reactions with alkyl or allylic halides. Although they are considered more reactive than Grignards reactions in alkylation, their use is still limited due to competing side reactions such as radical reactions or metal-halogen exchange. Most organolithium reagents used in S_N2 alkylations are more stabilized, less basic, and less aggregated, such as heteroatom stabilized, aryl- or allyllithium reagents. HMPA has been shown to increase reaction rate and product yields, and the reactivity of aryllithium reagents is often enhanced by the addition of potassium alkoxides. Organolithium reagents can also carry out nucleophilic attacks with epoxides to form alcohols.

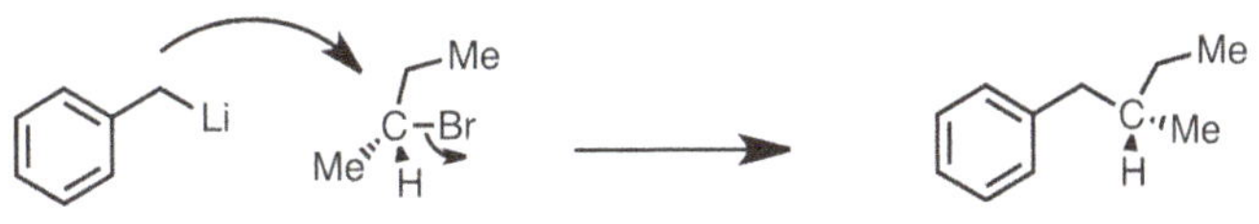

Alkylation of benzyllithium reagents with secondary alkylbmocle proceccd with inversion

Organolithium Reagent as Base

Organolithium reagents provide a wide range of basicity. *tert*-Butyllithium, with three weakly electron donating alkyl groups, is the strongest base commercially available (pKa = 53). As a result, the acidic protons on -OH, -NH and -SH are often protected in the presence of organolithium reagents. Some commonly used lithium bases are alkyllithium species such as *n*-butyllithium and lithium dialkylamides ($LiNR_2$). Reagents with bulky R groups such as lithium diisopropylamide (LDA) and lithium bis(trimethylsilyl)amide (LiHMDS) are often sterically hindered for nucleophilic addition, and are thus more selective toward deprotonation. Lithium dialkylamides ($LiNR_2$) are widely used in enolate formation and aldol reaction. The reactivity and selectivity of these bases are also influenced by solvents and other counter ions.

Metalation

Metalation with organolithium reagents, also known as lithiation or lithium-hydrogen exchange, is achieved when an organolithium reagent, most commonly an alkyllithium, abstracts a proton and forms a new organolithium species.

$$R-H+R'Li \rightarrow RLi+R'H$$

Common metalation reagents are the butyllithiums. *tert*-Butyllithium and *sec*-butyllithium are generally more reactive and have better selectivity than *n*-butyllithium, however, they are also more expensive and difficult to handle. Metalation is a common way of preparing versatile organolithium reagents. The position of metalation is mostly controlled by the acidity of the C-H bond. Lithiation often occurs at a position α to electron withdrawing groups, since they are good at stabilizing the electron-density of the anion. Directing groups on aromatic compounds and heterocycles provide regioselective sites of metalation, and directed ortho metelation is an important class of metalation reactions. Metalated sulfones, acyl groups and α-metalated amides are important intermediates in chemistry synthesis. Metalation of allyl ether with alkyllithium or LDA forms an anion α to the oxygen, and can proceed to 2,3-Wittig rearrangement. Addition of donor ligands such as TMEDA and HMPA can increase metalation rate and broaden substrate scope. Chiral organolithium reagents can be accessed through asymmetric metalation.

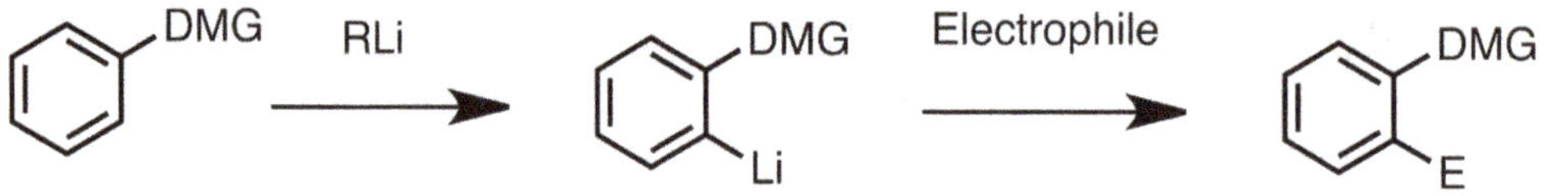

Directed ortho metalation is an important tool in the synthesis of regiospecific substituted aromatic compounds. This approach to lithiation and subsequent quenching of the intermediate lithium species with electrophile is often superior to electrophilic aromatic substitution due to its high regioselectivity. This reaction proceeds through deprotonation by organolithium reagents at the positions α to the direct metalation group (DMG) on the aromatic ring. The DMG is often a functional group containing a heteroatom that is Lewis basic, and can coordinate to the Lewis-acidic lithium cation. This generates a complex-induced proximity effect, which directs deprotonation at the α position to form an aryllithium species that can further react with electrophiles. Some of the most effective DMGs are amides, carbamates, sulfones and sulfonamides. They are strong electron-withdrawing groups that increase the acidity of alpha-protons on the aromatic ring. In the presence of two DMGs, metalation often occurs ortho to the stronger directing group, though mixed products are also observed. A number of heterocycles that contain acidic protons can also undergo ortho-metalation. However, for electron-poor heterocycles, lithium amide bases such as LDA are generally used, since alkyllithium has been observed to perform addition to the electron-poor heterocycles rather than deprotonation. In certain transition metal-arene complexes, such as ferrocenes, the transition metal attracts electron-density from the arene, thus rendering the aromatic protons more acidic, and ready for ortho-metalation.

Superbases

Addition of potassium alkoxide to alkyllithium greatly increases the basicity of organolithium species. The most common "superbase" can be formed by addition of KOtBu to butyllithium, often abbreviated as "LiCKOR"reagents. These "superbases" are highly reactive and often stereoselective reagents. In the example below, the LiCKOR base generates a stereospecific crotylboronate species through metalation and subsequent lithium-metalloid exchange.

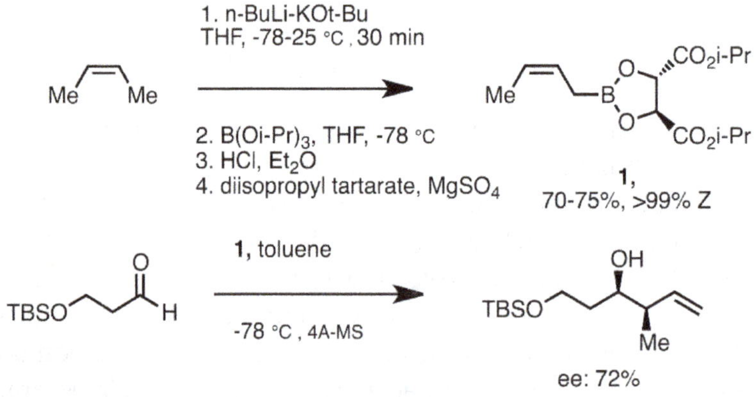

Superbase used to prepare (Z)-crotylboronate roagents

Asymmetric Metalation

Enantioenriched organlithium species can be obtained through asymmetric metalation of prochiral substrates. Asymmetric induction requires the presence of a chiral ligand such as (-)-sparteine. The enantiomeric ratio of the chiral lithium species is often influenced by the differences in rates of deprotonation. In the example below, treatment of *N*-Boc-*N*-benzylamine with *n*-butyllithium in the presence of (-)-sparteine affords one enantiomer of the product with high enantiomeric excess. Transmetalation with trimethyltinchlroride affords the opposite enantiomer.

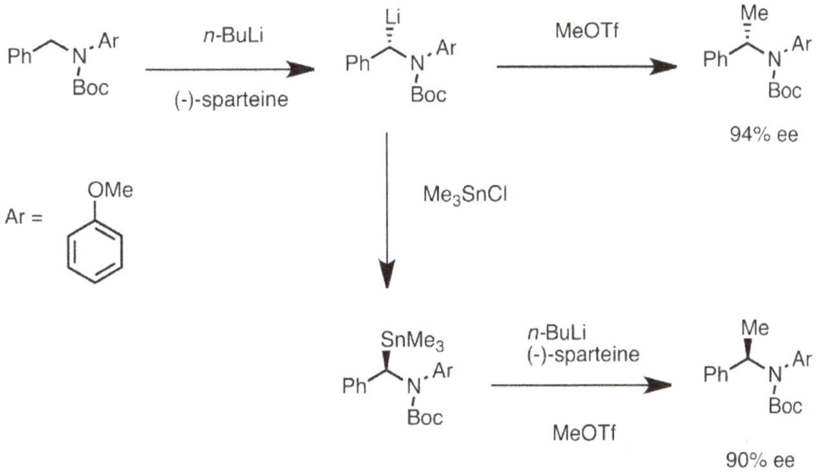

Enantioselective synthesis with n-BuLi in the presence of (-)-sparteine

Enolate Formation

Lithium enolates are formed through deprotonation of a C-H bond α to the carbonyl group by an organolithium species. Lithium enolates are widely used as nucleophiles in carbon-carbon bond formation reactions such as aldol condensation and alkylation. They are also an important intermediate in the formation of silyl enol ether.

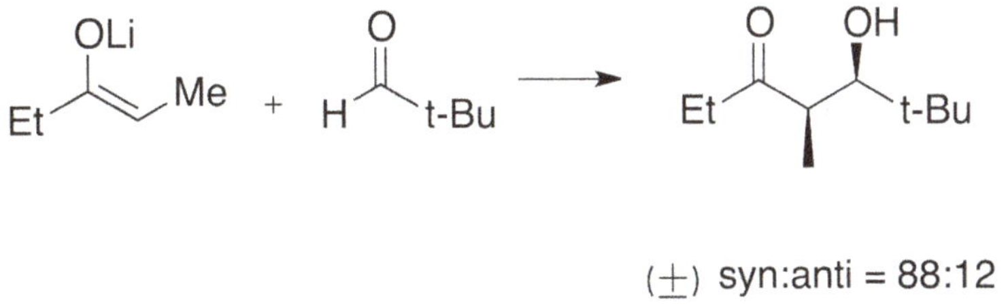

Lithium enolate formation can be generalized as an acid-base reaction, in which the relatively acidic proton α to the carbonyl group (pK =20-28 in DMSO) reacts with organolithium base. Generally, strong, non-nucleophilic bases, especially lithium amides such LDA, LiHMDS and LiTMP are used. THF and DMSO are common solvents in lithium enolate reactions.

The stereochemistry and mechanism of enolate formation have gained much interest in the chemistry community. Many factors influence the outcome of enolate stereochemistry, such as steric effects, solvent, polar additives, and types of organolithium bases. Among the many models used to explain and predict the selectivity in stereochemistry of lithium enolates is the Ireland model.

In this assumption, a monomeric LDA reacts with the carbonyl substrate and form a cyclic Zimmerman-Traxler type transition state. The (E)-enolate is favored due to an unfavorable *syn-pentane* interaction in the (Z)-enolate transition state.

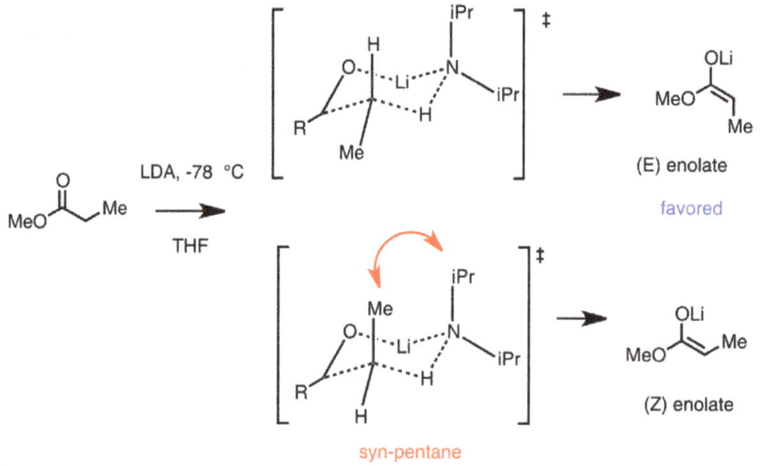

Addition of polar additives such as HMPA or DMPU favors the formation of (Z) enolates. The Ireland model argues that these donor ligands coordinate to the lithium cations, as a result, carbonyl oxygen and lithium interaction is reduced, and the transition state is not as tightly bound as a six-membered chair. The percentage of (Z) enolates also increases when lithium bases with bulkier side chains (such as LiHMDS) are used. However, the mechanism of how these additives reverse stereoselectivity is still being debated.

There have been some challenges to the Ireland model, as it depicts the lithium species as a monomer in the transition state. In reality, a variety of lithium aggregates are often observed in solutions of lithium enolates, and depending on specific substrate, solvent and reaction conditions, it can be difficult to determine which aggregate is the actual reactive species in solution.

Lithium-Halogen Exchange

Lithium halogen exchange is a metathesis reaction between an organohalide and organolithium species. Gilman and Wittig independently discovered this method in the late 1930s.

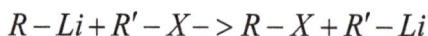

$$R-Li+R'-X->R-X+R'-Li$$

The mechanism of lithium-halogen exchange is still debated. One possible pathway involves a nucleophilic mechanism that generates a reversible "ate-complex" intermediate. Farnham and Calabrese were able to isolate "ate-complex" lithium bis(pentafluorophenyl) iodinate complexed with TMEDA and obtain an X-ray crystal structure. The "ate-complex" further reacts with electrophiles and provides pentafluorophenyl iodide and C_6H_5Li. A number of kinetic studies also support a nucleophilic pathway in which the carbanion on the lithium species attacks the halogen atom on the aryl halide. Another possible mechanism involves single electron transfer and the generation of radicals. In reactions of secondary and tertiary alkyllithium and alkyl halides, radical species were detected by EPR spectroscopy. However, whether these radicals are reaction intermediates is not definitive. The mechanistic studies of lithium -halogen exchange is also complicated by the formation of aggregates of organolithium species.

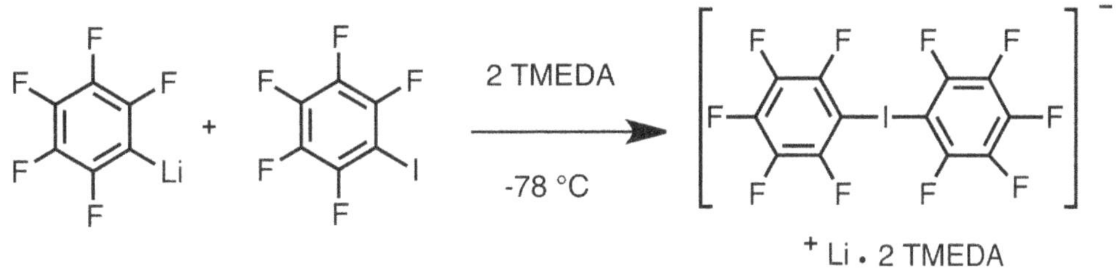

Proposed "atc-complex" in lithium halogen exchange

The rate of lithium halogen exchange is extremely fast. It is usually faster than nucleophilic addition and can sometimes exceed the rate of proton transfer. In the example below, the exchange between lithium and primary iodide is almost instantaneous, and outcompetes proton transfer from methanol to *tert*-butyllithium. The major alkene product is formed in over 90% yield.

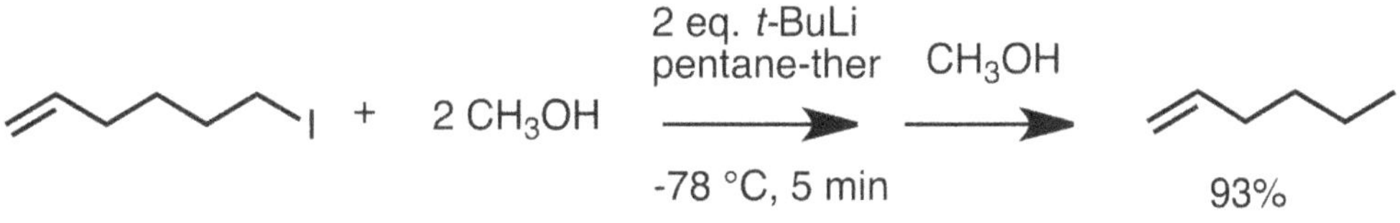

Lithium-halogen exchange is very useful in preparing new organolithium reagents. Exchange rates usually follow the trend I > Br > Cl. Alkyl- and arylfluoride are generally unreactive toward organolithium reagents. Lithium halogen exchange is kinetically controlled, and the rate of exchange is primarily influenced by the stabilities of the carbanion intermediates (sp > sp2 > sp3) of the organolithium reagents. For example, the more basic tertiary organolithium reagents (usually *n*-butyllithium, *sec*-butyllithium or *tert*-butyllithium) are the most reactive, and will react with primary alkyl halide (usually bromide or iodide) to form the more stable organolithium species. Therefore, lithium halogen exchange is most frequently used to prepare vinyl- aryl- and primary alkyllithium reagents. Lithium halogen exchange is also facilitated when alkoxy groups or heteroatoms are present to stabilize the carbanion, and this method is especially useful for the preparation of functionalized lithium reagents which cannot tolerate the harsher conditions required for reduction with lithium metal. Substrates such as vinyl halides usually undergo lithium-halogen exchange with retention of the stereochemistry of the double bond.

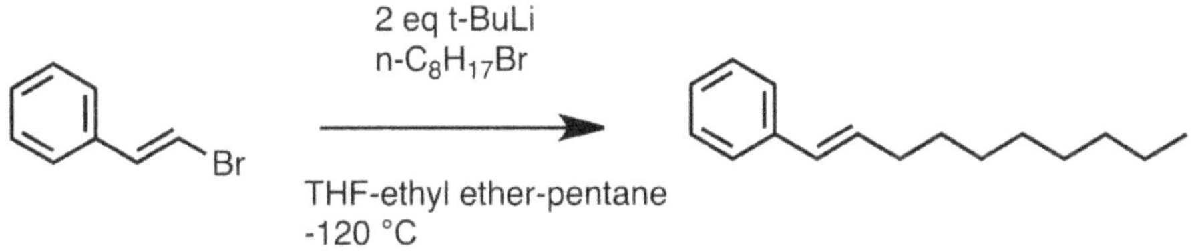

Double bond geomistry is preserved in lithium-halogen exchange

Below is an example of the use of lithium-halogen exchange in the synthesis of morphine. Here, *n*-butyllithium is used to perform lithium-halogen exchange with bromide. The nucleophilic carbanion center quickly undergoes carbolithiation to the double bond, generating an anion stabilized by the adjacent sulfone group. An intramolecular S_N2 reaction by the anion forms the cyclic backbone of morphine.

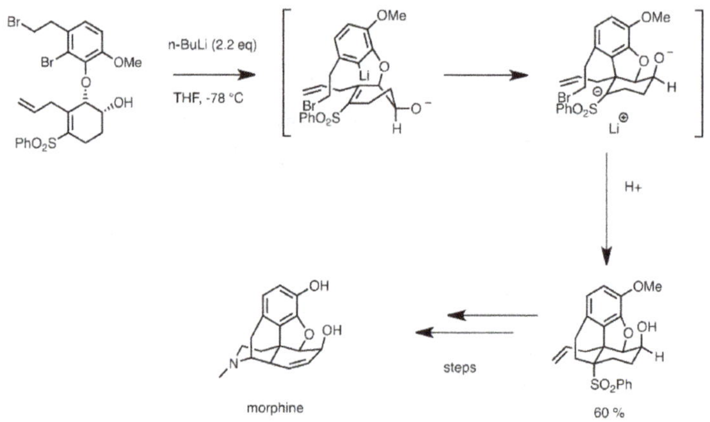

Lithium-halogen exchange in the synthesis of morphin to from tetracyclic core structure

Lithium halogen exchange is a crucial part of Parham cyclization. In this reaction, an aryl halide (usually iodide or bromide) exchanges with organolithium to form a lithiated arene species. If the arene bears a side chain with an electrophilc moiety, the carbanion attached to the lithium will perform intramolecular nucleophilic attack and cyclize. This reaction is a useful strategy for heterocycle formation. In the example below, Parham cyclization was used to in the cyclization of an isocyanate to form isoindolinone, which was then converted to a nitrone. The nitrone species further reacts with radicals, and can be used as "spin traps" to study biological radical processes.

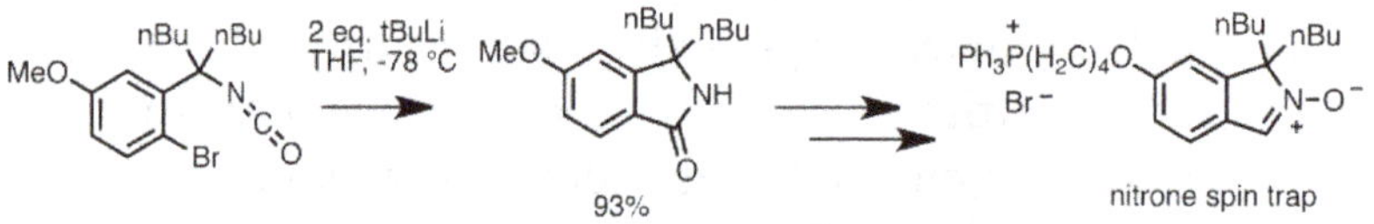

Parham cyclization in the synthesis of a nitrone "spin trap"

Transmetalation

Organolithium reagents are often used to prepare other organometallic compounds via transmetalation. Organocopper, organotin, organosilicon, organoboron, organophosphorus, organocerium and organosulfur compounds are frequently prepared by reacting organolithium reagents with appropriate electrophiles.

$$R-M+n-BuLi \rightarrow R-Li+n-BuM$$

Common types of transmetalation include Li/Sn, Li/Hg, and Li/Te exchange, which are fast at low temperature. The advantage of Li/Sn exchange is that the tri-alkylstannane precursors undergo few side reactions, as the resulting n-Bu_3Sn byproducts are unreactive toward alkyllithium reagents. In the following example, vinylstannane, synthesized from terminal alkyne, forms vinyllithium through transmetalation with n-BuLi.

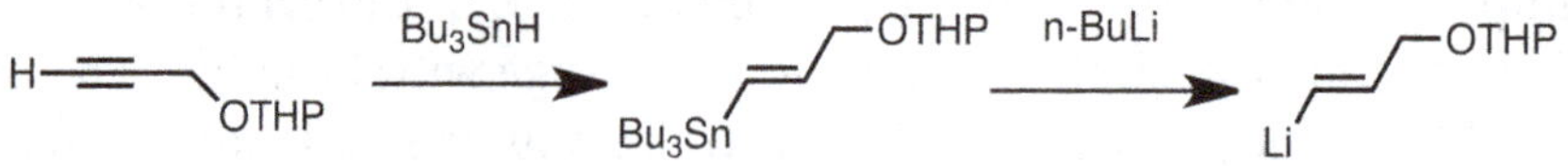

Organolithium can also be used in to prepare organozinc compounds through transmetalation with zinc salts.

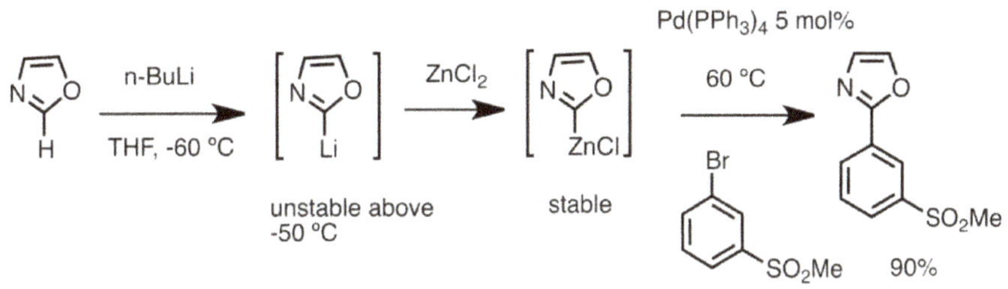

Llithium diorganocuprates can be formed by reacting alkyl lithium species with copper(I) halide. The resulting organocuprates are generally less reactive toward aldehydes and ketones than organolithium reagents or Grignard reagents.

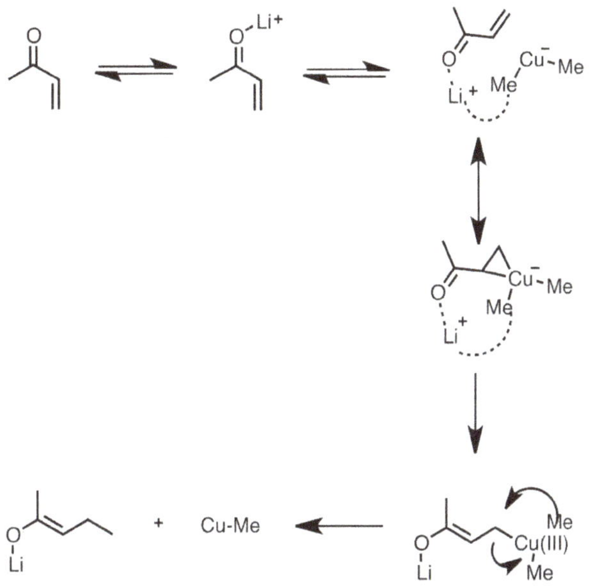

Reaction pathway of 1.4 conjugate addition of organocuprate

Preparation

Most simple alkyllithium reagents, and common lithium amides are commercially available in a variety of solvents and concentrations. Organolithium reagents can also be prepared in the laboratory.

Reaction with Lithium Metal

Reduction of alkyl halide with metallic lithium can afford simple alkyl and aryl organolithium reagents.

$$R-X+2Li \rightarrow R-Li+Li-X$$

Industrial preparation of organolithium reagents is achieved using this method by treating the alkyl chloride with metal lithium containing 0.5-2% sodium. The conversion is highly exothermic. The sodium initiates the radical pathway and increases the rate. The reduction proceeds via a radical pathway. Sometimes, lithium metal in the form of fine powders are used in the reaction with certain catalysts such as naphthalene or 4,4'-di-t-butylbiphenyl (DTBB). Another substrate that can be reduced with lithium metal to generate alkyllithium reagents is sulfides. Reduction of sulfides is useful in the formation of functionalized organolithium reagents such as alpha-lithio ethers, sulfides, and silanes.

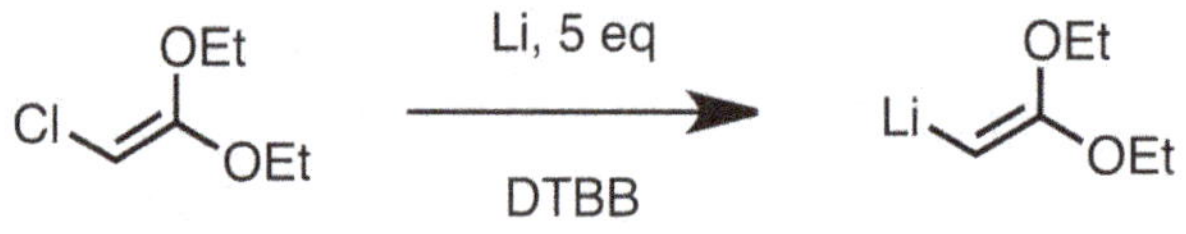

Preparation of organolithium reagents through reduction with lithium metal

Metalation

A second method of preparing organolithium reagents is through metalation (lithium hydrogen exchange). The relative acidity of hydrogen atoms controls the position of lithiation.

This is the most common method for preparing alkynyllithium reagents, because the terminal hydrogen bound to the *sp* carbon is very acidic and easily deprotonated. For aromatic compounds, the position of lithiation is also determined by the directing effect of substituent groups. Some of the most effective directing substituent groups are alkoxy, amido, sulfoxide, sulfonyl. Metalation often occurs at the position ortho to these substituents. In heteroaromatic compounds, metalation usually occurs at the position ortho to the heteroatom.

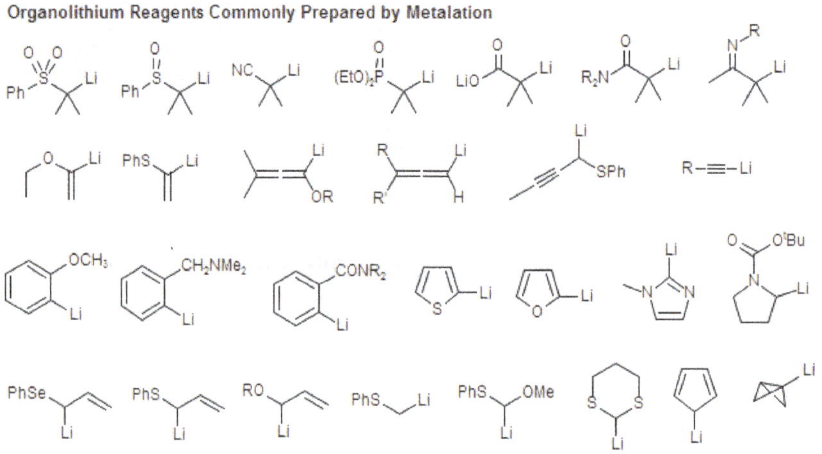

Lithium Halogen Exchange

A third method to prepare organolithium reagents is through lithium halogen exchange.

tert-butyllithium or *n*-butyllithium are the most commonly used reagents for generating new organolithium species through lithium halogen exchange. Lithium-halogen exchange is mostly used to convert aryl and alkenyl iodides and bromides with *sp2* carbons to the corresponding organolithium compounds. The reaction is extremely fast, and often proceed at -60 to -120 °C.

Transmetalation

The fourth method to prepare organolithium reagents is through transmetalation. This method can be used for preparing vinyllithium.

Shapiro Reaction

In the Shapiro reaction, two equivalents of strong alkyllithium base react with p-tosylhydrazone compounds to produce the vinyllithium, or upon quenching, the olefin product.

Handling

Organolithium compounds are highly reactive species and require specialized handling techniques. They are often corrosive, flammable, and sometimes pyrophoric (spontaneous ignition when exposed to oxygen or moisture). Alkyllithium reagents can also undergo thermal decomposition to form the corresponding alkyl species and lithium hydride. Organolithium reagents are typically stored below 10 °C. Reactions are conducted using air free techniques. The concentration of alkyllithium reagents is often determined by titration:

Organolithium reagents react with ethers, which are often used as solvents:

Approximate half-lives of common lithium reagents in typical solvents						
Solvent/Temp	n-BuLi	s-BuLi	t-BuLi	MeLi	CH_2=C(OEt)-Li	CH_2=C($SiMe_3$)-Li
THF/-20 °C			40 min, 360 min			
THF/20 °C					>15 hr	17 hr
THF/35 °C	10 min					
THF/TMEDA/-20 °C	55 hr					
THF/TMEDA/ 0 °C	340 min					
THF/TMEDA/20 °C	40 min					
Ether/-20 °C			480 min			
Ether/0 °C			61 min			
Ether/20 °C	153 hr		<30 min			17 days
Ether/35 °C	31 hr					
Ether/TMEDA/ 20 °C	603 min					
DME/-70 °C		120 min	11 min			
DME/-20 °C	110 min	2 min	<<2 min			
DME/0 °C	6 min					

Allylic Substitution

Allylic Substitution with Carbon Nucleophiles

The metal-catalyzed allylic substitution is one most of the important processes in organic synthesis. Figure represents the catalytic cycle of a transition metal based allylic substitution reaction. The reaction begins with the coordination of the low valent metal complex to the double bond of an allylic system. Subsequent oxidative addition by removal of the leaving group X gives a π -allyl complex as intermediate. The intermediate could be a neutral or cationic species, depending on the nature of the ligands and the counter ion X. The nucleophile typically adds to the terminal carbon with inversion of configuration rather than *via* the metal cation with retention.

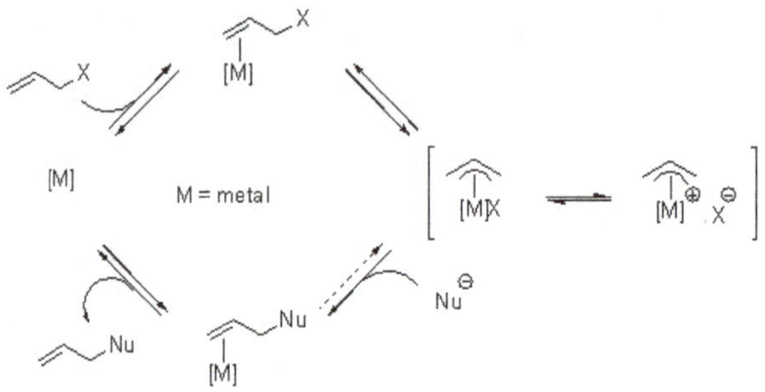

Palladium-Catalyzed Reactions

The palladium catalyzed allylic substitution reaction is a very powerful process. This topic covers some recent examples on the palladium catalyzed enantioselective allylic substitution with carbon nucleophiles. The use of azlactones as a soft stabilized pronucleophile is particularly important because they give rise to amino acids as products. Figure presents Trost's synthesis of spingofungins via alkylation of a geminal diacetate with an azlactone. The product is formed with good diastereo- and enantioselectivity.

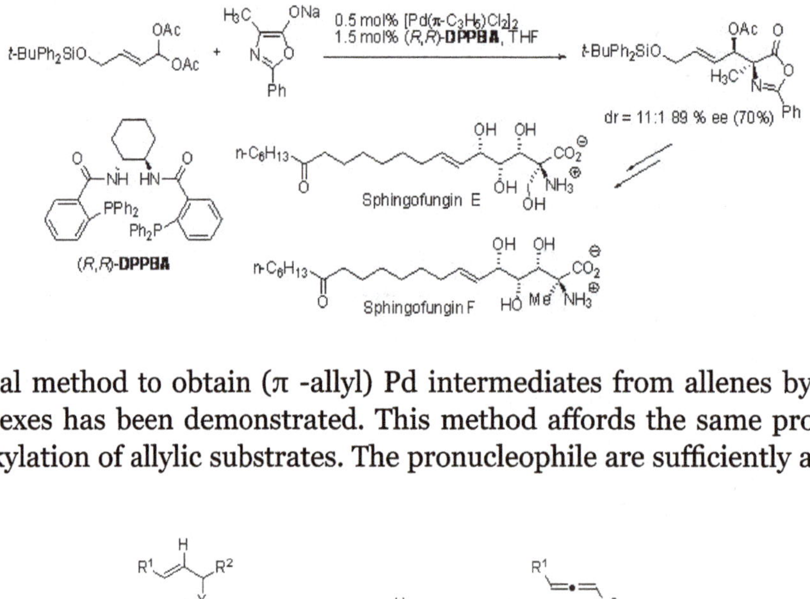

Atom economical method to obtain (π -allyl) Pd intermediates from allenes by addition of hydrido-Pd complexes has been demonstrated. This method affords the same products as that of the standard alkylation of allylic substrates. The pronucleophile are sufficiently acidic to produce $HPdL_2$ species.

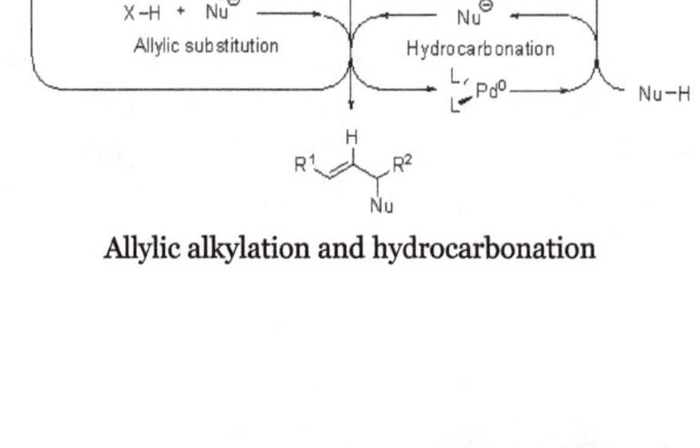

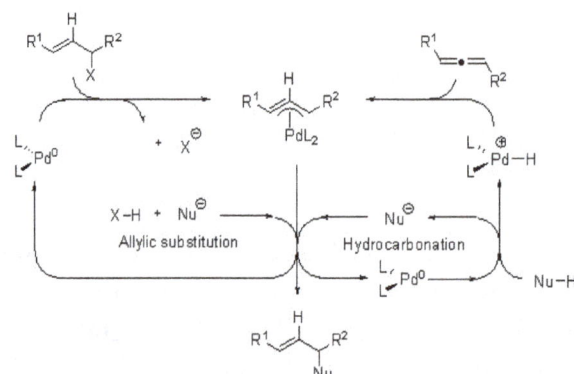

Allylic alkylation and hydrocarbonation

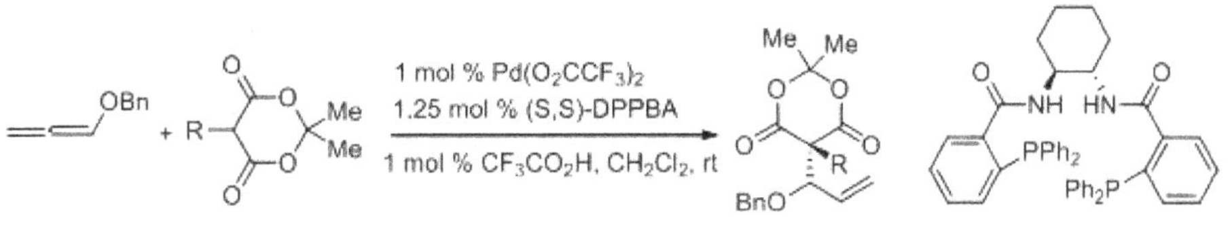

(S,S)-**DPPBA**

R	Yield [%]	ee [%]
CH_3	75	99
$(CH_3)_2CHCH_2$	61	88
CH_2=$CHCH_2$	82	96
$PhCH_2$	90	91
2-$C_4H_3OCH_2$	81	94
OH	63	82

The palladium catalyzed reaction of vinyl epoxide with nucleophiles provides branched products. This is due to interaction of the nucleophile with an alkoxy or OH moiety produced by reaction with the Pd(0) species. For example, the reaction of isoprene monoepoxide with β-keto esters preferentially gives the branched alkylation products in the form of the hemiacetals. The nature of the β-ketoester and optimization of the reaction conditions are crucial for the success of this process.

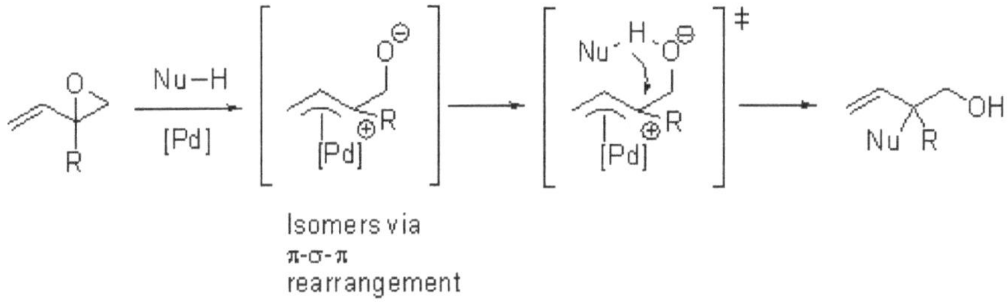

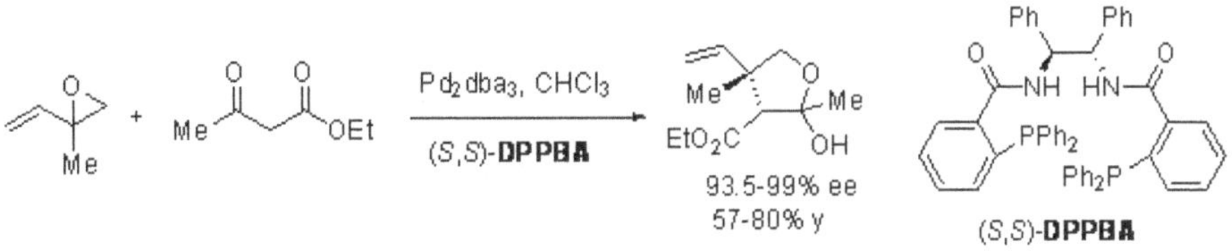

Bimetallic system having $Rh(acac)(CO)_2$, $Pd(Cp)(\pi\text{-}C_3H_5)$ and the ligand Anis Trap has been used for the allylic alkylation with α -cyanopropionic acid derivative as pronucleophile. The control of the stereochemistry is believed to take place via the nucleophile with a chiral Rh complex coordinating to the cyano group.

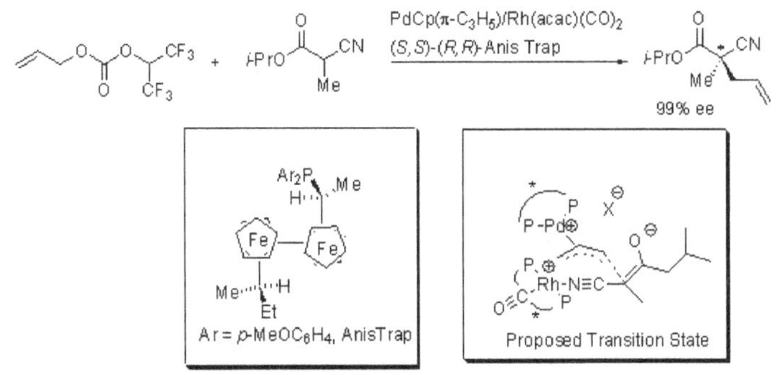

Recently, allylic alkylation has been realized by enolate generated in situ by decarboxylation. Both allylic β -keto carboxylates and allyic enol carbonates undergo facile decarboxylation after oxidative addition of a Pd(0) species.

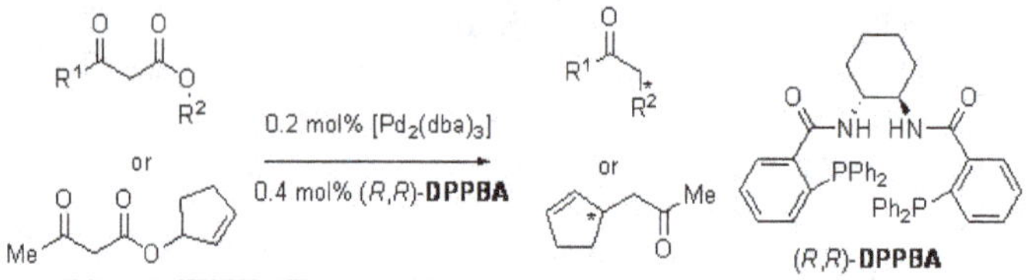

R1	R2	Yield [%]	ee [%]
Me	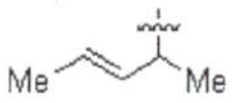	82	86
Me	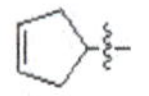	85	86
Me	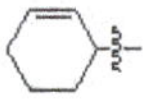	75	94
PhCH2	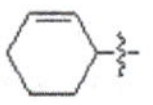	71	90
i-Pr	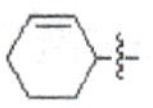	94	80
Ph	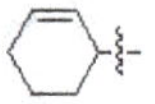	69	92

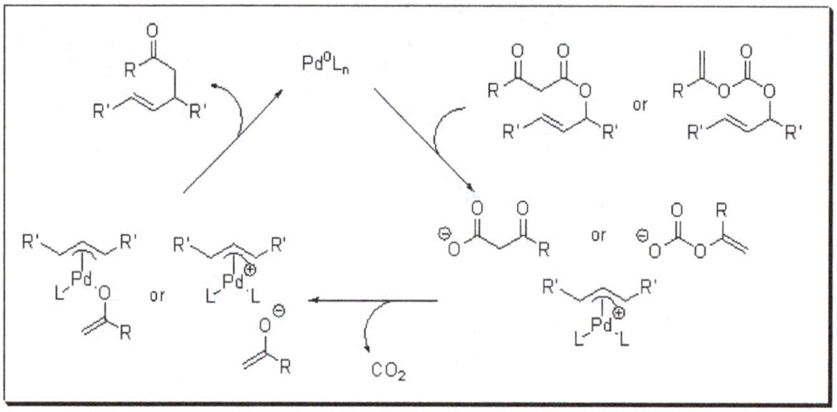

Nickel-Catalyzed Reactions

In comparison to the palladium catalyzed reactions, the nickel based chemistry is less explored. In addition, the nickel based chemistry less popular with the reactions of soft nucleophiles and few examples only so far investigated. For example, the reaction of allylic acetates has been studied with soft nucleophiles such as dimethyl malonate using a wide range of phosphine ligands. Linear allylic substrates give a mixture of regioisomers, whereas in cyclohexenyl acetate, the regioselectivity does not play any role affording the alkylated product with moderate enantioselectivity in the presence of chiral phosphine L1 .

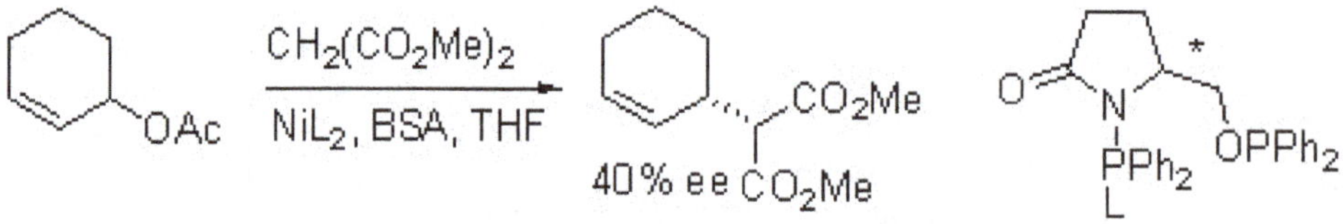

However, the nickel based systems are very popular with the reactions of hard nucleophiles such as boronic acids, borates and Grignard reagents. For example, the reaction of 1,3-disubstituted allyl ethers with Grignard reagents can be accomplished using nickel phosphine complex with good enantioselectivity. The reaction of methyl ether gave better results compared to phenyl ethers. In this reaction, if the reaction is quenched before complete consumption of the staring material, a significant kinetic resolution is observed.

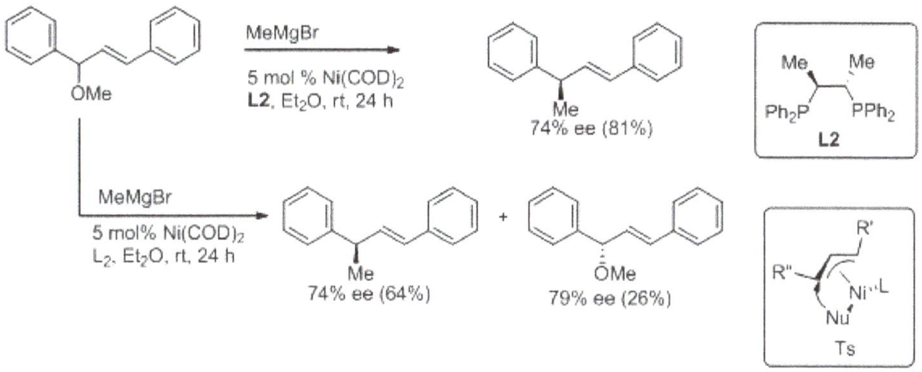

Molybdenum-Catalyzed Reactions

Although the palladium catalyzed systems dominate in π -allyl chemistry, analogues Mo-catalyzed reactions have also emerged as powerful reactions in organic synthesis. The Mo-based reactions are the one first showed different regioselectivity compared to the palladium catalyzed systems. Figure illustrates the mechanism for the asymmetric Mo-catalyzed allylic alkylation.

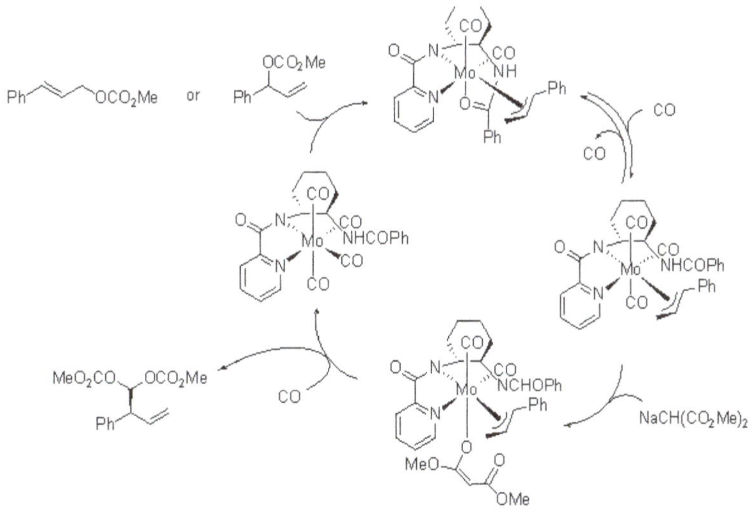

Copper-Catalyzed Reactions

In case of the nonsymmetrical allylic substrates, the palladium catalyzed allylic alkylation reactions show poor regioselectivity. In this context, the copper based chemistry is an interesting alternative and lots of efforts have been made on this topic during last years. The copper based systems tolerate a wide range of hard and nonstabilized nucleophiles. Figure presents the regioselectivity in copper-catalyzed allylation reactions. In unsymmetrical substrates, nucleophile may attack directly at the leaving group (S_N2) or at the allylic position (S_N2') under migration of the double bond depending on the reaction parameters as well as the substrate and nucleophile.

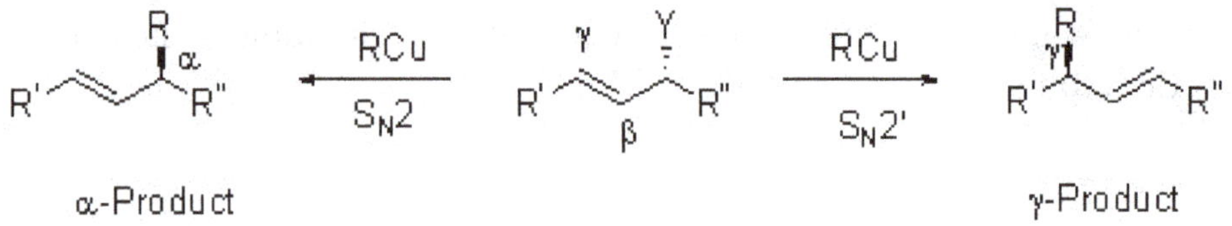

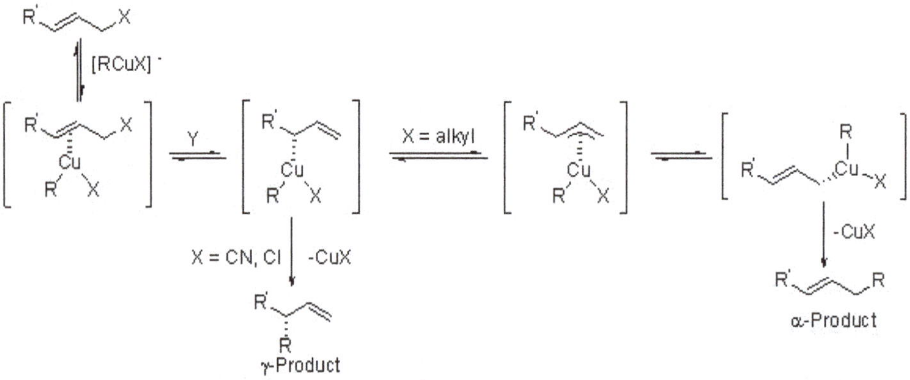

The observed results suggest that the regioselectivity and stereoselectivity are established at different stages. For example, the reaction of chiral carbamates with achiral copper reagent gives S_N2' product with excellent enantioselectivity.

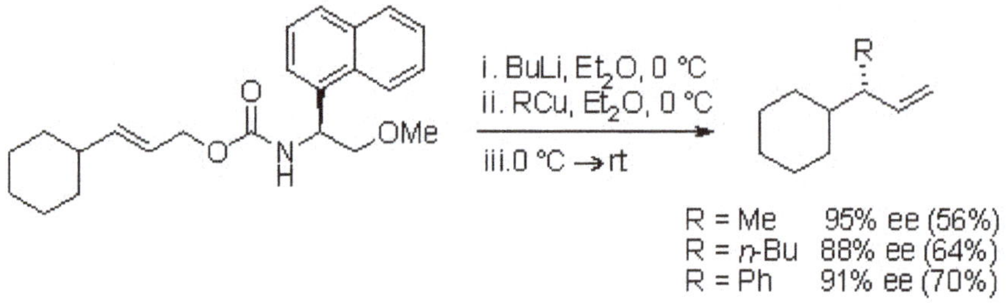

References

- Fernandez, I.; Bickelhaupt, F. M. (2012). "Alder-ene reaction: Aromaticity and activation-strain analysis". Journal of Computational Chemistry. 33 (5): 509–16. PMID 22144106. doi:10.1002/jcc.22877
- Smith, Michael B.; March, Jerry (2007), Advanced Organic Chemistry: Reactions, Mechanisms, and Structure (6th ed.), New York: Wiley-Interscience, ISBN 0-471-72091-7
- Inagaki, S.; Fujimoto, H; Fukui, K. J. (1976). "Orbital interaction in three systems". J. Am. Chem. Soc. 41 (16): 4693. doi:10.1021/ja00432a001
- Hein, Sara M. (June 2006). "An Exploration of a Photochemical Pericyclic Reaction Using NMR Data". Journal of Chemical Education. 83 (6): 940–942. Bibcode:2006JChEd..83..940H. doi:10.1021/ed083p940
- Clayden, Jonathan; Greeves, Nick (2005). Organic chemistry. Oxford: Oxford Univ. Press. p. 212. ISBN 978-0-19-850346-0
- McGarrity, J. F.; Ogle, C.A. (1985). "High-field proton NMR study of the aggregation and complexation of n-bu-

tyllithium in tetrahydrofuran". J. Am. Chem. Soc. 107: 1805–1810. doi:10.1021/ja00293a001

- Goebel, M. T.; Marvel, C. S. (1933). "The Oxidation of Grignard Reagents". Journal of the American Chemical Society. 55 (4): 1693–1696. doi:10.1021/ja01331a065
- Wardell, J.L. (1982). "Chapter 2". In Wilinson, G.; Stone, F. G. A.; Abel, E. W. Comprehensive Organometallic Chemistry, Vol. 1 (1st ed.). New York: Pergamon. ISBN 0080406084
- Smith, David H. (1999). "Grignard Reactions in "Wet" Ether". Journal of Chemical Education. 76 (10): 1427. Bibcode:1999JChEd..76.1427S. doi:10.1021/ed076p1427
- Farnham, W. B.; Calabrese, J. C. (1986). "Novel hypervalent (10-I-2) iodine structures". J. Am. Chem. Soc. 108: 2449–2451. doi:10.1021/ja00269a055
- The Preparation of Organolithium Reagents and Intermediates Leroux.F., Schlosser. M., Zohar. E., Marek. I., Wiley, New York. 2004. ISBN 978-0-470-84339-0
- Gellert, H; Ziegler, K. (1950). "Organoalkali compounds. XVI. The thermal stability of lithium alkyls.". Liebigs Ann. Chem. 567: 179–185. doi:10.1002/jlac.19505670110

3

Carbon-Heteroatom Bonds and Overman Rearrangement in Asymmetric Synthesis

The Overman rearrangement converts readily available allylic alcohols to allylic amines by a two-step process. Carbon-heteroatom formation is an important part of asymmetric synthesis. Asymmetric synthesis is best understood in confluence with the major topics listed in the following chapter.

Carbon-heteroatom Bond Reactions

Asymmetric carbon-heteroatom bond formation is among the fundamentally important reactions. This topic covers the carbon-heteroatom bond-forming reactions using transition-metal-complex as well as the chiral Lewis acid catalyzed protocols.

Allylic Substitution Reactions

Much effort has been devoted on controlling the regioselectivity and enantioselectivity in allylic substitution of substrates 1 or 2. The palladium-catalyzed allylic substitution is versatile, however, the (E)-linear product 3 is often formed. Thus, the control of regioselectivity has been recently the main focus to provide product 4.

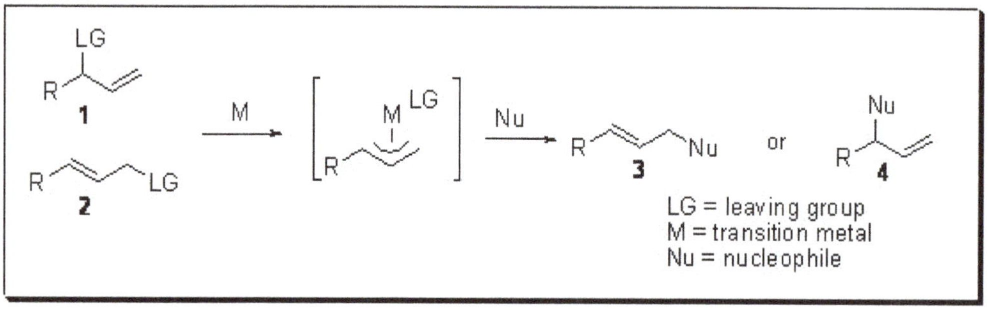

Allylic Amination and Etherification of Allylic Alcohol Derivatives

Chiral iridium complex having phosphoramidate 4a or 5a has been shown to catalyze the allylic amination of carbonate to give branched product with excellent enantioselectivity. An activated form of the iridium complex by in situ C-H activation at CH_3 group of a hindered ligand 4a has been identified.

The direct reaction of allylic alcohols has been studied to give allylic amines in the presence of chiral iridium complex derived from $[Ir(COD)Cl]_2$ and ligand 6. In this reaction, sulfamic acid serves not only as a nitrogen source but also as an in situ activator of the hydroxyl group of the allylic alcohol

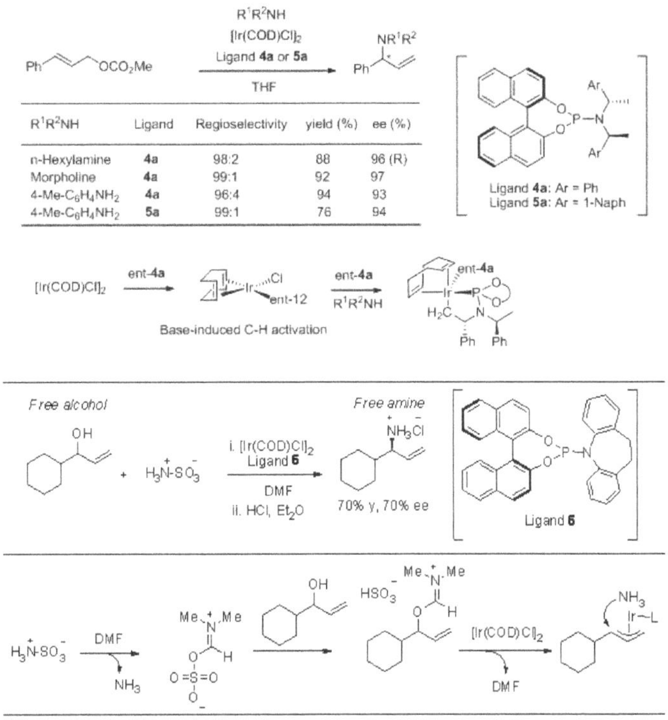

R¹R²NH	Ligand	Regioselectivity	yield (%)	ee (%)
n-Hexylamine	4a	98:2	88	96 (R)
Morpholine	4a	99:1	92	97
4-Me-C₆H₄NH₂	4a	96:4	94	93
4-Me-C₆H₄NH₂	5a	99:1	76	94

Allylic amination is important for the construction of nitrogen-based heterocyclic compounds. The enantioselective intramolecular allylic amination has been accomplished using chiral iridium complex derived from $[Ir(CDD)Cl_2]_2$ and ligand 7. Good enantioselectivity has been obtained upon activation using 1,5,7-triazabicylo[4.4.0]undec-5-ene (TBD) as base. The catalytic system has also been used for the sequential aminations of bis -allylic carbonate via an inter- followed by an intra-molecular reactions.

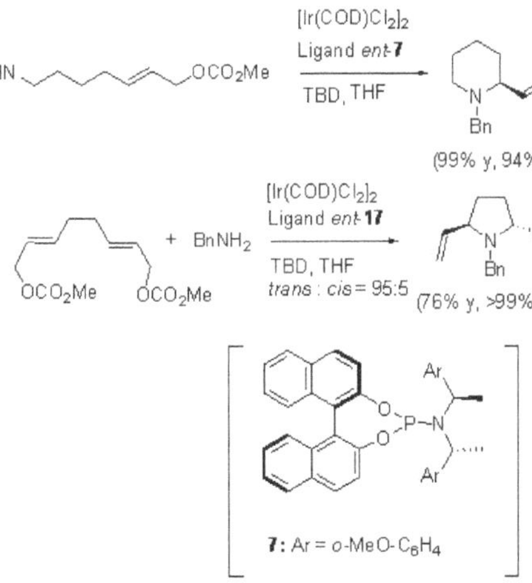

Enantioselective allylic amination is also a powerful tool for the construction of natural products. For example, asymmetric desymmetrization of meso -diol with p -tosylisocyanate using chiral palladium complex gives easy access to chiral nitrogen-substituted heterocycles which are precursor for the synthesis of (-)-swainsonine.

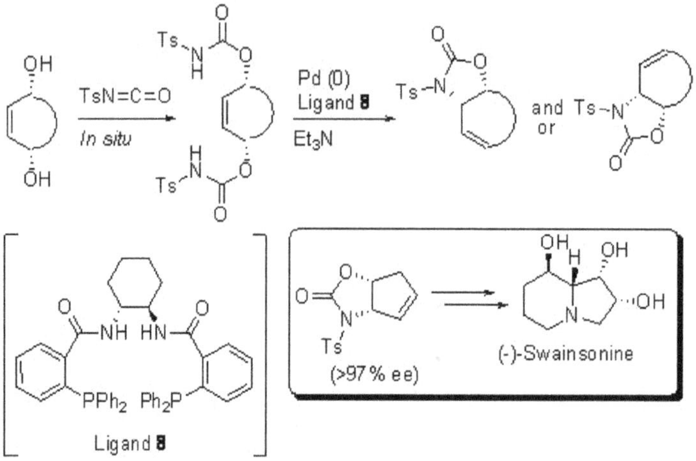

The chiral palladium catalyzed enantioselective allylic amination has also been utilized for the total synthesis of (-)-tubifoline, (-)-dehydrotubifoline and (-)-strychnine.

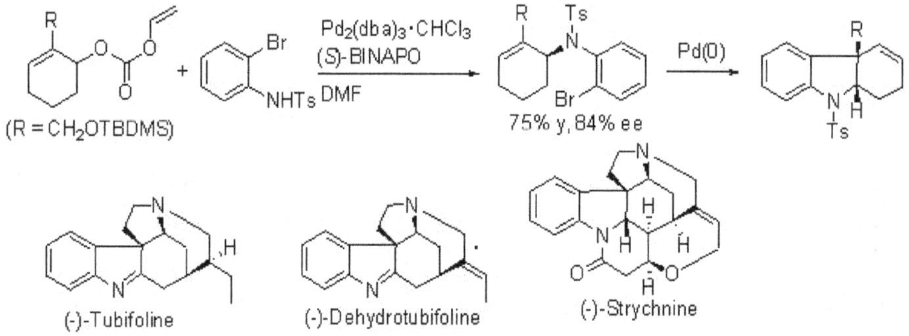

The one-pot enantioselective synthesis of azacycle has been shown using a ruthenium-catalyzed ene-yne addition followed by a palladium-catalyzed asymmetric allylic amination.

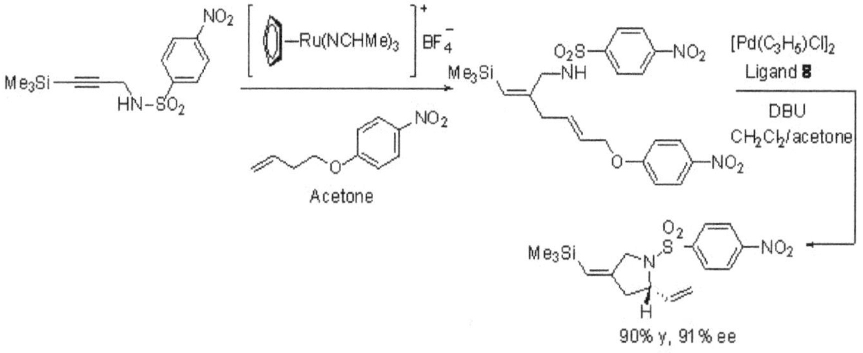

The regio- and enantioselective allylic etherification has been studied using chiral ruthenium complex. For example, planar-chiral cyclopentadienyl ruthenium complex 9 catalyzes efficiently the reaction of cinnamoyl chloride with 3-methylphenol with high enantioselectivity and yield.

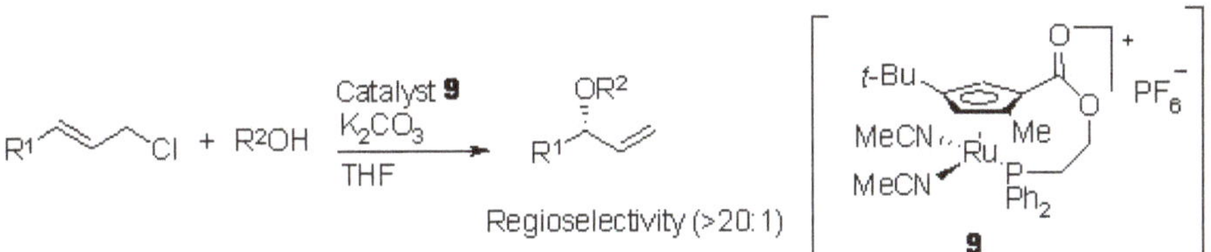

R^1	R^2	Yield [%]	ee [%]
Ph	Ph	99	92
Ph	4-$CF_3C_6H_4$	99	91
Ph	4-ClC_6H_4	98	93
4-$CF_3C_6H_4$	Ph	90	94
4-ClC_6H_4	Ph	96	86
Me	Ph	36	80
1-napthyl	Ph	98	82

Enantioselective allylic substitutions of carbonates with a diboron using copper(I)-based catalysts has been demonstrated. For example, Cu(I)-phosphine complex generated in situ from Cu(O-t-Bu) with ligand 10 has been shown to catalyze the reaction of allylboronate with carbonate in excellent regioselectivity and enantioselectivity. Addition-elimination mechanism having the generation of Cu-alkene π -complex and borylalkylcopper intermediate has been suggested.

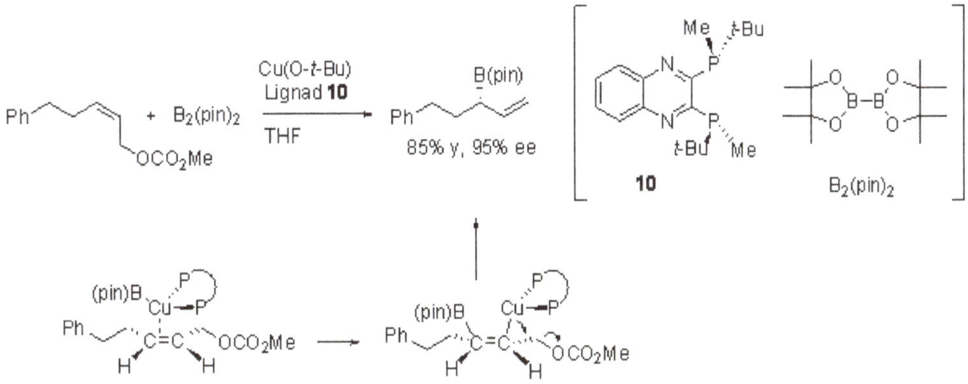

Reaction of π -Allyl Intermediates

Nucleophilic attack of an amine to a π -allyl intermediate can afford an allylic amine derivative. For example, palladium complex derived from $[Pd(C_3H_5)Cl]_2$ and ligand 11 catalyzes the reaction of racemic vinyloxirane with phthalimide in nearly quantitative yield. Involvement of the hydrogen bond of the nucleophile to the oxygen leaving group is proposed to deliver the nucleophile to the adjacent carbon to provide the target molecule. The process has been utilized for the synthesis of (+)-broussonetine G.

Palladium based systems has also been utilized for the cycloaddition reaction of epoxides and aziridines with heterocumulenes.

Enantioselective copper(I)-catalyzed substitution reactions of propargylic acetates with amines has been explored. For examples, copper complexes derived from copper(I) salts and ligands 12 and 13 catalyze the reaction of propargylic amination with 85% ee.

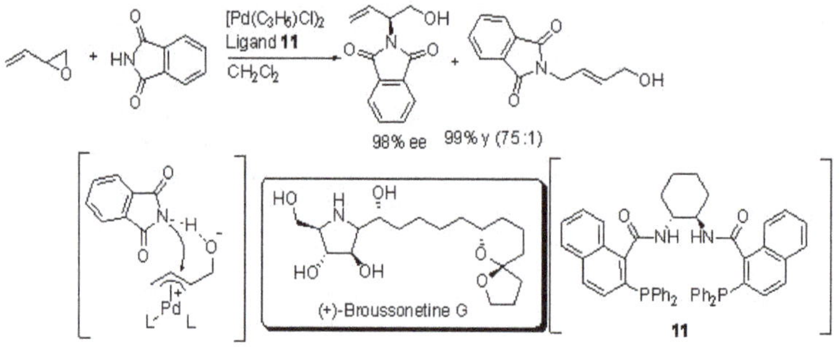

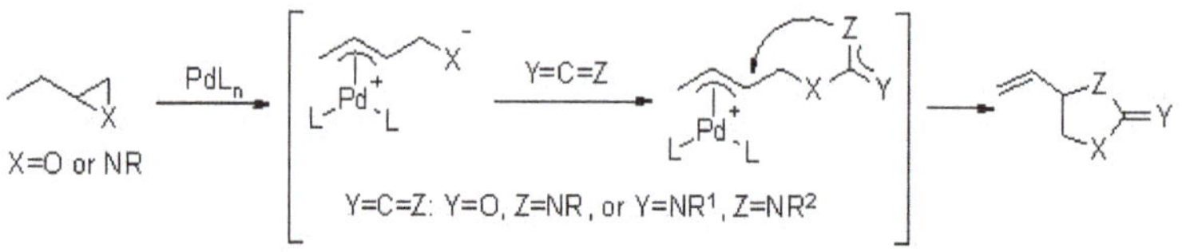

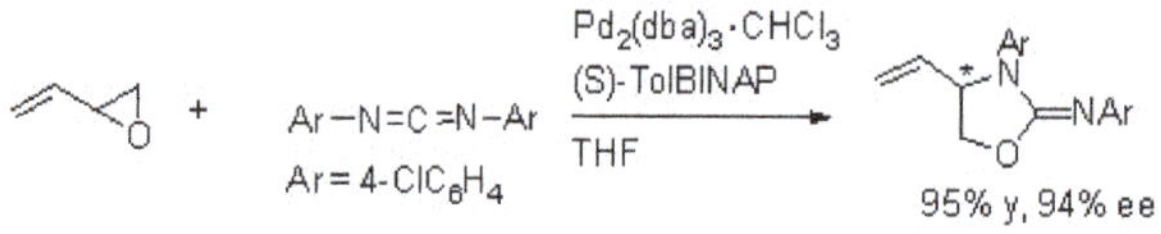

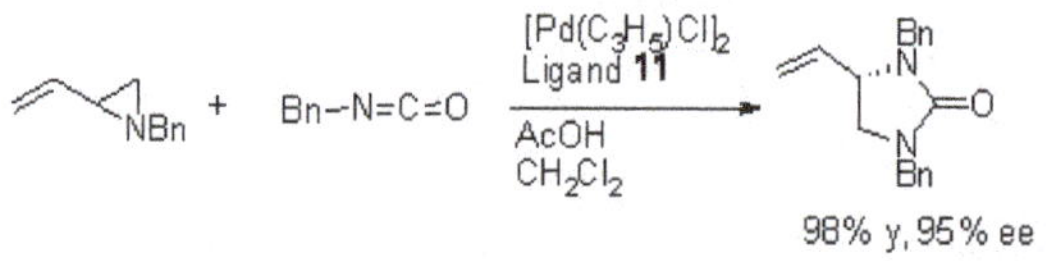

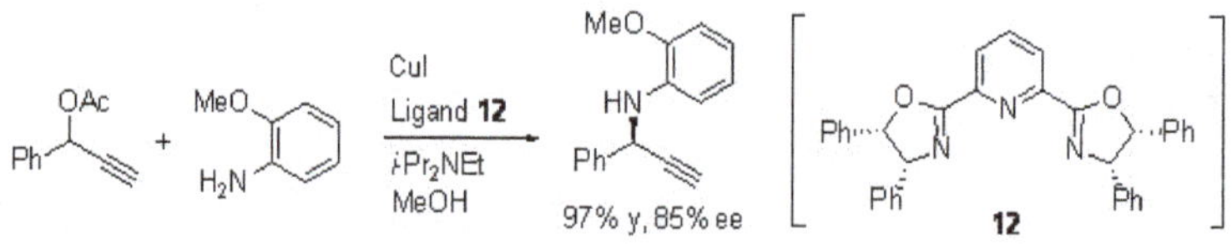

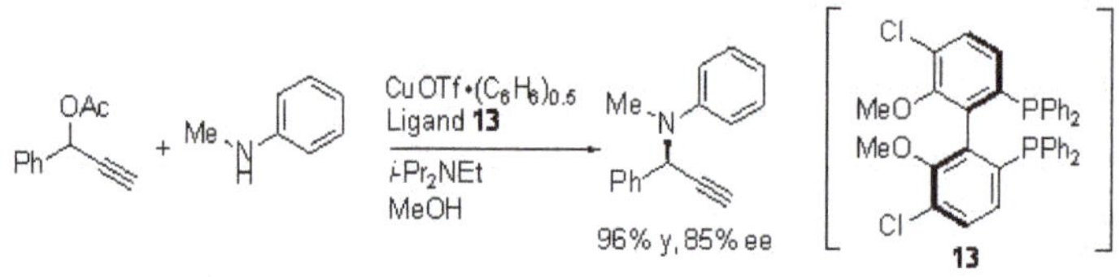

Overman Rearrangement

The Overman rearrangement is a chemical reaction that can be described as a Claisen rearrangement of allylic alcohols to give allylic trichloroacetamides through an imidate intermediate. The Overman rearrangement was discovered in 1974 by Larry Overman.

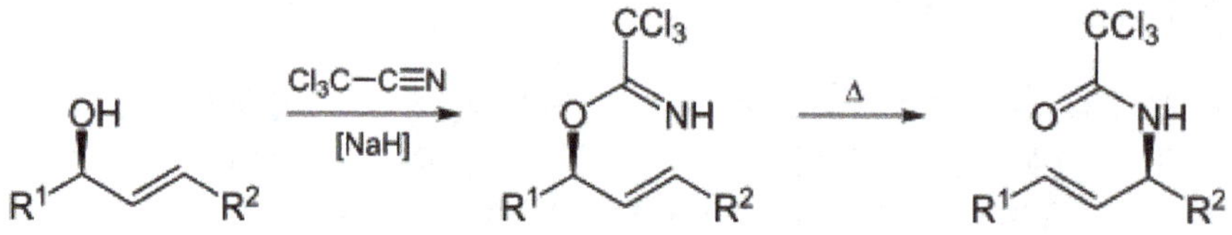

The [3,3]-sigmatropic rearrangement is diastereoselective and requires heating or the use of Hg(II) or Pd(II) salts as catalysts. The resulting allylamine structures can be transformed into many chemically and biologically important natural and un-natural amino acids (like (1-adamantyl)glycine).

The Overman rearrangement may also be used for asymmetric synthesis.

Aza-Claisen Rearrangement and Related Reactions

Aza-Claisen rearrangement, known as the Overman rearrangement, has been extensively studied that allows us to synthesize chiral allylic amines from achiral allylic imidates with excellent enantioselectivity. For example, prochiral N -arylbenzimidates can be converted into chiral N- a rylbenzamides in the presence of ferrocenyloxazoline palladacycle , FOP-TFA.

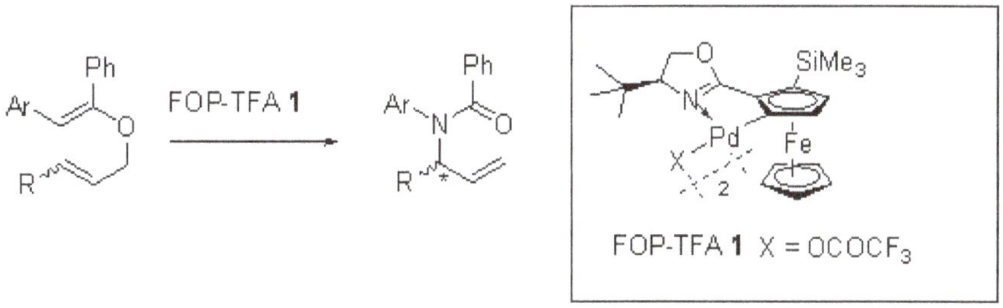

This catalytic system has also been shown to promote the cyclization of allylic N -arylsulfonyl carbamates to give five-membered nitrogen containing heterocycles. An involvement of aminopalladation of the alkene followed by insertion of the alkene into the Pd-N has been proposed.

This procedure has also been extended for the allylic etherification reaction. For example, the reaction of (Z)-allylic trichloroacetimidates with carboxylic acids in the presence of COP-OAc 2 gives chiral allylic esters in high enantiopurity. Under these reaction conditions, E -stereoisomer show inferior results. In these reactions, the COP-OAc activates the carbon-carbon double bond for attack by external oxygen nucleophile and the trichloroacetimidate group serves as a leaving group along with templating the catalyst to the double bond.

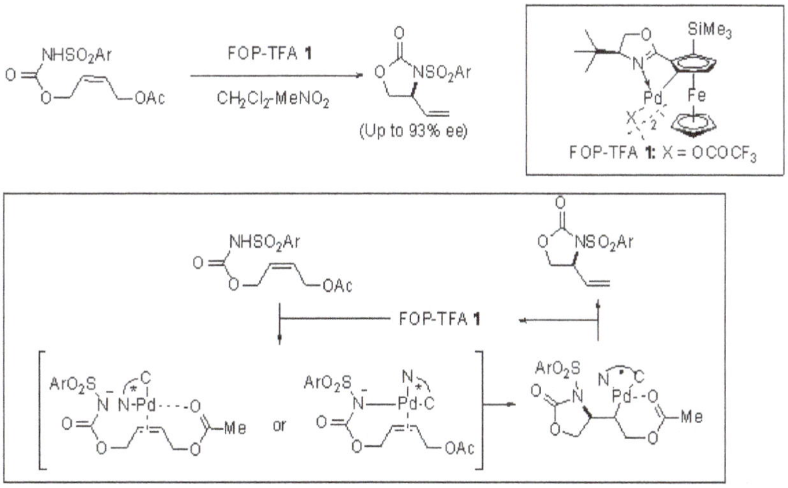

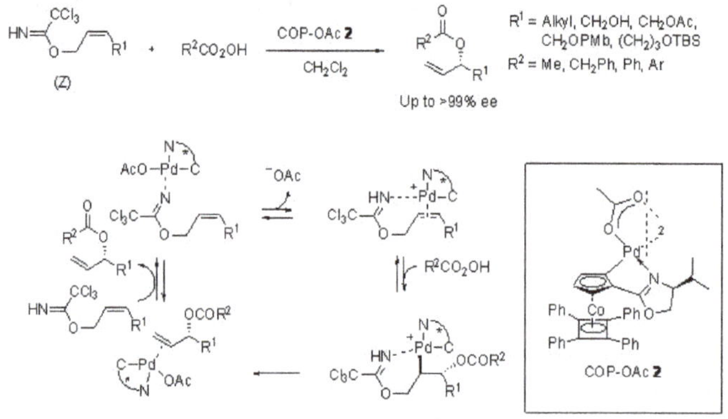

Hydroamination

Hydroamination is the addition of an N-H bond of an amine across a carbon-carbon multiple bond of an alkene, alkyne, diene, or allene. In the ideal case, hydroamination is atom economical and green. Amines are common in fine-chemical, pharmaceutical, and agricultural industries.

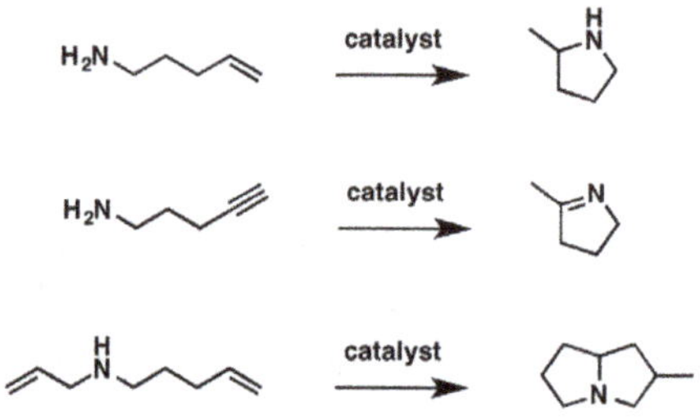

Examples of intramolecular hydroamination

Hydroamination can be used intramolecularly to create heterocycles or intermolecularly with a separate amine and unsaturated compound. The development of catalysts for hydroamination remains an active area, especially for alkenes.

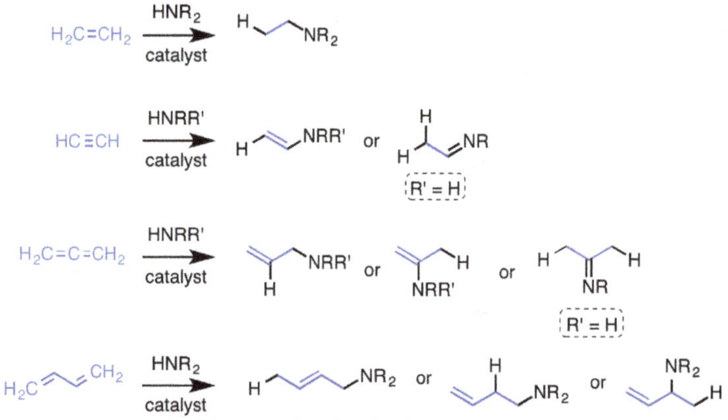

Prototypical intermolecular hydroamination reactions.

History

The first intramolecular hydroaminations were reported by Tobin J. Marks in 1989 using metallocene derived from rare-earth metals such as lanthanum, lutetium, and samarium. Catalytic rates correlated inversely with the ionic radius of the metal, perhaps as a consequence of steric interference from the ligands. In 1992, Marks developed the first chiral hydroamination catalysts by using a chiral auxiliary, which were the first hydroamination catalysts to favor only one specific stereoisomer. Chiral auxiliaries on the metallocene ligands were used to dictate the stereochemistry of the product. The first non-metallocene chiral catalysts were reported in 2003, and used bisarylamido and aminophenolate ligands to give higher enantioselectivity.

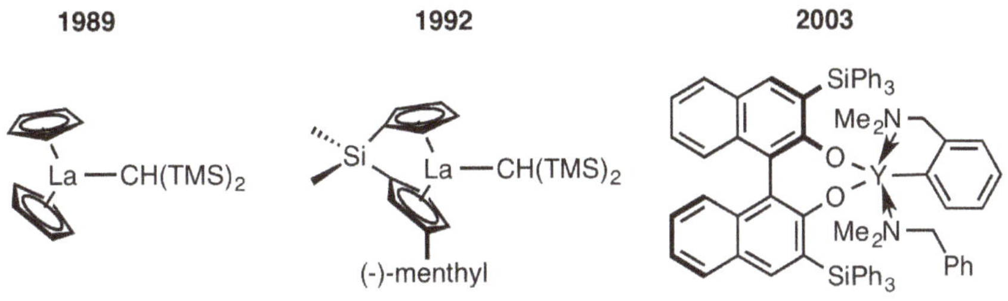

Notable hydroamination catalysts by year of publication

Reaction Scope

Hydroamination has been examined with a variety of amines, unsaturated substrates, and vastly different catalysts. Amines that have been investigated span a wide scope including primary, secondary, cyclic, acyclic, and anilines with diverse steric and electronic substituents. The unsaturated substrates that have been investigated include alkenes, dienes, alkynes, and allenes. For intramolecular hydroamination, various aminoalkenes have been examined.

Products

Addition across the unsaturated carbon-carbon bond can be Markovnikov or anti-Markovnikov depending on the catalyst. Interestingly, when considering the possibly of R/S chirality, four products can be obtained: Markovnikov with R or S and anti-Markovnikov addition with R or S. Although there have been many reports of catalytic hydroamination with a wide range of metals, there are far fewer describing enantioselective catalysis to selectively make one of the four possible products. Recently, there have been reports of selectively making the thermodynamic or kinetic product, which can be related to the racemic Markovnikov or anti-Markovnikov structures.

Catalysts and Catalytic Cycle

Catalysts

Many metal-ligand combinations have been reported to catalyze hydroamination, including main group elements including alkali metals such as lithium, group 2 metals such as calcium, as well as group 3 metals such as aluminum, indium, and bismuth. In addition to these main group examples, extensive research has been conducted on the transition metals with reports of early, mid, and late metals, as well as first, second, and third row elements. Finally the lanthanides have been thoroughly investigated. Zeolites have also shown utility in hydroamination.

Catalytic Cycles

The mechanism of metal-catalyzed hydroamination has been well studied. Particularly well studied is the organolanthanide catalyzed intramolecular hydroamination of alkenes. First, the catalyst is activated by amide exchange, generating the active catalysis (i). Next, the alkene inserts into the Ln-N bond (ii). Finally, protonolysis occurs generating the cyclized product while also regenerating the active catalyst (iii). Although this mechanism depicts the use of a lanthanide catalyst, it is the basis for rare-earth, actinide, and alkali metal based catalysts.

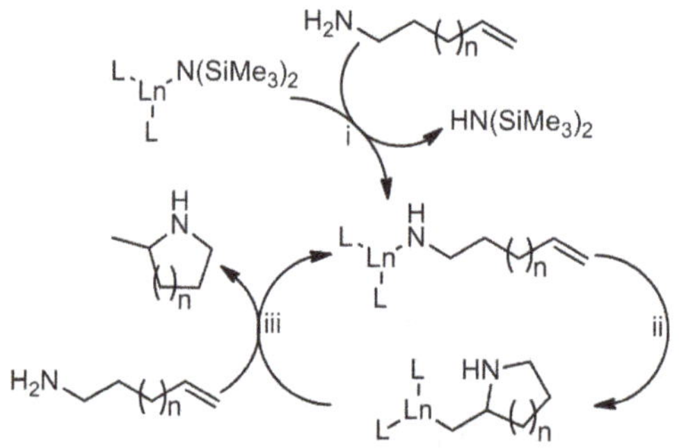

Proposed catalytic cycle for intramolecular hydroamination

Late transition metal hydroamination catalysts have multiple models based on the regioselective determining step. The four main categories are (1) nucleophilic attack on an alkene alkyne, or allyl ligand and (2) insertion of the alkene into the metal-amide bond. Generic catalytic cycles appear below. Mechanisms are supported by rate studies, isotopic labeling, and trapping of the proposed intermediates.

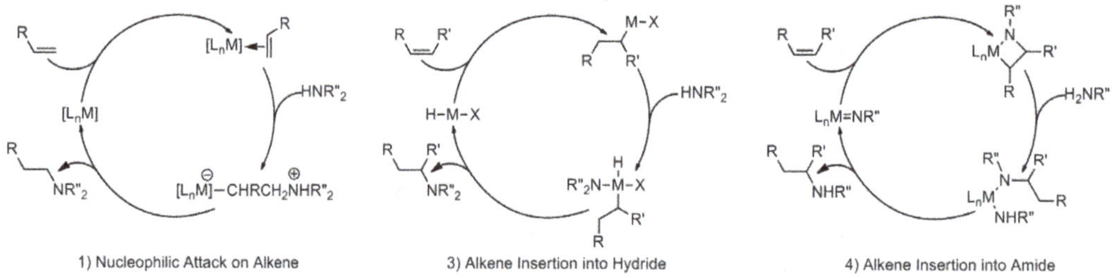

Common catalytic cycles for hydroamination

Thermodynamics and Kinetics

The hydroamination reaction is approximately thermochemically neutral. The reaction however suffers from a high activation barrier, perhaps owing to the repulsion of the electron-rich substrate and the amine nucleophile. The intermolecular reaction also is accompanied by highly negative changing entropy, making it unfavorable at higher temperatures. Consequently, catalysts are necessary for this reaction to proceed. As usual in chemistry, intramolecular processes occur at faster rates than intermolecular versions.

Thermodynamic vs Kinetic Product

In general, most hydroamination catalysts require elevated temperatures to function efficiently, and as such, only the thermodynamic product is observed. The isolation and characterization of the rarer and more synthetically valuable kinetic allyl amine product was reported when allenes was used at the unsaturated substrate. One system utilized temperatures of 80 °C with a rhodium catalyst and aniline derivatives as the amine. The other reported system utilized a palladium catalyst at room temperature with a wide range of primary and secondary cyclic and acyclic amines.

Both systems produced the desired allyl amines in high yield, which contain an alkene that can be further functionalized through traditional organic reactions.

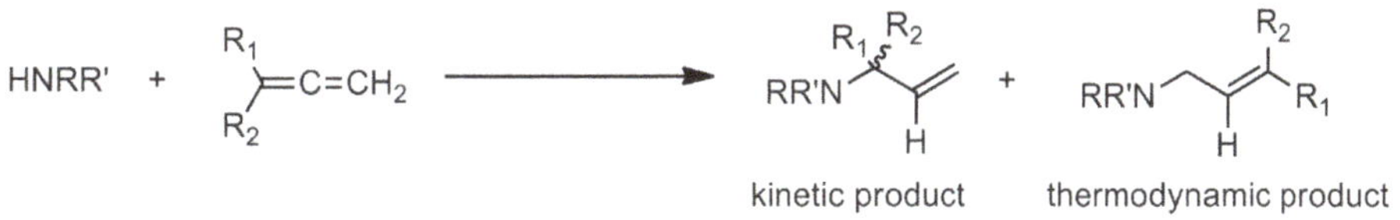

Possible thermodynamic and kinetic products when utilizing an allene

Base Catalyzed Hydroamination

Strong bases catalyze hydroamination, an example being the ethylation of piperidine using ethylene.

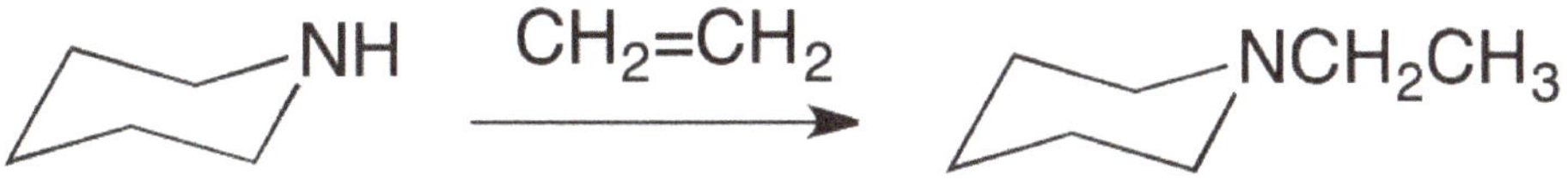

Hydroamination of ethylene with piperidine proceeds with no transition metal catalyst, but requires a strong base.

Such base catalyzed reactions proceed well with ethylene but higher alkenes are less reactive.

Hydroamination Catalyzed by Group (IV) Complexes

Certain titanium and zirconium complexes catalyze intermolecular hydroamination of alkynes and allenes. Both stoichiometric and catalytic variants were initially examined with zirconocene bis(amido) complexes. Titanocene amido and sulfonamido complexes catalyze the intra-molecular hydroamination of aminoalkenes via a [2+2] cycloaddition that forms the corresponding azametallacyclobutane, as illustrated in Figure. Subsequent protonolysis by incoming substrate gives the α-vinyl-pyrrolidine (1) or tetrahydropyridine (2) product. Experimental and theoretical evidence support the proposed imido intermediate and mechanism with neutral group IV catalysts.

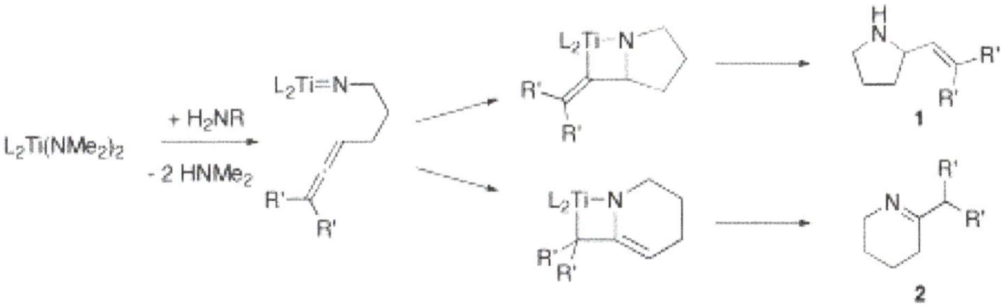

The catalytic hydroamination of aminoallenes to form chiral α-vinyl-pyrrolidine (1) and tetrahydropyridine (2) products. L_2 = Cp_2 or bis(amide).

Formal Hydroamination

The addition of hydrogen and an amino group (NR_2) using reagents other than the amine HNR_2 is known as a "formal hydroamination" reaction. Although the advantages of atom economy and/or ready available of the nitrogen source are diminished as a result, the greater thermodynamic driving force, as well as ability to tune the aminating reagent are potentially useful. In place of the amine, hydroxylamine esters and nitroarenes have been reported as nitrogen sources.

Applications

Hydroamination could find applications due to the valuable nature of the resulting amine, as well as the greenness of the process. Functionalized allylamines, which can be produced through hydroamination, have extensive pharmaceutical application, although presently such species are not prepared by hydroamination. Hydroamination has been utilized to synthesize the allylamine Cinnarizine in quantitative yield. Cinnarizine treats both vertigo and motion sickness related nausea.

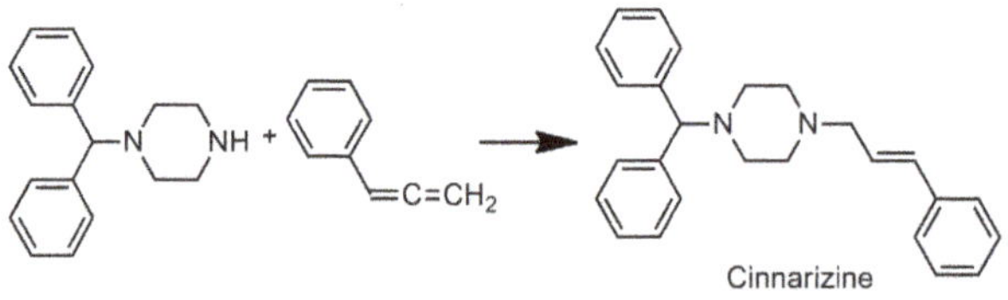

Synthesis of cinnarizine via hydroamination

Hydroamination is also promising for the synthesis of alkaloids. An example was the total synthesis of (-)-epimyrtine.

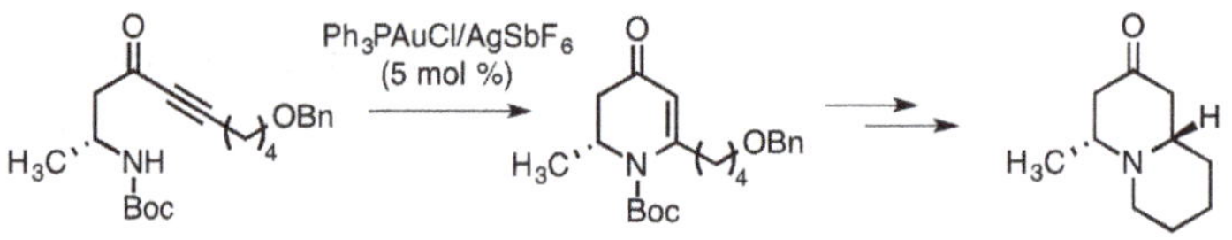

Gold-catalyzed hydroamination used for the total synthesis of (-)-epimyrtine

Hydroalkoxylation of Allenes

Hydroalkoxylation of allenes has been accomplished using 1:2 mixture of the dppm$(AuCl)_2$ and chiral silver phosphonate to give furan with 97% ee.

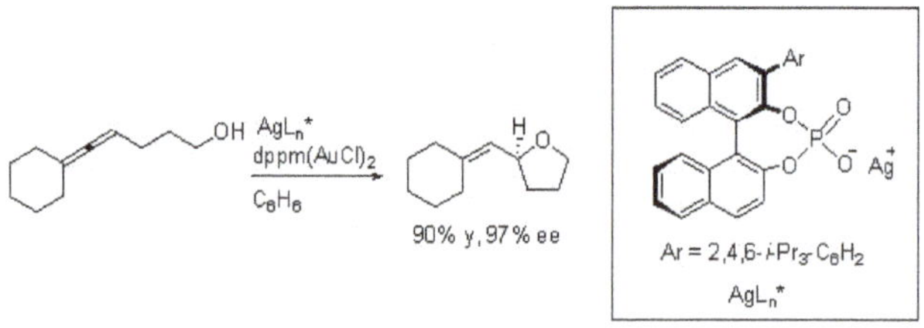

Oxidation Reactions

Wacker-type tandem cyclization reaction of alkenyl alcohol is reported using chiral palladium(II)-spirobis(isoxazoline) with excellent enantioselectivity. In this reaction, benzoquinone reoxidizes the reduced palladium(0) to palladium(II) species to complete the catalytic cycle.

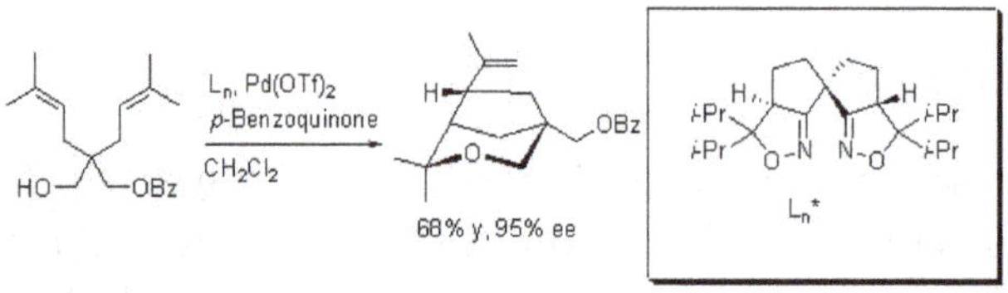

Palladium complex derived from $Pd(TFA)_2$ and (S,S)-BOXAX has been found to be effective for the synthesis of chiral chroman framework in the presence of benzoquinone.

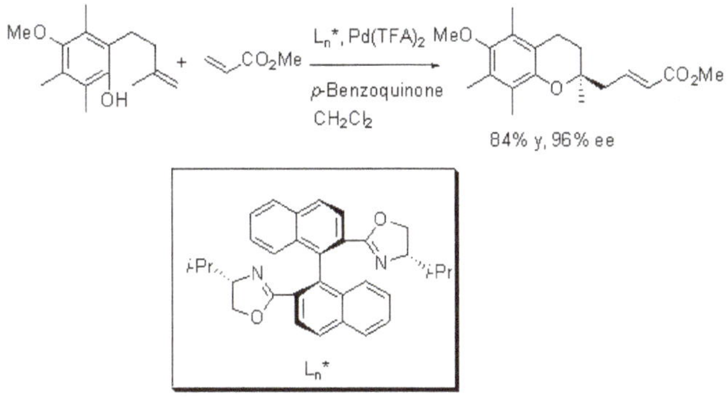

The mercury(II) complex derived from $Hg(TFA)_2$ and bisoxazoline has been used for the mercurio-cyclization with high enantioselectivity.

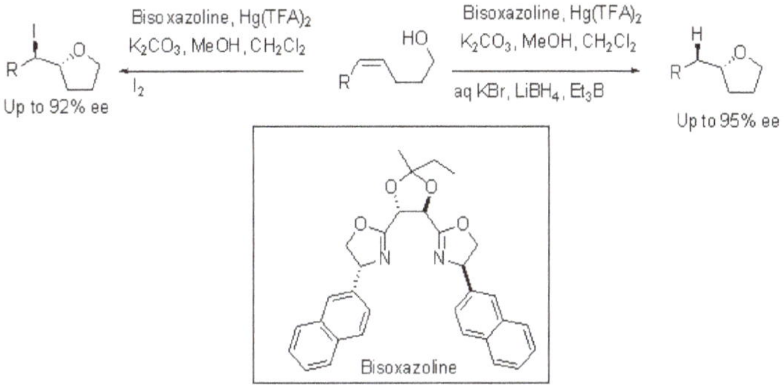

Chiral cobalt(II)-salen has been used for the enantioselective intramolecular iodoetherification to procure 2-substituted tetrahydrofurans with up to 90% ee.

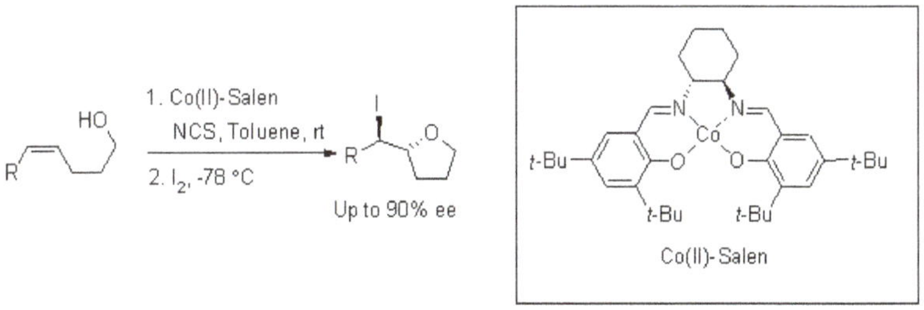

R	Yield [%]	ee [%]
$(CH_2)_3Ph$	94	84
Me	96	67
Et	89	82
n-Pr	85	85
i-Pr	83	73
$(CH_2)_3OTr$	89	90

References

- Brunet, Jean-Jacques; Neibecker, Denis; Niedercorn, Francine (1 February 1989). "Functionalisation of alkenes: catalytic amination of monoolefins". Journal of Molecular Catalysis. 49 (3): 235–259. doi:10.1016/0304-5102(89)85015-1
- Mapp CE (2001). "Agents, old and new, causing occupational asthma". Occup. Environ. Med. 58 (5): 354–60. PMC 1740131. PMID 11303086. doi:10.1136/oem.58.5.354

- Grützmacher, ed. by Antonio Togni; Hansjörg (2001). Catalytic heterofunctionalization: from hydroanimation to hydrozirconation (1. ed.). Weinheim [u.a.]: Wiley-VCH. ISBN 3527302344
- Johns, Adam M.; Sakai, Norio; Ridder, André; Hartwig, John F. (1 July 2006). "Direct Measurement of the Thermodynamics of Vinylarene Hydroamination". Journal of the American Chemical Society. 128 (29): 9306–9307. PMID 16848446. doi:10.1021/ja062773e
- Beller, edited by M.; Bolm, C. (2004). Transition metals for organic synthesis : building blocks and fine chemicals (2nd rev. and enl. ed.). Weinheim ;[Chichester]: Wiley-VCH. ISBN 9783527306138
- Odom, A. L. (2005). "New C–N and C–C bond forming reactions catalyzed by titanium complexes". Dalton Trans. 2 (2): 225–233. PMID 15616708. doi:10.1039/b415701j
- Crabtree, Robert H. (2005). The organometallic chemistry of the transition metals (4th ed.). Hoboken, N.J.: John Wiley. ISBN 0-471-66256-9
- Ryan Hili; Andrei K. Yudin (2006). "Readily Available Unprotected Amino Aldehydes". J. Am. Chem. Soc. 128 (46): 14772–3. doi:10.1021/ja065898s
- Weast, Robert C.; et al. (1978). CRC Handbook of Chemistry and Physics (59th ed.). West Palm Beach, FL: CRC Press. ISBN 0-8493-0549-8
- Crimmin, Mark R.; Casely, Ian J.; Hill, Michael S. (1 February 2005). "Calcium-Mediated Intramolecular Hydroamination Catalysis". Journal of the American Chemical Society. 127 (7): 2042–2043. PMID 15713071. doi:10.1021/ja043576n
- Aravinda B. Pulipaka; Stephen C. Bergmeier (2008). "A Synthesis of 6-Azabicyclo[3.2.1]octanes. The Role of N-Substitution". J. Org. Chem. 73 (4): 1462–7. doi:10.1021/jo702444c
- Some Aziridines, N-, S- and O-Mustards and Selenium (PDF). IARC Monographs on the Evaluation of Carcinogenic Risks to Humans. 9. 1975. ISBN 92-832-1209-6
- Hong, Sukwon; Marks, Tobin J. (1 September 2004). "Organolanthanide-Catalyzed Hydroamination". Accounts of Chemical Research. 37 (9): 673–686. doi:10.1021/ar040051r

4

A Comprehensive Study of Hydrosilylation

The addition of Si-H bonds across unsaturated bonds is called hydrosilation. Alkyl and vinyl silanes are the products of alkenes and alkynes while silyl ethers are produced by aldehydes and ketones. The topics discussed in the chapter are of great importance to broaden the existing knowledge on asymmetric synthesis.

Hydrosilylation

Hydrosilylation, also called catalytic hydrosilation, describes the addition of Si-H bonds across unsaturated bonds. Ordinarily the reaction is conducted catalytically and usually the substrates are unsaturated organic compounds. Alkenes and alkynes give alkyl and vinyl silanes; aldehydes and ketones give silyl ethers. The process was first reported in academic literature in 1947. Hydrosilylation has been called the "most important application of platinum in homogeneous catalysis."

Scope and Mechanism

The catalytic transformation represents an important method for preparing organosilicon compounds. An idealized transformation is illustrated by the addition of triethylsilane to diphenylacetylene:

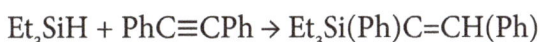

$$Et_3SiH + PhC{\equiv}CPh \rightarrow Et_3Si(Ph)C{=}CH(Ph)$$

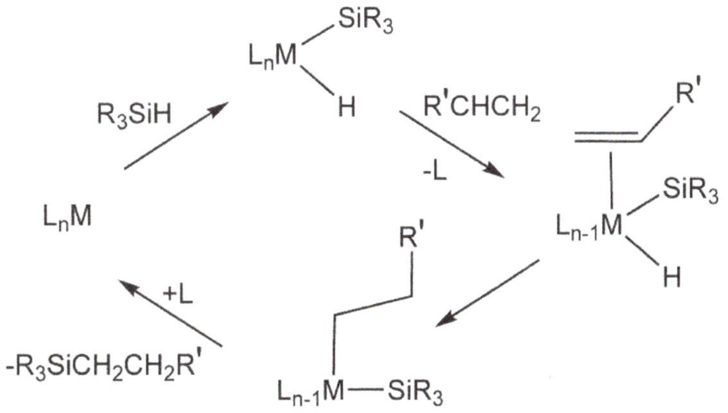

Idealized mechanism for metal-catalysed hydrosilylation of an alkene.

The reaction is related mechanistically to hydrogenation, and similar catalysts are sometimes employed for the two catalytic processes. Popular industrial catalysts are "Speier's catalyst," H_2PtCl_6, and Karstedt's catalyst (an alkene-stabilized platinum(0) catalyst. One prevalent mechanism, called the Chalk-Harrod mechanism, assumes an intermediate metal complex that contains a hydride, a silyl ligand (R_3Si), and the alkene substrate. The reaction usually produces anti-Markovnikov addition alkane, i.e., silicon on the terminal carbon. Variations of the Chalk-Harrod mechanism. Some cases involve insertion of alkene into M-Si bond followed by reductive elimination. In certain cases, hydrosilylation results in vinyl or allylic silanes resulting from beta-hydride elimination.

These reactions can also be catalyzed using nanomaterial-based catalysts.

Asymmetric Hydrosilylation

Using chiral phosphines as spectator ligands, catalysts have been developed for catalytic asymmetric hydrosilation. A well studied reaction is the addition of trichlorosilane to styrene to give 1-phenyl-1-(trichlorosilyl)ethane:

$$Cl_3SiH + PhCHCH_2 \rightarrow (Ph)(CH_3)CHSiCl_3$$

Nearly perfect enantioselectivities (ee's) can be achieved using palladium catalysts supported by binaphthyl-substituted monophosphine ligands.

Surface Hydrosilylation

Silicon wafers can be etched in hydrofluoric acid (HF) to remove the native oxide and form a hydrogen-terminated silicon surface. The hydrogen-terminated surfaces undergo hydrosilation with unsaturated compounds (such as terminal alkenes and alkynes), to form a stable monolayer on the surface. For example:

$$\text{Si-}H + H_2C{=}CH(CH_2)_7CH_3 \rightarrow \text{Si-}CH_2CH H\text{-}(CH_2)_7CH_3$$

The hydrosilylation reaction can be initiated with UV light at room temperature or with heat (typical reaction temperature 120-200 °C), under moisture- and oxygen-free conditions. The resulting monolayer, which is stable and inert, inhibits oxidation of the base silicon layer, relevant to various device applications.

Unsaturated Hydrocarbon

Unsaturated hydrocarbons are hydrocarbons that have double or triple covalent bonds between adjacent carbon atoms. Those with at least one carbon-to-carbon double bond are called alkenes and those with at least one carbon-to-carbon triple bond are called alkynes. The position of the double or triple bond is shown by a number written either at the start of the name, or just before the -ene or -yne suffix (e.g. pent-2-ene and 2-butyne). The number represents the position of the first of the two carbons making the bond, in the longest carbon chain. Alkenes and alkynes with more than one double or triple bond respectively are named with a prefix preceding the -ene or -yne (e.g. 2,4-pentadiene).

Unsaturated hydrocarbons with both double and triple bonds have the suffix -enyne and are handled in a similar manner.

The physical properties of unsaturated hydrocarbons are very similar to those of the corresponding saturated compounds. They are slightly soluble in water.

Except for aromatic compounds, unsaturated hydrocarbons are highly reactive and undergo addition reactions to their multiple bonds. Typical reagents added are hydrogen halides, water, sulfuric acid, elemental halogens and alcohols.

Testing

To test whether the hydrocarbon is unsaturated one would typically use the bromine test, which involves the addition of bromine water to the hydrocarbon in question; if the bromine water is decolourised by the hydrocarbon it can then be concluded that the hydrocarbon is unsaturated.

Hydrosilylation of Alkenes

Asymmetric hydrosilylation and hydroboration of carbon-carbon double bonds followed by oxidative cleavage of the C-Si and C-B bonds give effective methods for the construction of optically active alcohols.

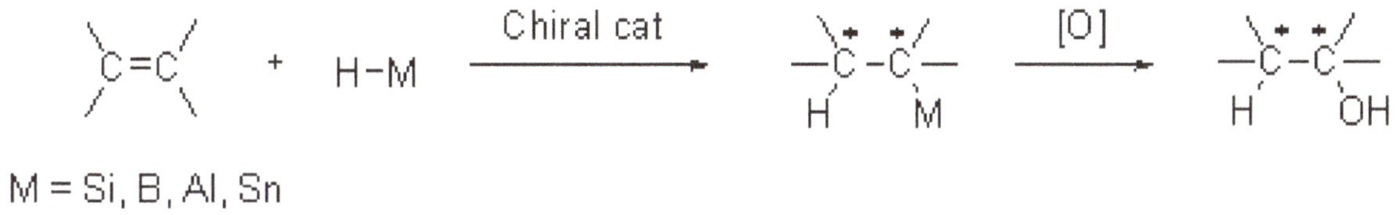

Asymmetric hydrosilylation of carbon-carbon unsaturated substrates provides effective methods for the synthesis of optically active organosilanes, which are versatile intermediates in organic synthesis. Chiral alkyl and aryl silanes can be converted into optically active alcohols with retention configuration by oxidative cleavage of a carbon-silicon bond into carbon-oxygen bond, while the diastereoselective reaction of chiral allyl- and allenyl silanes with C=O bond can give homoallylic and homopropargylic alcohols.

Reactions of Styrene and its Derivatives

Chiral Palladium-catalyzed asymmetric hydrosilylation of styrene with trichlorosilane has been extensively studied. The reaction proceeds with excellent regioselectivity to give 1-phenyl-1-silylethane via a stable π -benzyl palladium intermediate.

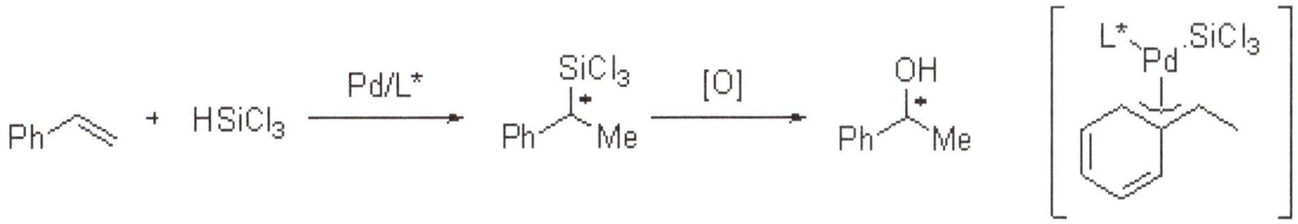

Figure illustrates the possible mechanism. Deuterium-labeling studies suggest that the β -hydrogen elimination is found to be much faster compared to the reductive elimination from the intermediate II . The involvement hydropalladation in the catalytic cycle has been revealed by the side product analysis from the reaction of o -allylstyrene.

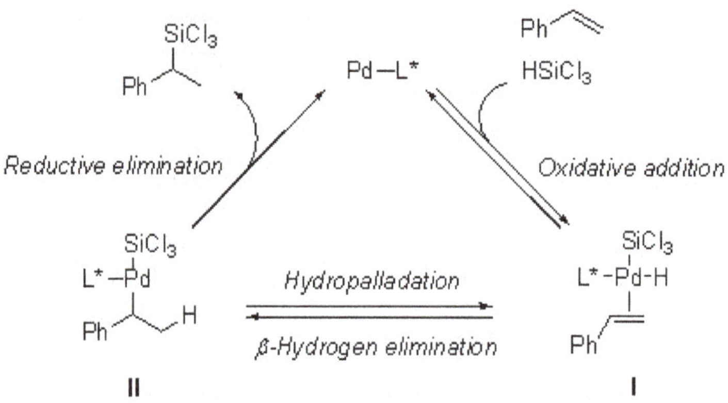

The reaction has been utilized in the synthesis of 1-aryl-1,2-diols from arylacetylenes. Platinum-catalyzed hydrosilylation of arylacetylene gives (E)-1-aryl-2-(trichlorosilyl)ethanes that could be further reacted with trichlorosilane in the presence of chiral palladium complex to afford optically active 1-aryl-1,2-bis(trichlorosilyl)ethanes. The latter could be transformed into optically active 1,2-diol via oxidative cleavage of the carbon-silicon bond into carbon oxygen bond.

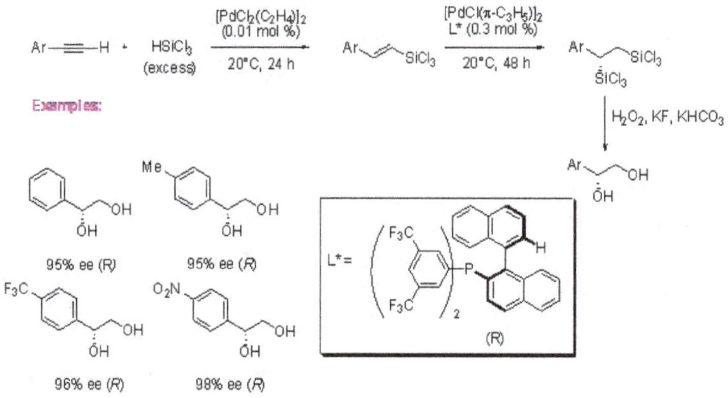

Other chiral catalysts have also been employed for the asymmetric hydrosilylation of alkenes. The chiral bis (oxazolinyl) phenylrhodium complex catalyzes the asymmetric hydrosilylation of styrenes with hydro(alkoxy) silanes in high enantioselectivity, although the regioselectivity is found to be somewhat moderate.

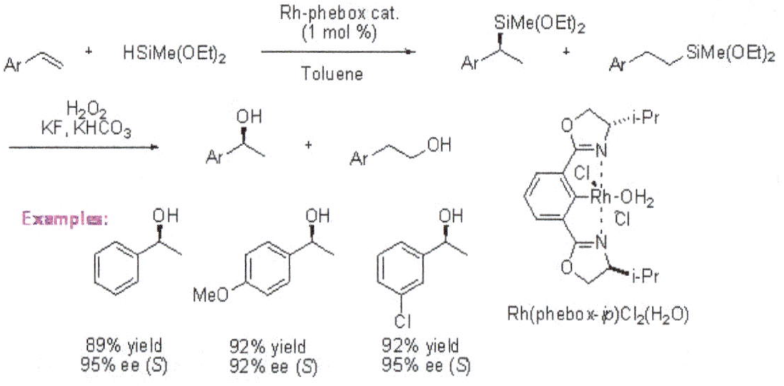

α -Substituted styrenes proceed reaction with phenylsilane to afford benzylic tert-alkylsilanes in the presence chiral organolanthanide as catalyst in moderate enantioselectivity.

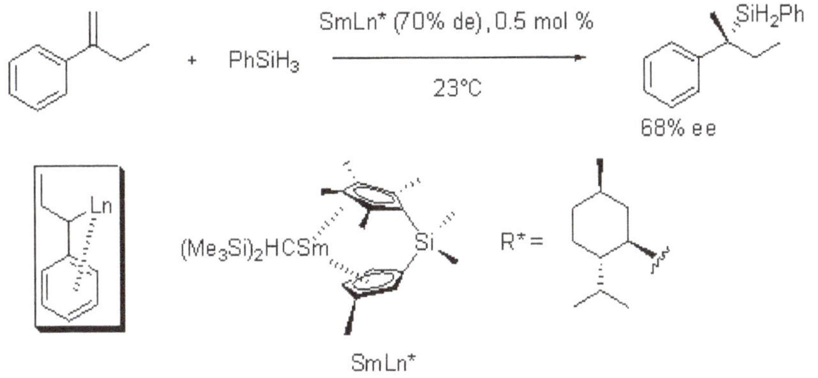

Reactions of 1,3-Dienes

The reaction of 1,3-dienes with hydrosilanes having electron-withdrawing groups on silicon affords synthetically useful optically active silanes in the presence of chiral palladium complex. The reaction proceeds in a 1,4-fashion providing chiral allylsilanes that could be converted into homoallylic alcohols on the reaction with aldehydes.

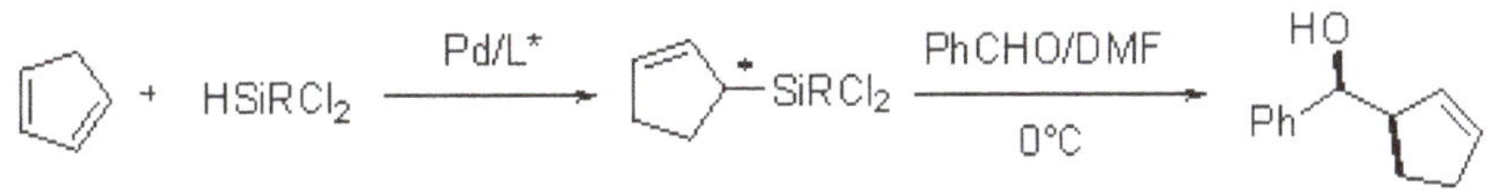

The use of ferrocenylphosphine and mop-phen ligands has been demonstrated for the hydrosilylation of cyclo-1,3-hexadiene in the presence of palladium salts. The reaction with phenyldifluorosilane afforded the highest enantioselectivity compared to that with trichlorsilane or methyldichlorosilane. Based on the reaction of with deuterium-labeled silane the involvement of π-allylpalladium intermediate and 1,4-cis-addition has been proposed.

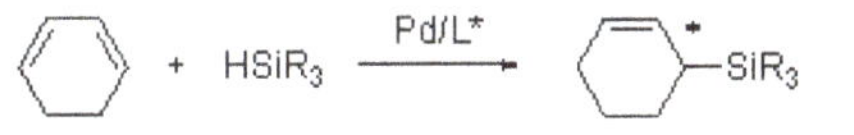

Ligand	$HSiR_3$	Yield (%)	ee (%)
(*R*)-(*S*)-ppfa	$HSiMeCl_2$	95	2 (*S*)
(*R*)-(*S*)-ppfOAc	$HSiCl_3$	44	38 (*S*)
(*R*)-(*S*)-ppfOAc	$HSiF_2Ph$	58	77 (*S*)
(*R*)-(*S*)-ppfOMe	$HSiF_2Ph$	50	54 (*S*)
(*R*)-mop-phen	$HSiCl_3$	80	72 (*S*)

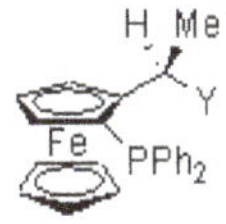

(*R*)-(*S*)-ppfa: Y = NMe_2
(*R*)-(*S*)-ppfOAc: Y = OAc
(*R*)-(*S*)-ppfOMe: Y = OMe

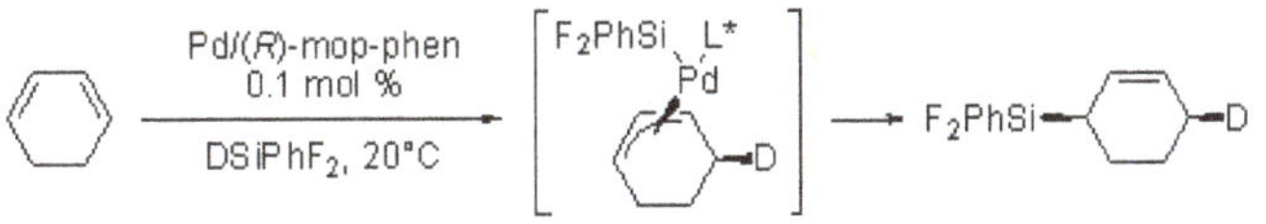

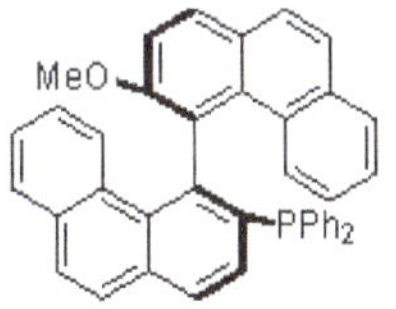

(*R*)-mop-phen

In case of linear 1,3-dienes, the regioselectivity has become an issue. In the reaction of 1-phenyl-1,3-butadiene using ferrocenyl ligand, (R)-(S)-ppfa, the formation of a mixture of regioisomeric allylsilanes is observed. However, in the reaction of alkyl substituted 1,3-dienes, 1,3-hexadiene and 1,3-decadiene, a single regioisomer is obtained with moderate enantioselectivity. Improvement in the enantioselectivity is observed employing the bis (ferrocenyl)monophophine ligands a-d having two planar chiral ferrocenyl moieties on phosphorus atom.

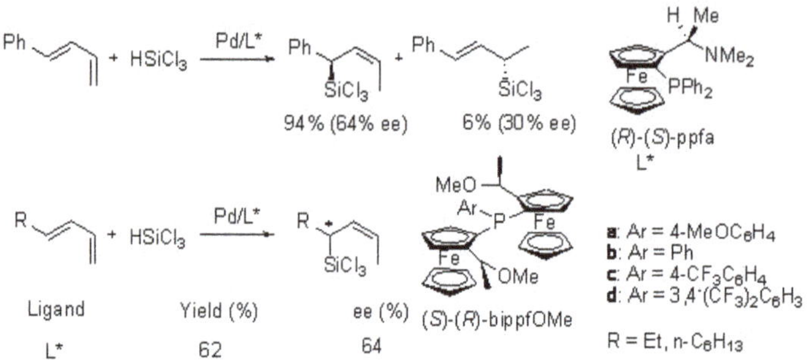

Ligand	Yield (%)	ee (%)
L*	62	64
a	43	68
b	78	76
c	89	78
d	91	87

Reactions of 1,3-Dienes

The reaction of 1-buten-3-ynes substituted with bulky groups at the alkyne terminus affords enantiomerically enriched allenylsilanes in the presence of palladium complex. For example, the reaction of 5,5-dimethyl-1-hexen-3-yne using (S)-(R)-bisppfOMe a proceeds in a 1,4-fashion to give allenyl(trichloro)silanes in high regio-and enantioselectivity. Further enhancement in the enantioselectivity is shown employing chiral phosphametallocene b having a sterically demanding η^5 -C_5 Me_5 moiety.

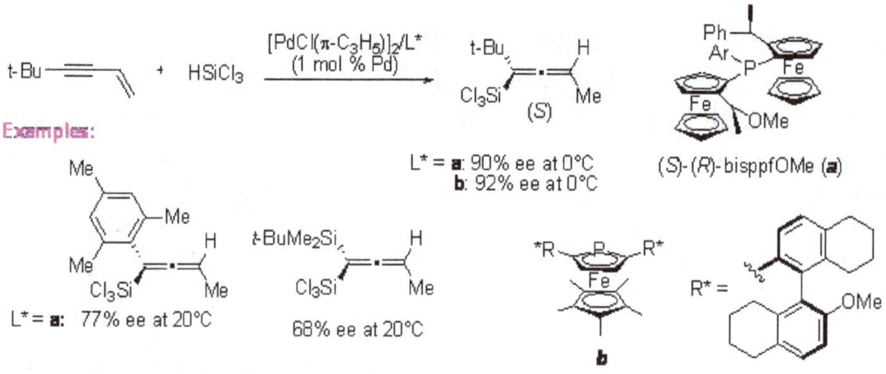

Reactions of Alkyl Substituted Alkenes

Hydrosilylation of simple terminal alkenes give branched products with high regioselectivity. The palladium systems show exceptional catalytic system compared to Pt, Ni and Rh based systems. For example, the hydrosilylation of 1-octene with trichlorosilane using palladium-(S)—MeO-mop gives a 93:7 mixture of 1-octylsilane and 2-octylsilane with 95% ee.

The above catalytic system is also effective for the hydrosilylation of cyclic alkenes, such as norbornene and bicyclo[2.2.2]octane, 2,5-dihydrofuran and norbornadiene. For example, the reaction of norbornene gives exo adduct exclusively. The hydrosilylated product can be transformed into exo -2-norbornanol or endo -2-bromonorbornane via the corresponding pentafluorosilicate. In addition, chiral ferrocenylmonophosphines a-d are too found to be effective for this process with excellent enantioselectivity.

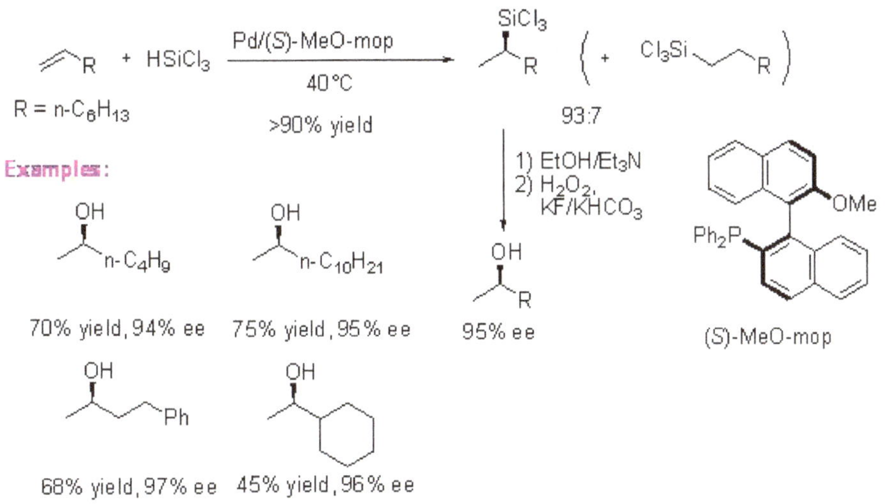

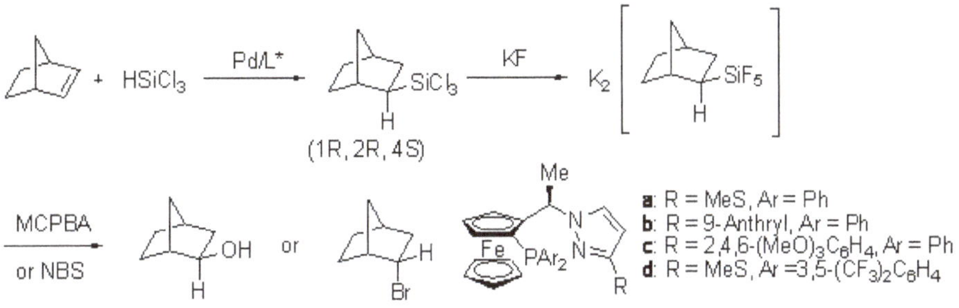

Ligand	Yield (%)	ee (%)
(*R*)-Meo-mop	99	96
a	56	91
b	54	81
c	30	92
d	59	99.5

Chiral yttrium hydride complex (d_0 metal complex) bearing non-Cp ligand catalyzes the hydrosilylation of norbornene with phenylsilane to produce exo -adduct with 90% ee. More recently, the first chirality transfer from silicon to carbon in a reagent-controlled reaction of norbornene is reported in the presence of achiral palladium complex. The hydrosilylation of norbornene with chiral silane A having 85% ee is found to form the hydrosilylated product B with 93% ee exhibiting asymmetric amplification.

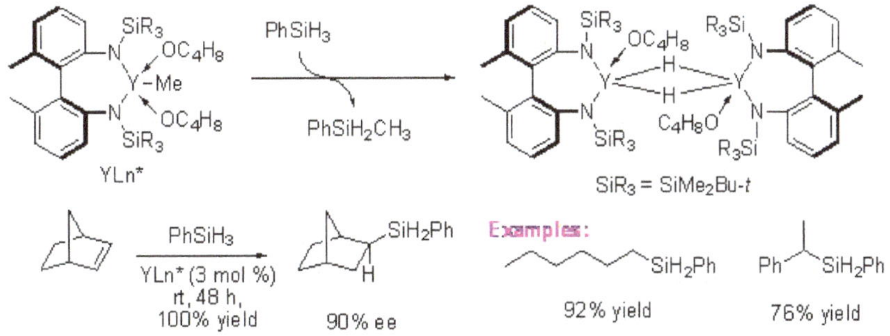

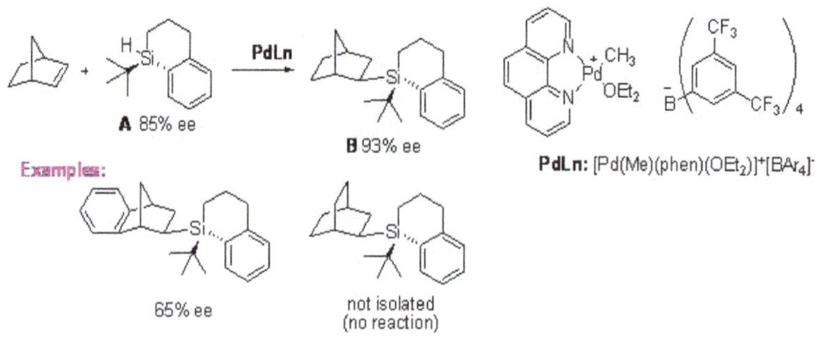

Intramolecular Hydrosilylation

Synthesis of optically active polyols from allylic alcohols can be achieved using chiral Rh-catalyzed intramolecular hydrosilylation followed by oxidation of allyloxy hydrosilanes. For example, hydrosilyl ether of di(2-propenyl)methanol can be converted into optically active 1,3-diol using intramolecular hydrosilylation in the presence of chiral rhodium-(R,R)-diop followed by oxidation. Rh-BINAP is also found to be effective catalyst for the intramolecular hydrosilylation of hydrosilyl ethers of allyl alcohols.

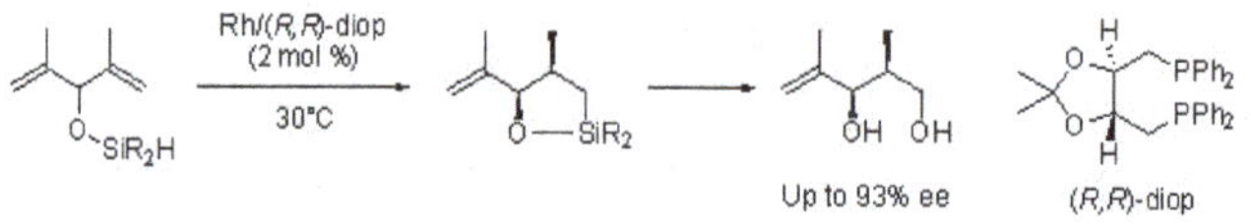

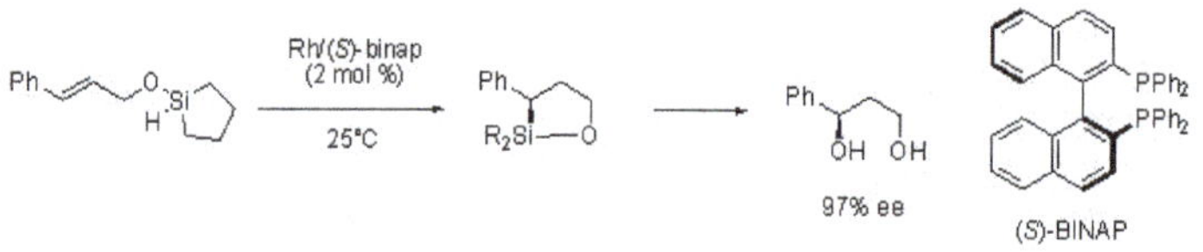

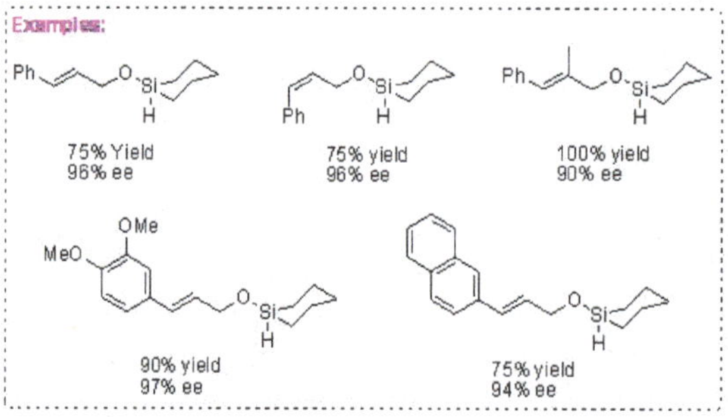

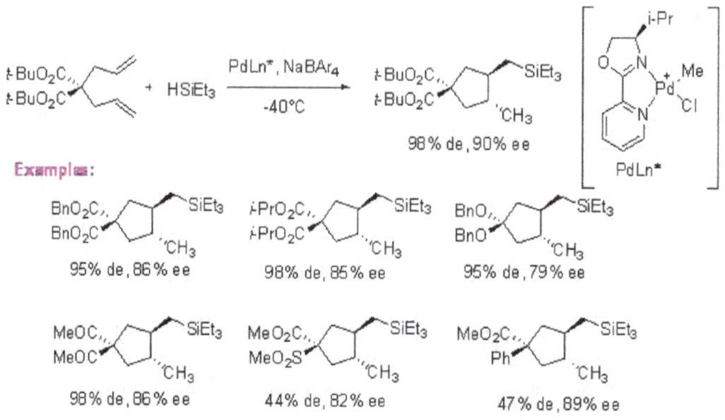

Hydroacylation

Hydroacylation is a type of organic reaction in which an alkene is inserted into the a formyl C-H bond. The product is a ketone. The reaction requires a metal catalyst. It is almost invariably practiced as an intramolecular reaction using homogeneous catalysts, often based on rhodium phosphines.

$$RCHO + CH_2{=}CHR' \rightarrow RC(O)CH_2CH_2R'$$

With an alkyne in place of alkenes, the reaction produce an α,β-unsaturated ketone.

Examples

The reaction was discovered as part of a synthetic route to certain prostanoids. The reaction required tin tetrachloride and a stoichiometric amount of Wilkinson's catalyst. An equal amount of a cyclopropane was formed as the result of decarbonylation.

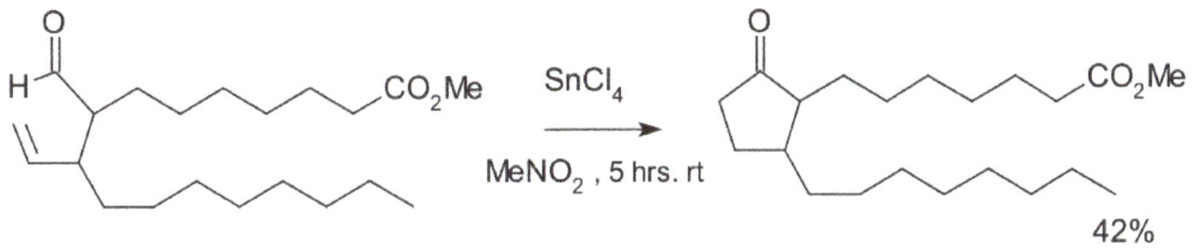

The first catalytic application involved cycization of 4-pentenal to cyclopentanone using with Wilkinson's catalyst. In this reaction the solvent was saturated with ethylene.

$$CH_2{=}CHCH_2CH_2CHO \rightarrow (CH_2)_4CO$$

Reaction Mechanism

Labeling studies establish the following regiochemistry:

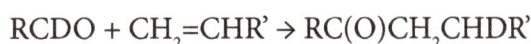

$$RCDO + CH_2{=}CHR' \rightarrow RC(O)CH_2CHDR'$$

In terms of the reaction mechanism, hydroacylation begins with oxidative addition of the aldehydic carbon-hydrogen bond. The resulting acyl hydride complex next binds the alkene. The sequence of oxidative addition and alkene coordination is often unclear. Via migratory insertion, the alkene inserts into either the metal-acyl or the metal-hydride bonds. In the final step, the resulting alkyl-acyl or beta-ketoalkyl-hydride complex undergoes reductive elimination. A competing side-reaction is decarbonylation of the aldehyde. This process also proceeds via the intermediacy of the acyl metal hydride:

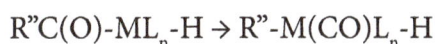

$$R''C(O)\text{-}ML_n\text{-}H \rightarrow R''\text{-}M(CO)L_n\text{-}H$$

This step can be followed by reductive elimination of the alkane:

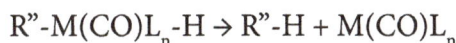

$$R''\text{-}M(CO)L_n\text{-}H \rightarrow R''\text{-}H + M(CO)L_n$$

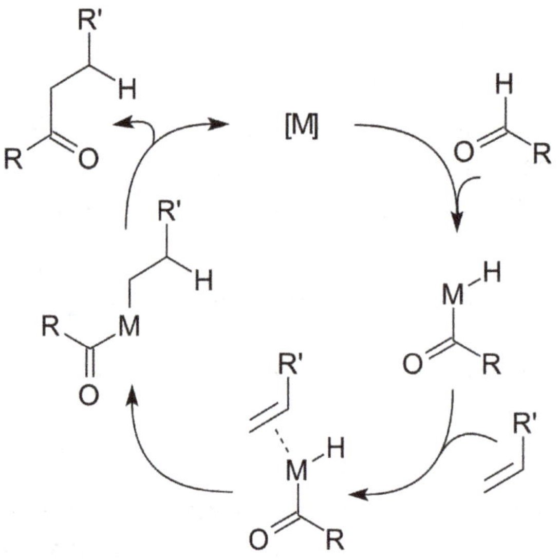

Asymmetric Hydroacylation

Hydroacylation as an asymmetric reaction was demonstrated in the form of a kinetic resolution. A true asymmetric synthesis was also described. Both conversions employed rhodium catalysts and a chiral diphosphine ligand. In one application the ligand is Me-DuPhos:

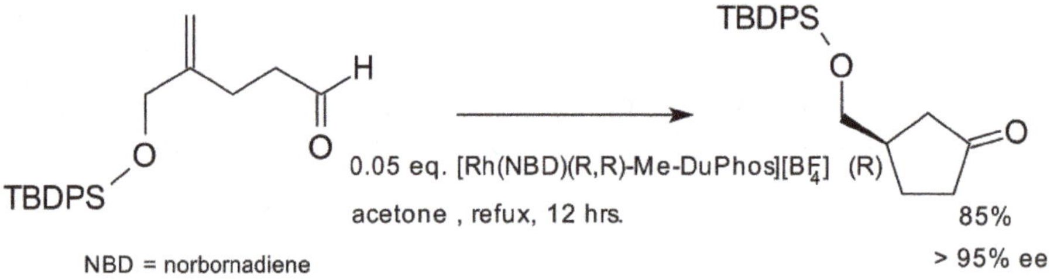

Asymmetric Cyclization

Cyclization/Hydrosilylation

Asymmetric cyclization and hydrosilylation of a ,ω-diunsaturated compounds such as 1,6-dienes and 1,6-enynes affords powerful tool for the construction of optically active functionalized carbocycles. For example, the tandem reaction of diallylmalonate in the presence of cationic Pd complex bearing a chiral pyridine-oxazoline proceeds with high diastereoselectivity to yield the corresponding trans-substituted cyclopentane with 90% ee.

The reactions of 1,6-diynes using cationic Rh complexes bearing chiral bisphosphine gives the hydrosilylated alkylidenecyclopentanes with high enantioselectivity. For example, the 1,6-enyne proceeds reaction with triethylsilane in the presence of cationic Rh and (*R*)-biphemp to give hydrosilylated alkylidene cyclopentane in 92% ee. Subsequently, chiral Rh complex containing spiro diphosphine (*R*)-sdp is found to be effective for this process.

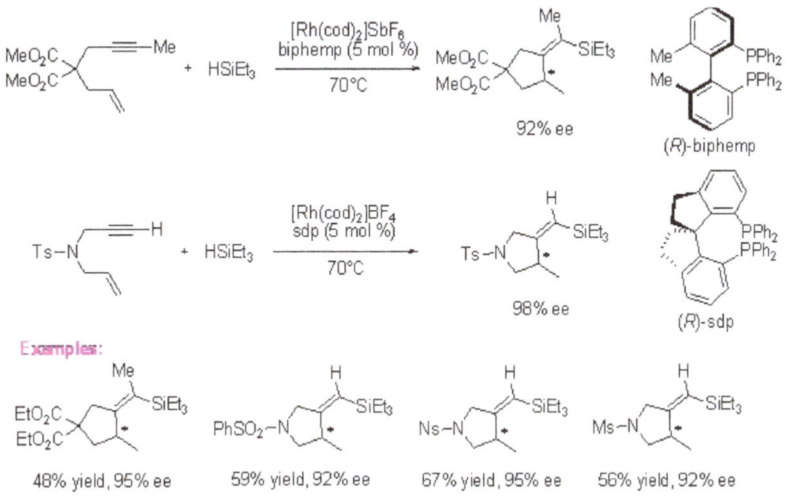

The synthesis of carbocycles can also be accomplished by the cyclization of ω-formyl-1,3-dienes in the presence of hydrosilanes and chiral nickel complex. For example, zerovalent nickel complex of (2 R ,5 R)-2,5-dimethyl-1-phenylphospholane catalyzes the cyclization of 1,3-dienes with a tethered formyl group in the presence of triethoxysilane to give five-membered carbocycle with 73% ee.

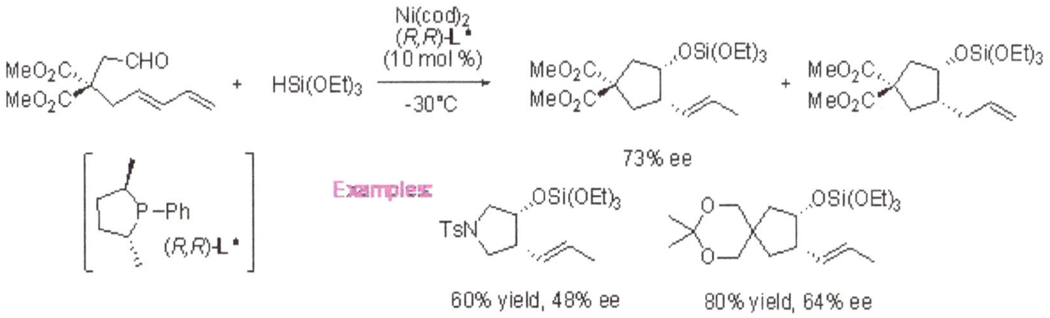

Hydroboration

In chemistry, hydroboration refers to the addition of a hydrogen-boron bond to C-C, C-N, and C-O double bonds, as well as C-C triple bonds. This chemical reaction is useful in the organic synthesis of organic compounds. The development of this technology and the underlying concepts were recognized by the Nobel Prize in Chemistry to Herbert C. Brown. He shared the Nobel prize in chemistry with Georg Wittig in 1979 for his pioneering research on organoboranes as important synthetic intermediates.

Hydroboration produces organoborane compounds that react with a variety of reagents to produce useful compounds, such as alcohols, amines, alkyl halides. The most widely known reaction of the organoboranes is oxidation to produce alcohols typically by hydrogen peroxide. This type of reaction has promoted research on hydroboration because of its mild condition and a wide scope of olefins tolerated. Another research subtheme is metal-catalysed hydroboration.

Addition of a H-B Bond to C-C Double Bonds

Hydroboration is typically anti-Markovnikov, i.e. the hydrogen adds to the most substituted carbon of the double bond. That the regiochemistry is reverse of a typical HX addition reflects the polarity of the $B^{\delta+}$-$H^{\delta-}$ bonds. Hydroboration proceeds via a four-membered transition state: the hydrogen and the boron atoms added on the same face of the double bond. Granted that the mechanism is concerted, the formation of the C-B bond proceeds slightly faster than the formation of the C-H bond. As a result, in the transition state, boron develops a partially negative charge while the more substituted carbon bears a partially positive charge. This partial positive charge is better supported by the more substituted carbon.

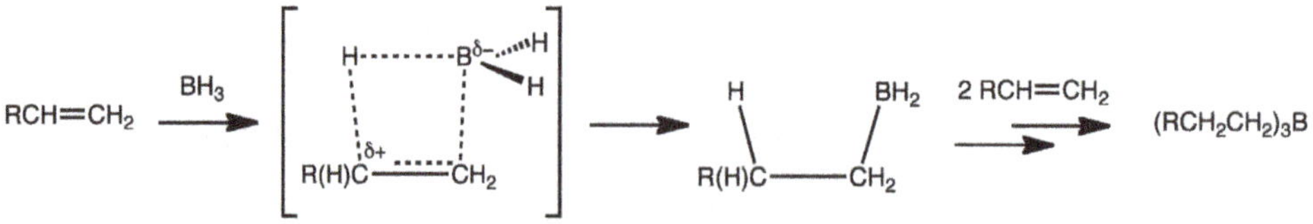

Hydroboration of a terminal alkene to a trialkylborane, showing idealized image of the cyclic transition state.

If BH_3 is used as the hydroborating reagent, reactions typically proceed beyond the monoalkyl borane compounds, especially for less sterically hindered small olefins. Trisubstituted olefins can rapidly produce dialkyl boranes, but further alkylation of the organoboranes is slowed because of steric hindrance. This significant rate difference in producing di- and tri-alkyl boranes is useful in the synthesis of bulky boranes that can enhance regioselectivity.

Reactions Involving Substituted Alkenes

For trisubstituted alkenes such as 1, boron is predominantly placed on the less substituted carbon. The minor product, in which the boron atom is placed on the more substituted carbon, is usually produced in less than 10%. A notable case with lower regioselectivity is styrene, and the selectivity is strongly influenced by the substituent on the para position.

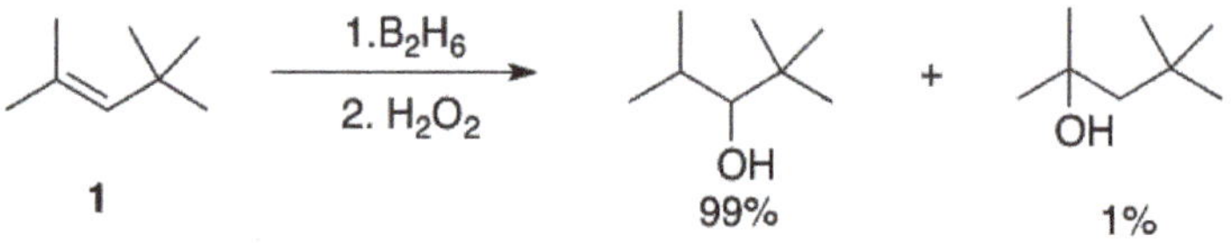

Hydroboration of 1,2-disubstituted alkenes, such as a *cis* or *trans* olefin, produces generally a mixture of the two organoboranes of comparable amounts, even if the substituents are very different in terms of steric bulk. For such 1,2-disubstituted olefins, regioselectivity can be observed only when one of the two substituents is a phenyl ring. In such cases, such as *trans*-1-phenylpropene, the boron atom is placed on the carbon adjacent to the phenyl ring. The observations above indicate that the addition of H-B bond to olefins is under electronic control rather than steric control.

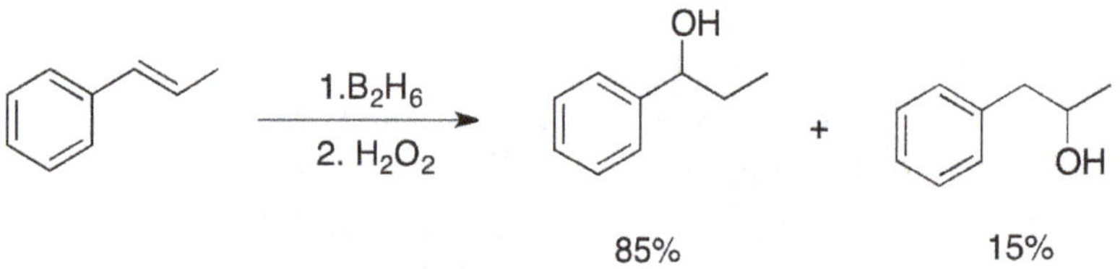

(2)

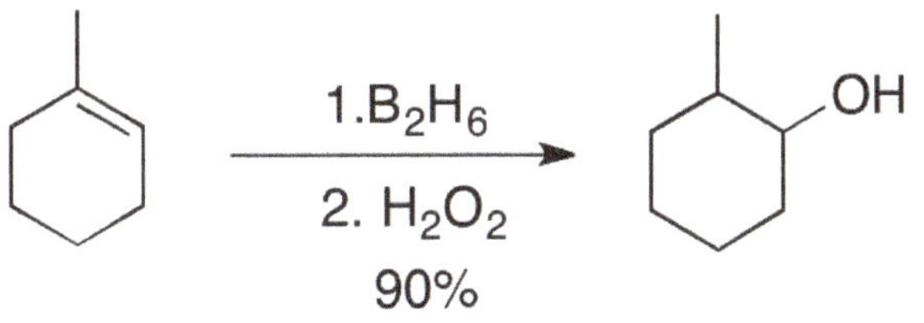

Reactions of Organoboranes

The C-B bonds generated by hydroboration are reactive with various reagents, the most common one being hydrogen peroxide. Because the addition of H-B to olefins is stereospecific, this oxidation reaction will be diastereoselective when the alkene is tri-substituted. Hydroboration-oxidation is thus an excellent way of producing alcohols in a stereospecific and anti-Markovnikov fashion.

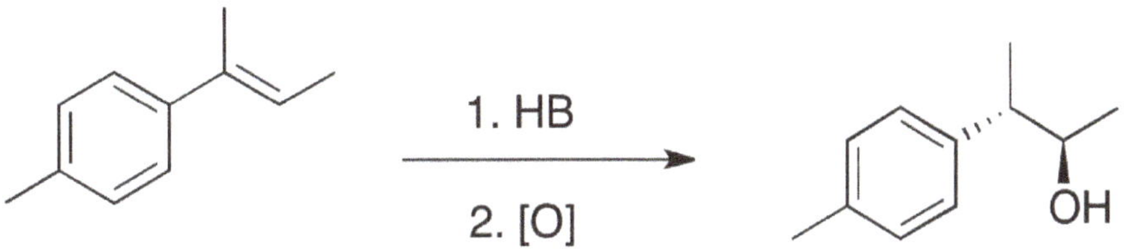

Hydroboration can also lead to amines by treating the intermediate organoboranes with chloramine or O-hydroxylaminesulfonic acid (HSA).

Terminal olefins are converted to the corresponding alkyl bromides and alkyl iodides by treating the organoborane intermediates with bromine or iodine. Such reactions have not however proven very popular, because succinimide-based reagents such as NIS and NBS are more versatile and do not require rigorous conditions as do organoboranes. etc.

Borane Sources

Of the many hydroborating reagents available, borane (BH_3) is commercially available as THF solutions wherein it exists as the adduct BH_3(THF). Long term storage of BH_3/THF requires stabilization by a small amount of sodium borohydride and storage at 0 °C. The concentration of BH_3 usually cannot exceed 2M. The related borane dimethylsulfide complex $BH_3S(CH_3)_2$ (BMS) is comparatively more convenient because it is more stable and it can be obtained in highly concentrated forms. Less volatile sulfides have also been developed for odor control. These borane sulfide adducts are stable at room temperature and soluble in ethers and dichloromethane.

Borane adducts with phosphines and amines are also available. Borane makes a strong adduct with triethylamine; using this adduct requires harsher conditions in hydroboration. This can be advantageous for cases such as hydroborating trienes to avoid polymerization. More sterically hindered tertiary and silyl amines can deliver borane to alkenes at room temperature. Another advantage of these borane complexes is that it is possible to recover the amine carriers.

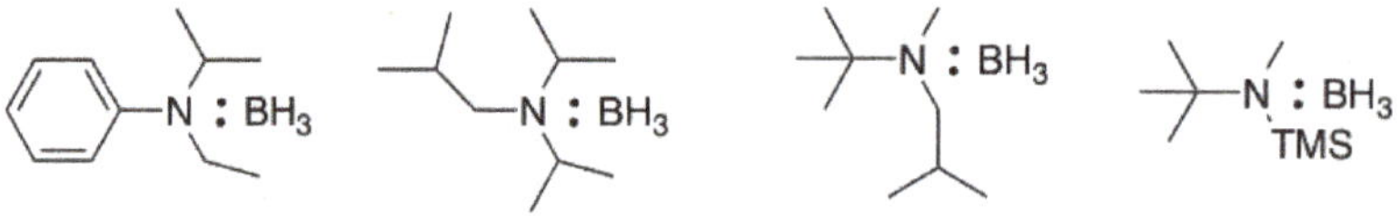

A way of producing borane in situ is to reduce BF_3 with $NaBH_4$.

Monosubstituted Boranes

Monosubstituted boranes of the form RBH_2 are available for R = alkyl and halide. One important example of monoalkyl boranes is $IpcBH_2$, monoisopinocampheylborane. It is available in both enantiomeric forms. Monobromo- and monochloro-borane can be prepared from BMS and the corresponding boron trihalides. The stable complex of monochloroborane and 1,4-dioxane is a superior for selective hydroboration of terminal alkenes.

Disubstituted Boranes

Hydroboration can be highly regio- and diastereoselective by using bulky dialkylborane compounds such as bis-3-methyl-2-butylborane (disiamylborane). Another dialkylborane that exhibits excellent selectivity is 9-borabycyclo[3,3,1]nonane, commonly abbreviated as 9-BBN. Relative to disiamylborane, 9-BBN is more sensitive to electronic influences. Additionally, 9-BBN allows shorter reaction time and higher regioselectivity.

Disiamylborane

Among hindered dialkylboranes, disiamylborane (abbreviated Sia_2BH) is well known for selective hydroboration of less hindered, usually terminal alkenes in the presence of more substituted alkenes. Disiamylborane must be freshly prepared as its solutions can only be stored at 0 °C for a few hours. Dicyclohexylborane, Chx_2BH, is another example that has improved thermal stability than Sia_2BH.

9-BBN

The most versatile among dialkylboranes is 9-BBN. It exists predominantly as a dimer. It can be distilled without decomposition at 195 °C (12mm Hg). Such property allows 9-BBN to react at 60–80 °C, and most alkenes react within one hour in such temperature range. Even tetrasubstituted alkenes undergo hydroboration with 9-BBN at elevated temperature. As mentioned before, 9-BBN has excellent regioselectivity in hydroboration of alkenes. It is more sensitive to subtle steric differences than Sia_2BH, because the rigid hetereocyclic substituents prevent internal rotation to relieve steric hindrance in the transition state. Reflecting its greater sensitivity to electronic factors, 9-BBN is more reactive towards alkenes than alkynes.

Other Secondary Boranes

Simple, unhindered dialkylboranes are reactive at room temperature towards most alkenes and terminal alkynes but are difficult to prepare in high purity, since they exist in equilibrium with mono- and trialkylboranes. One common way of preparing them is the reduction of dialkylhalogenoboranes with metal hydrides. An important synthetic application using such dialkylboranes, such as diethylborane, is the transmetallation of the organoboron compounds to form organozinc compounds. Dimesitylborane $(C_6H_2Me_3)_2BH$ is a particularly bulky secondary borane. Because of severe steric hindrance, it does not react readily even with simple terminal alkenes. Prolonged reaction time is required at room temperature. On the other hand, alkynes undergo monohydroboration with Mes_2BH easily to produce alkenylboranes.

For catalytic hydroboration, Pinacol pinacolborane and catecholborane are widely used. They also demonstrate higher reactivity toward alkynes.

Hydroboration of Alkenes

Chiral Rh catalyzed hydroboration of alkenes provides effective method for the synthesis of optically active organoboranes, which are versatile intermediates in organic synthesis. The carbon-boron bond can be converted into several functional group by subsequent carbon-carbon, carbon-oxygen, boron-carbon or carbon-nitrogen bond-forming reactions with retention of stereochemistry.

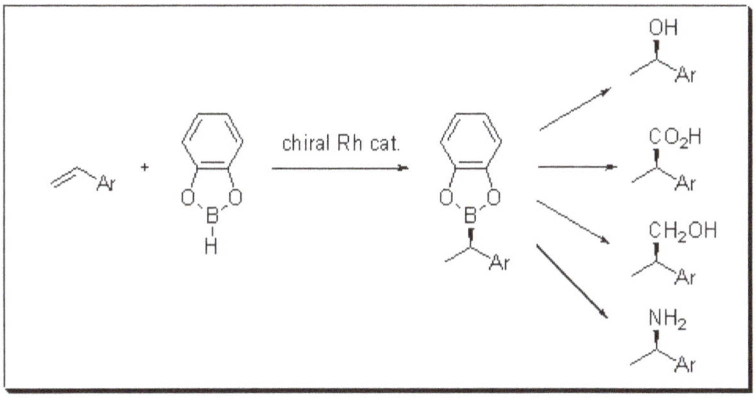

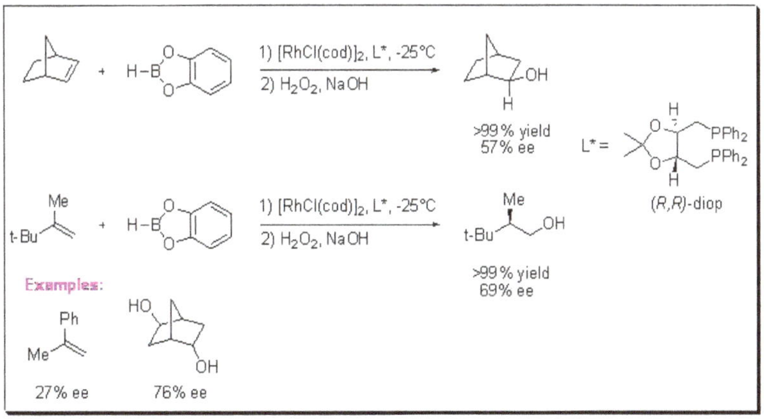

The first catalytic asymmetric hydroboration of norbornene and 2-tert-butylpropene with catecholborane appeared in the presence of Rh-(R, R)-diop complex. The products, 2-hydroxynorbornane and 2,3,3-trimethylbutanol are obtained after the treatment with alkaline hydrogen peroxide solution.

The use of the combination of chiral borane and achiral catalyst has been demonstrated for the asymmetric hydroboration. For example, the hydroboration of 4-methoxystyrene proceeds with chiral borane derived from pseudoephedrine in the presence of achiral rhodium complex to the corresponding secondary alcohol with 76% ee after the oxidation.

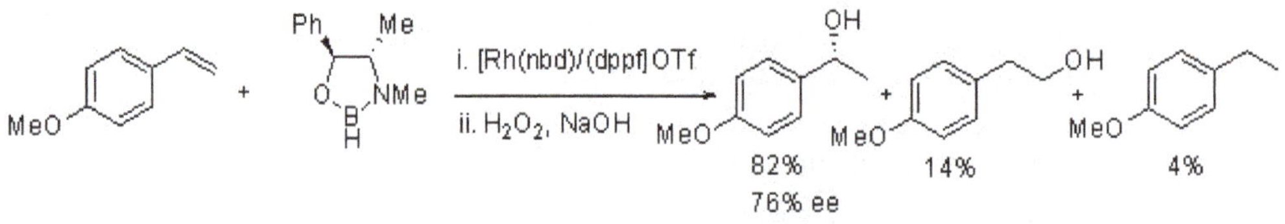

The reaction of vinylarenes with catecholborane has been extensively studied using chiral Rh complex. For example, the cationic Rh-(R)-BIANP catalyzes the hydroboration of styrene with complete branch selectivity to afford 1-phenylethanol with 96% ee after oxidation. The regioselectivity is opposite to that observed with uncatalyzed reactions.

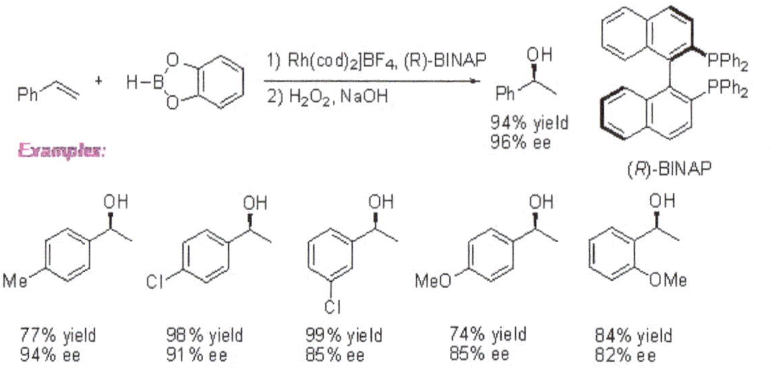

Asymmetric desymmetrization of meso -bicyclic hydrazines has been shown with catecholborane using chiral Rh and Ir-based complexes. A reversal of enantioselectivity is observed between the Rh and Ir catalysts.

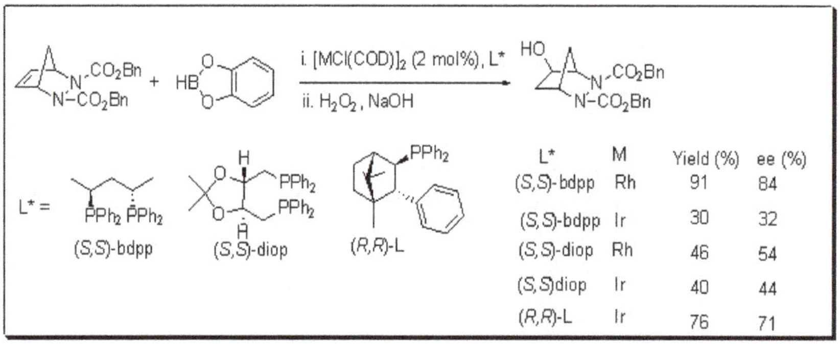

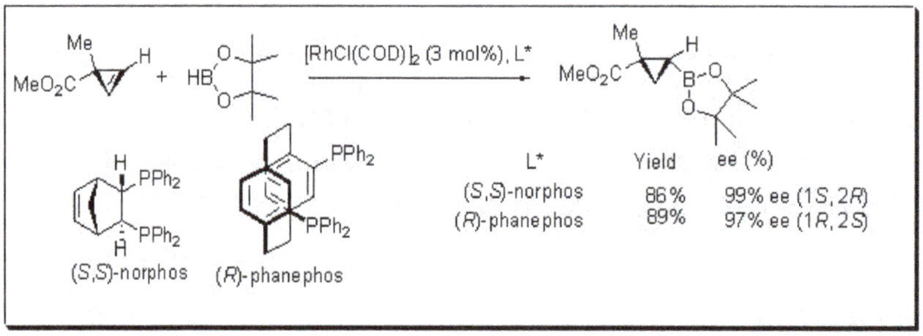

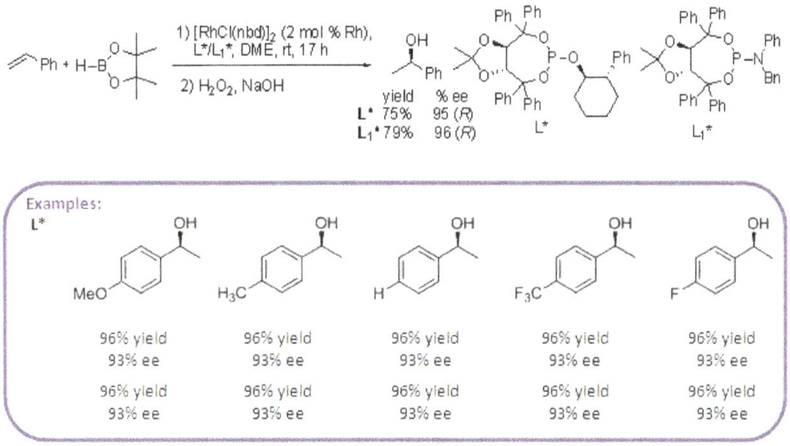

The reaction of cyclopropene is studied with pinacolborane as a new hydroborating agent in the presence of a series of chiral Rh phosphine complexes. The reaction using pinacolborane showed enhanced selectivity compared to that with catecholborane due to steric control between the substrates and the hydroborating agent.

Rh complexes with chiral monodentate phosphate and phosphoramidite derived from taddol are studied for the hydroboration of vinylarenes with pinacolborane. The reactions of a series of vinylarenes having electron withdrawing- and donating substituted proceed with high enantioselectivity.

Hydroalumination and Hydrostannation of Alkenes

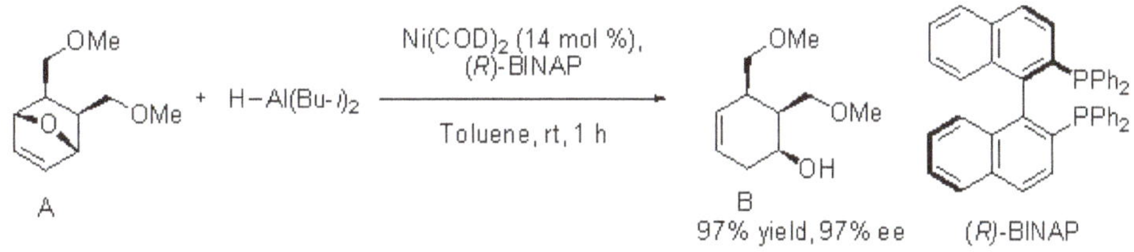

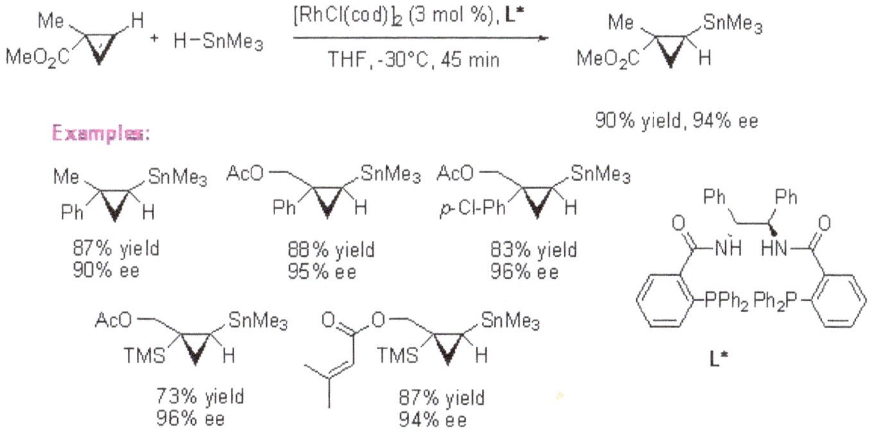

While the catalytic asymmetric hydrosilylation and hydroboration reactions are well known, the catalytic hydroalumination and hydrostannation of alkenes are rare. Chiral nickel complex is used for the asymmetric hydroalumination of oxabicyclic alkenes. For example, Ni-(R)-BINAP catalyzes the reaction of A with iso -Bu_2 AlH to give B with 97% ee.

The first example for the asymmetric hydrostannation of cyclopropenes is appeared using Rh-complex bearing chiral diphenylphosphinobenzoic acid-derived L*. The product trans- cyclopropylstannane is obtained with 94% ee. The procedure is general and the reaction a series of substituted cyclopropenes is demonstrated.

Reactions of Alkenyl- and Alkynylaluminium Compounds

Reactions of alkenyl- and alkynylaluminium compounds involve the transfer of a nucleophilic alkenyl or alkynyl group attached to aluminium to an electrophilic atom. Stereospecific hydroalumination, carboalumination, and terminal alkyne metalation are useful methods for generation of the necessary alkenyl- and alkynylalanes.

Introduction

Aluminium, like its congener boron, is less electronegative than carbon (Al, 1.61; C, 2.55); thus, aluminium-bound carbons in organoalanes possess partial negative charge and are nucleophilic as a result. Generally, however, organoalanes are not nucleophilic enough to transfer an organic group on their own (the exception being when carbonyl and enone acceptors are used, due to the high oxophilicity of aluminium). In most cases, nucleophilic activation of organoalanes is necessary for group transfer to take place. Like organoboranes, organoalanes possess an empty p orbital on the aluminium center that can receive electron density from an added nucleophile. The resulting negatively charged aluminate is much more nucleophilic than the neutral alane.

This concept has been applied to methods for the synthesis of organic compounds from alkenyl- and alkynylalanes. The most notable applications are methods for the stereospecific synthesis of olefins. Alkenylalanes, which are easily synthesized with complete stereocontrol through alkyne hydroalumination, transfer the alkenyl unit to a variety of electrophiles. Alkynylalanes are less commonly used because alkali metal acetylides can be used for many of the same transformations as alkynylalanes. However, alkynylalanes are useful for the coupling of tertiary halides and alkynes (a reaction difficult to effect with alkali metal alkynes) and for conjugate addition and epoxide opening reactions.

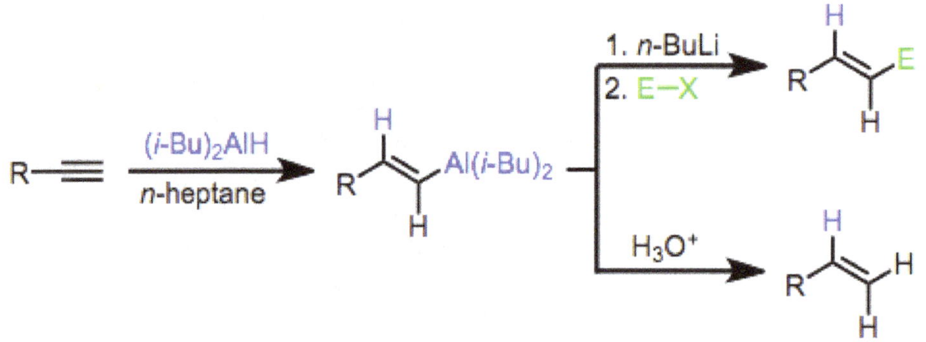

Mechanism and Stereochemistry

Hydroalumination of Alkynes

Hydroalumination of alkynes may be either stereospecifically *cis* or *trans* depending on the conditions employed. When a dialkylalane such as di(isobutyl)aluminium hydride (DIBAL-H) is used, the hydrogen and aluminium delivered from the reagent end up *cis* in the resulting alkenylalane. This stereospecificity can be explained by invoking a concerted addition of the H–Al bond across the triple bond. In the transition state, partial positive charge builds up on the carbon forming a bond to hydrogen; thus, the carbon better able to stabilize a positive charge becomes attached to hydrogen in the product alkenylalane. Hydroaluminations of terminal alkynes typically produce terminal alkenylalanes as a result. Selectivity in hydroaluminations of internal alkynes is typically low, unless an electronic bias exists in the substrate (such as a phenyl ring in conjugation with the alkyne).

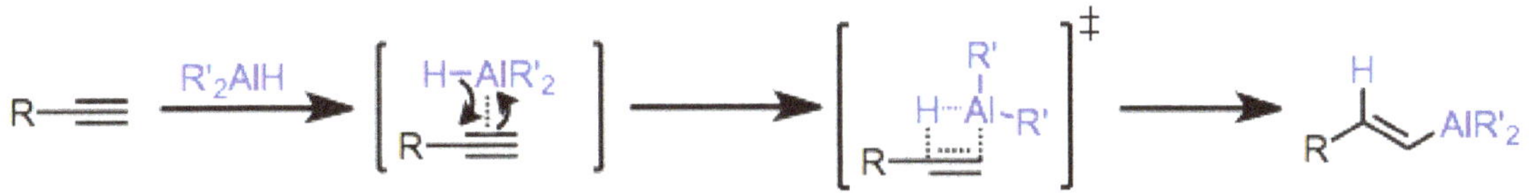

Stereospecific *trans* hydroalumination is possible through the use of lithium aluminium hydride. The mechanism of this transformation involves the addition of hydride to the carbon less able to stabilize the developing negative charge (viz., in the β position to an electron-withdrawing group). Coordination of aluminium to the resulting *trans* vinyl carbanion leads to the observed *trans* configuration of the product.

Reactions of Alkenyl- and Alkynylalanes and Aluminates

Neutral alanes are not nucleophilic enough to deliver organic groups to electrophilic substrates. However, upon activation by a nucleophile, the resulting aluminates are highly nucleophilic and add to electrophiles with retention of configuration at the migration carbon. Thus, stereospecific hydroalumination followed by nucleophilic attack provides a method for the stereospecific synthesis of olefins from alkynes.

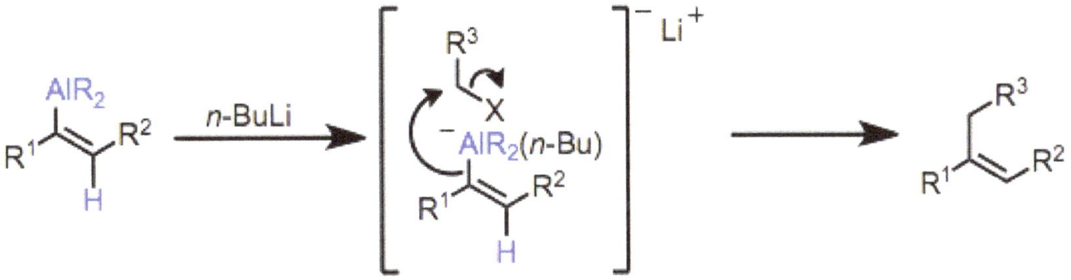

Scope and Limitations

Because unsaturated alanes are oxygen- and moisture-sensitive, they are most often prepared for immediate use without isolation. However, the method of preparation determines the configuration of the intermediate unsaturated alane, which is directly related to the configuration of the product (transfer of the alkenyl group occurs with retention of configuration). Thus, an understanding of available hydroalumination methods is important for the study of reactions of unsaturated alanes. This second describes the most common methods of hydroalumination, and subsequent chemical reactions that the resulting alkenylalanes may undergo.

Preparation of Alkenylalanes

Stereospecific *cis*-hydroalumination is possible through the use of dialkylalanes. The most common reagent used for this purpose is di(isobutyl)aluminium hydride (DIBAL-H). Analogous to hydroboration reactions with R_2BH, hydroalumination with R_2AlH leads to the attachment of aluminium at the carbon less able to stabilize developing positive charge (anti-Markovnikov selectivity). Metalation of terminal alkynes is a significant side reaction that occurs under these conditions. If metalation is desired, tertiary amine complexes of DIBAL-H are useful.

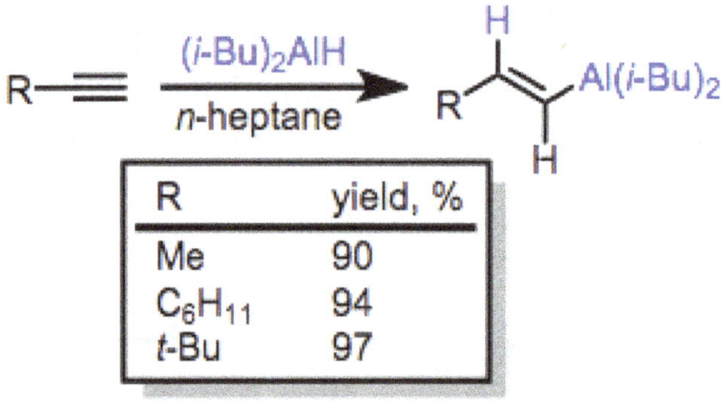

R	yield, %
Me	90
C_6H_{11}	94
t-Bu	97

The use of silyl acetylenes avoids the problem of competitive metalation of terminal alkenes. The stereoselectivity of hydroalumination can be altered through a change in solvent: tertiary amine solvents provide the *trans* alkenylalane and hydrocarbon solvents provide the *cis* isomer.

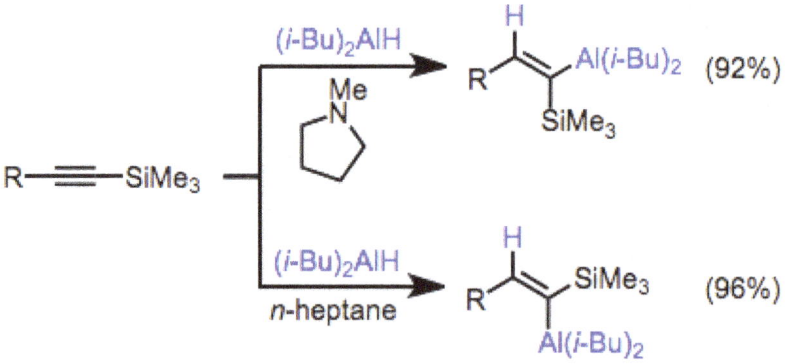

Lithium aluminium hydride hydroaluminates alkynes to afford *trans* alkenylalanes. In equation

hydride adds to the terminal carbon, which places the developing negative charge next to the stabilizing phenyl substituent.

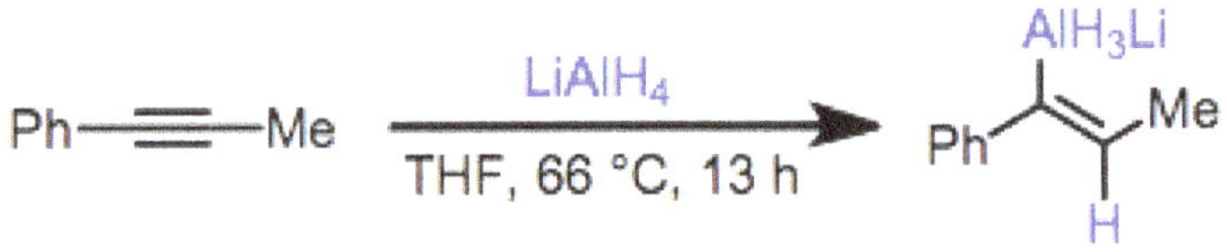

Reactions of Alkenylalanes

Alkenyl- and alkynylaluminates are most commonly generated through the addition of *n*-butyllithium to the alkenylalane. The alkenyl and alkynyl groups, which are better able to stabilize negative charge, are transferred in preference to the alkyl group. When these intermediates react with alkyl halides, functionalized olefins are produced.

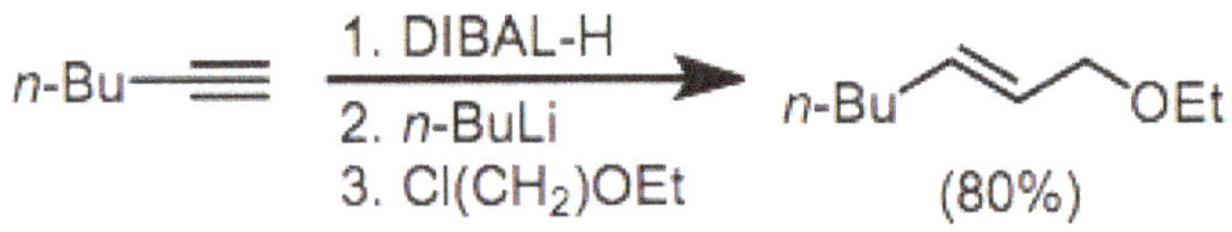

Treatment of alkenylaluminates with halogen electrophiles such as *N*-bromosuccinimide (NBS) and iodine leads to the formation of halogenated olefins. These products are useful for cross-coupling reactions.

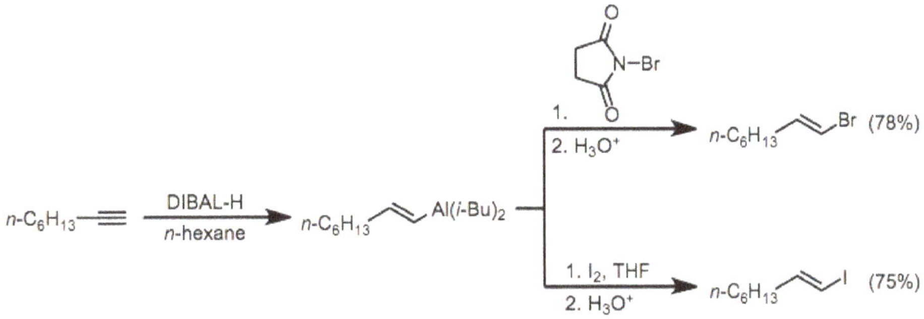

Zirconium-catalyzed carboalumination of alkynes by trimethylalane is a convenient method for accessing substituted alkenylalanes stereoselectively. Upon exposure to aldehydes and ketones, alkenylalanes form secondary or tertiary allylic alcohols. Formaldehyde is a useful reagent in this context for the introduction of a hydroxymethyl unit.

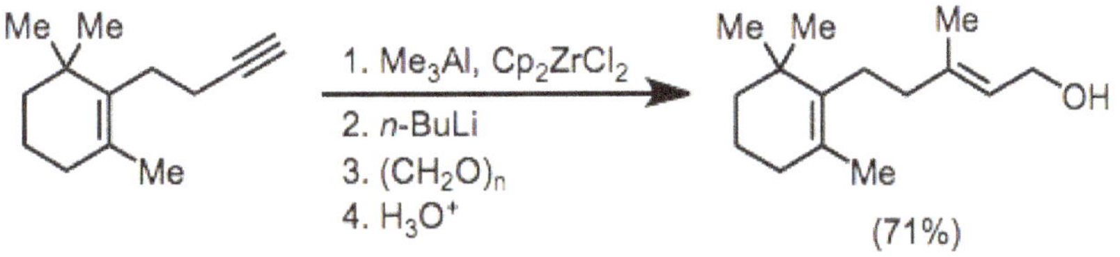

Alkynylalanes are primarily used in place of the corresponding alkali metal acetylides when the latter reagents are ineffective. The coupling of an acetylide and tertiary alkyl halide is an example of a reaction that cannot be accomplished with alkali metal acetylides, which displace halides in an S_N2 fashion.

The corresponding alkynylalanes are able to couple to tertiary halides via an S_N1-like mechanism.

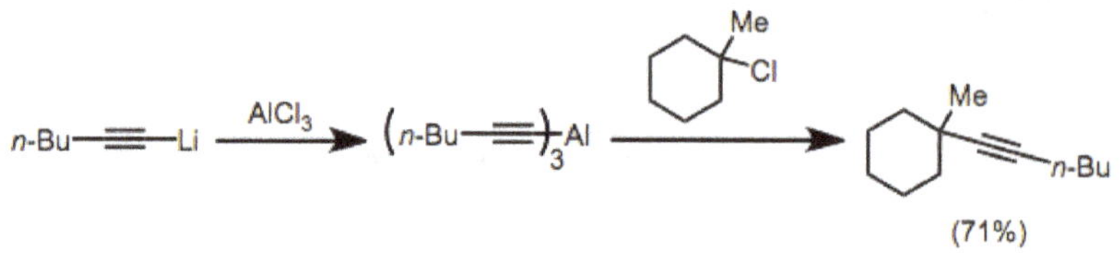

Alkynyl- and alkenylalanes add in a conjugate fashion to enones in the *s-cis* conformation without nucleophilic activation. Enones locked in an *s-trans* conformation, such as cyclohexenone, are unreactive. The coordination of oxygen to aluminium is believed to be necessary for this reaction.

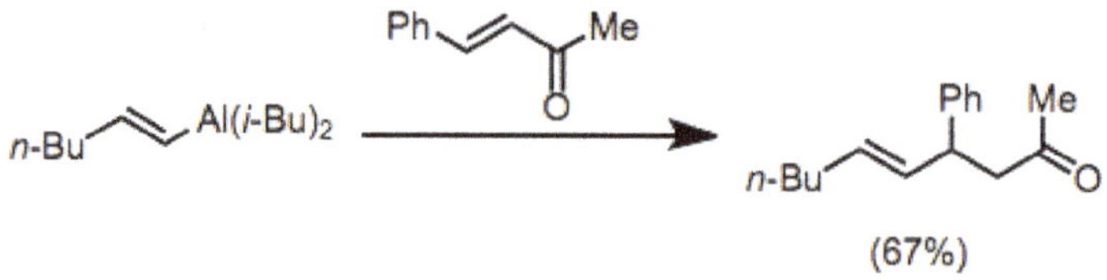

When alkynes and dialkylalanes are combined in a 2:1 ratio, 1,3-dienes result. The aluminium-carbon bond of the initially formed alkenylalane adds across a second molecule of alkyne, forming a conjugated dienylalane. Protonolysis provides the metal-free diene product.

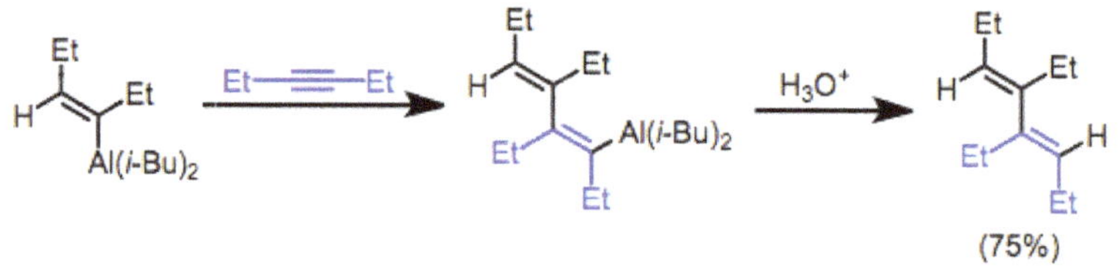

Alkenyl- and alkynylalanes undergo transmetalation to a variety of metals, including boron, zirconium, and mercury.

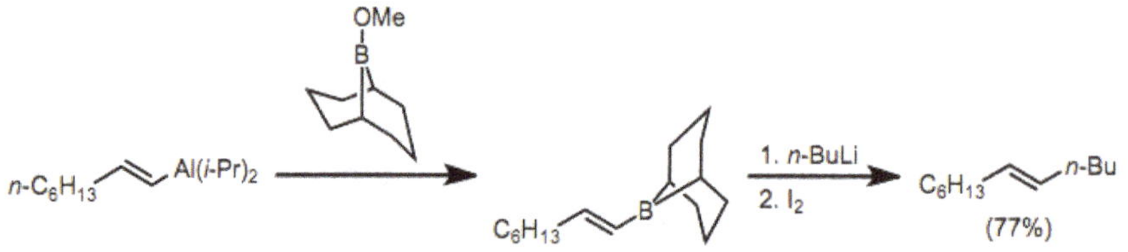

Experimental Conditions and Procedure

Typical Conditions

Organoaluminium compounds are extremely pyrophoric and should be handled only under inert atmosphere. Dialkylaluminium hydrides and lithium aluminium hydride are both available commercially; the former is available either neat or in solutions that may be standardized using known procedures. Solvents and other reagents should be scrupulously dried. Workup of these reactions should employ extreme pH's (10% hydrochloric acid or 6 N sodium hydroxide), as moderate pH encourages the formation of gelatinous aluminium hydroxide, which renders product separation difficult.

Example Procedure

$$C_6H_{13}\text{———} \xrightarrow[\text{2. NBS, -30 °C to rt, 1 h}]{\text{1. DIBAL-H, 50 °C}} C_6H_{13}\diagup\diagdown Br$$

(78%)

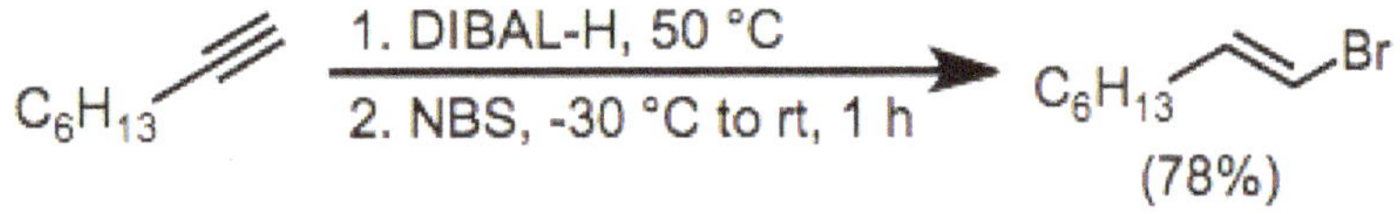

To 2.76 g (25.0 mmol) of 1-octyne was added 25.0 mL of a 1.07 M solution of diisobutylaluminium hydride (26.8 mmol) in n-hexane while the temperature was maintained at 25–30° by means of a water bath. The solution was stirred at room temperature for 30 minutes and then was heated at 50° for 4 hours. The resultant alkenylalane was cooled to −30°, diluted with 15 mL of dry ether, and treated with 5.35 g (30.1 mmol) of N-bromosuccinimide while keeping the temperature below −15°. The reaction mixture was gradually warmed to room temperature and stirred for 1 hour before being poured slowly into a mixture of 6 N hydrochloric acid (50 mL), n-pentane (10 mL), and ice (10 g). The layers were separated, and the aqueous phase was extracted with pentane. The combined organic extract was washed successively with 1 N sodium hydroxide, 10% sodium sulfite, and saturated aqueous sodium chloride and then was treated with a few crystals of BHT to inhibit isomerization of the alkenyl bromide. After drying over magnesium sulfate, distillation afforded 3.72 g (78%) of (E)-1-bromo-1-octene, bp 67° (5 mm), n_D^{26} 1.4617. This compound, which contained 4% of 1-bromo-1-octyne, was stored over a few crystals of BHT. ^1H NMR (CDCl$_3$): δ 0.92 (m, 3 H), 1.1-1.7 (m, 8 H), 1.90-2.34 (m, 2 H), 5.85-6.30 (m, 2 H).

References

- Brown, H. C.; Zwefei, G. (1960). "Isomerization of Organoboranes Derived Addition Mechanism of Isomerization from Branched-Chain and Ring Olefins- Further Evidence for the Elimination-Addition Mechanism of Isomerizaton". Journal of the American Chemical Society. 82: 1504. doi:10.1021/ja01491a058

- Sommer, L.; Pietrusza, E.; Whitmore, F. "Peroxide-catalyzed addition of trichlorosilane to 1-octene". J. Amer. Chem. Soc. 69 (1): 188. doi:10.1021/ja01193a508

- Hayashi, T.; Yamasaki, K. (2007). "C−E Bond Formation through Asymmetric Hydrosilylation of Alkenes". In Crabtree, Robert H.; Mingos, D. Michael P. Comprehensive Organometallic Chemistry III. Amsterdam: Elsevier. ISBN 978-0-08-045047-6. doi:10.1016/B0-08-045047-4/00140-0

- Allred, E. L.; Sonnenbcrg, J.; Winstcin S. (1960). "Preparation of Homobenzyl and Homoallyl Alcohols by the Hydroboration Method". Journal of Organic Chemistry. 25: 25. doi:10.1021/jo01071a007

- Transition-Metal-Promoted Aldehyde-Alkene Addition Reactions Charles F. Lochow, Roy G. Miller J. Am. Chem. Soc., 1976, 98 (5), pp 1281–1283 doi:10.1021/ja00421a050

- Brown, H. C.; Kulkarni, S. U. (1981). "Organoboranes: XXV. Hydridation of dialkylhaloboranes. New practical syntheses of dialkylboranes under mild conditions". Journal of Organometallic Chemistry. 218: 299. doi:10.1016/S0022-328X(00)81001-3

- Brown, H.C.; Zaidlewicz, M. (2001). Organic Syntheses Via Boranes, Vol. 2. Milwaukee, WI: Aldrich Chemical Co.,. ISBN 978-0-9708441-0-1

- Pelter, A.; Singaram, S.; Brown, H. C. (1983). "The dimesitylboron group in organic chemistry. 6 Hydroborations with dimesitylborane". Tetrahedron Letters. 24: 1433. doi:10.1016/S0040-4039(00)81675-5

- Dodd, D.S.; Ochlschlager, A. C. (1992). "Synthesis of inhibitors of 2,3-oxidosqualene-lanosterol cyclase: conjugate addition of organocuprates to N-(carbobenzyloxy)-3-carbomethoxy-5,6-dihydro-4-pyridone". Journal of Organic Chemistry. 57: 2794. doi:10.1021/jo00036a008

- Marek Zaidlewicz, Ofir Baum, Morris Srebnik, "Borane Dimethyl Sulfide" Encyclopedia of Reagents for Organic Synthesis doi:10.1002/047084289X.rb239.pub2

5

Hydrogenation: An Overview

When a chemical reaction occurs between molecular hydrogen and another element, usually in the presence of palladium or nickel as a catalyst, the resulting reaction is known as hydrogenation. Catalysts make the reactions feasible as non-catalytic reactions only take place in high temperatures. The aspects elucidated in this chapter are of vital importance, and provide a better understanding of hydrogenation.

Hydrogenation

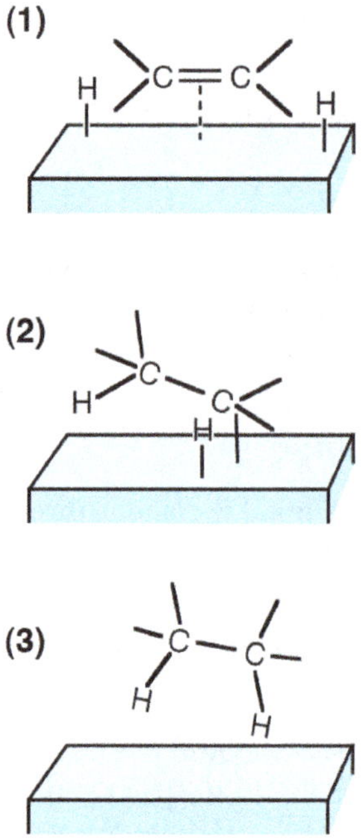

Steps in the hydrogenation of a C=C double bond at a catalyst surface, for example Ni or Pt :
(1) The reactants are adsorbed on the catalyst surface and H_2 dissociates.
(2) An H atom bonds to one C atom. The other C atom is still attached to the surface.
(3) A second C atom bonds to an H atom. The molecule leaves the surface.

Hydrogenation – to treat with hydrogen – is a chemical reaction between molecular hydrogen (H_2) and another compound or element, usually in the presence of a catalyst such as nickel, palladium or platinum. The process is commonly employed to reduce or saturate organic compounds. Hydrogenation typically constitutes the addition of pairs of hydrogen atoms to a molecule, often an alkene. Catalysts are required for the reaction to be usable; non-catalytic hydrogenation takes place only at very high temperatures. Hydrogenation reduces double and triple bonds in hydrocarbons.

Process

It has three components, the unsaturated substrate, the hydrogen (or hydrogen source) and, invariably, a catalyst. The reduction reaction is carried out at different temperatures and pressures depending upon the substrate and the activity of the catalyst.

Related or Competing Reactions

The same catalysts and conditions that are used for hydrogenation reactions can also lead to isomerization of the alkenes from cis to trans. This process is of great interest because hydrogenation technology generates most of the trans fat in foods. A reaction where bonds are broken while hydrogen is added is called hydrogenolysis, a reaction that may occur to carbon-carbon and carbon-heteroatom (oxygen, nitrogen or halogen) bonds. Some hydrogenations of polar bonds are accompanied by hydrogenolysis.

Hydrogen Sources

For hydrogenation, the obvious source of hydrogen is H_2 gas itself, which is typically available commercially within the storage medium of a pressurized cylinder. The hydrogenation process often uses greater than 1 atmosphere of H_2, usually conveyed from the cylinders and sometimes augmented by "booster pumps". Gaseous hydrogen is produced industrially from hydrocarbons by the process known as steam reforming. For many applications, hydrogen is transferred from donor molecules such as formic acid, isopropanol, and dihydroanthracene. These hydrogen donors undergo dehydrogenation to, respectively, carbon dioxide, acetone, and anthracene. These processes are called transfer hydrogenations.

Substrates

An important characteristic of alkene and alkyne hydrogenations, both the homogeneously and heterogeneously catalyzed versions, is that hydrogen addition occurs with "syn addition", with hydrogen entering from the least hindered side. Typical substrates are listed in the table

Substrates for and products of hydrogenation		
Substrate	**Product**	**Comments**
$R_2C{=}CR'_2$ (alkene)	$R_2CHCHR'_2$ (alkane)	many catalysts one application is margarine
RC≡CR (alkyne)	*cis*-RHC=CHR' (alkene)	over-hydrogenation to alkane can be problematic
RCHO (aldehyde)	RCH_2OH (primary alcohol)	easy substrate
R_2CO (ketone)	R_2CHOH (secondary alcohol)	more challenging than RCHO prochiral for unsymmetrical ketones
RCO_2R' (ester)	RCH_2OH + R'OH (two alcohols)	challenging substrate

RR'CNR" (imine)	RR'CHNHR" (amine)	easy substrate often use transfer hydrogenation actual precursor is N-protonated
RC(O)NR'$_2$ (amide)	RCH$_2$NR'$_2$ (amine)	challenging substrate
RCN (nitrile)	RCH$_2$NH$_2$ (primary amine)	product amine reactive toward precursor nitrile in some cases
RNO$_2$ (nitro)	RNH$_2$ (amine)	commercial applications use heterogeneous Ni and Ru catalysts major application is aniline

Catalysts

With rare exceptions, H_2 is unreactive toward organic compounds in the absence of metal catalysts. The unsaturated substrate is chemisorbed onto the catalyst, with most sites covered by the substrate. In heterogeneous catalysts, hydrogen forms surface hydrides (M-H) from which hydrogens can be transferred to the chemisorbed substrate. Platinum, palladium, rhodium, and ruthenium form highly active catalysts, which operate at lower temperatures and lower pressures of H_2. Non-precious metal catalysts, especially those based on nickel (such as Raney nickel and Urushibara nickel) have also been developed as economical alternatives, but they are often slower or require higher temperatures. The trade-off is activity (speed of reaction) vs. cost of the catalyst and cost of the apparatus required for use of high pressures. Notice that the Raney-nickel catalysed hydrogenations require high pressures:

Catalysts are usually classified into two broad classes: homogeneous catalysts and heterogeneous catalysts. Homogeneous catalysts dissolve in the solvent that contains the unsaturated substrate. Heterogeneous catalysts are solids that are suspended in the same solvent with the substrate or are treated with gaseous substrate.

Homogeneous Catalysts

Some well known homogeneous catalysts are indicated below. These are coordination complexes that activate both the unsaturated substrate and the H_2. Most typically, these complexes contain platinum group metals, especially Rh and Ir.

Homogeneous Hydrogenation Catalysts and their Precursors

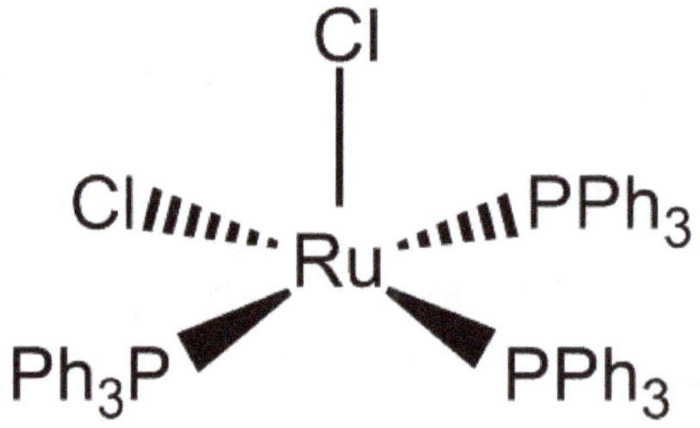

Dichlorotris(triphenylphosphine)ruthenium(II) is a precatalyst based on ruthenium.

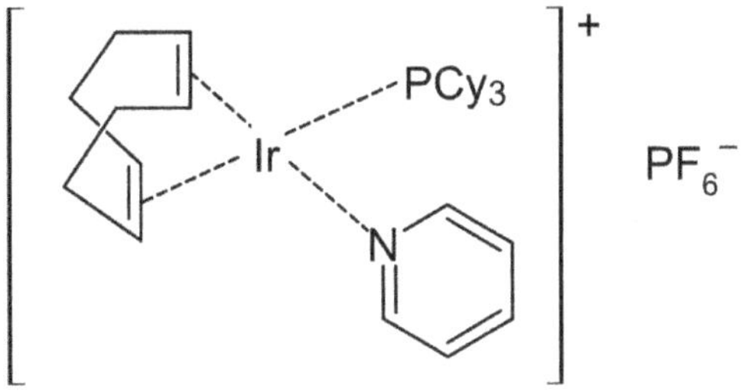

Crabtree's catalyst is a highly active catalyst featuring iridium.

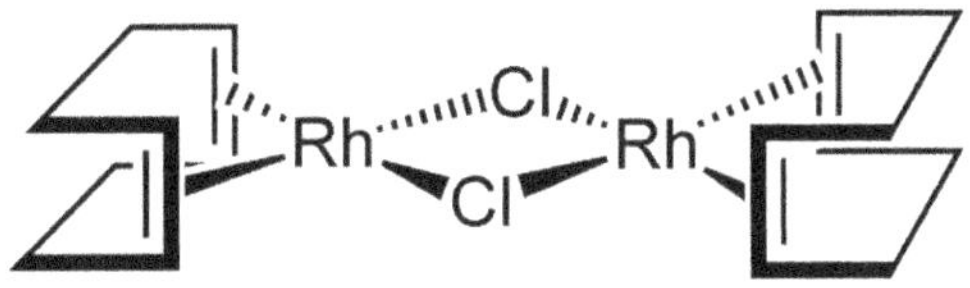

$Rh_2Cl_2(cod)_2$ is a precursor to many homogeneous catalysts.

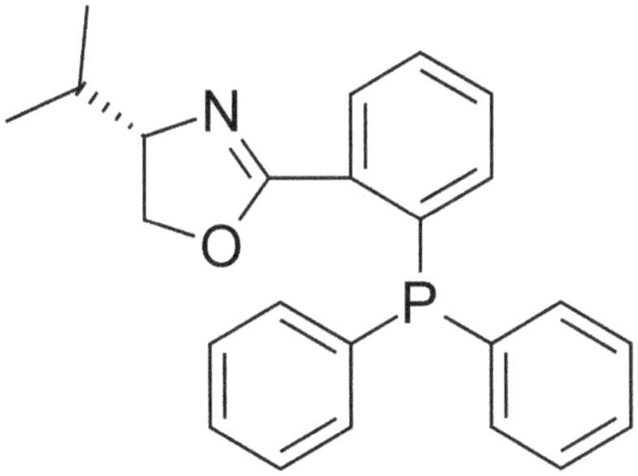

(S)-iPr-PHOX is a typical chelating phosphine ligand used in asymmetric hydrogenation.

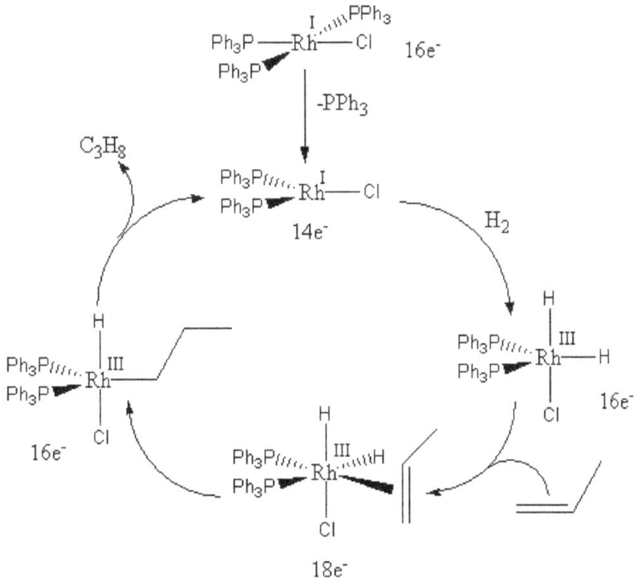

Catalytic hydrogenation of propylene

hydrogenation of propylene with Wilkinson's catalyst

Homogeneous catalysts are also used in asymmetric synthesis by the hydrogenation of prochiral substrates. An early demonstration of this approach was the Rh-catalyzed hydrogenation of enamides as precursors to the drug L-DOPA. To achieve asymmetric reduction, these catalyst are made chiral by use of chiral diphosphine ligands. Rhodium catalyzed hydrogenation has also been used in the herbicide production of S-metolachlor, which uses a Josiphos type ligand (called Xyliphos). In principle asymmetric hydrogenation can be catalyzed by chiral heterogeneous catalysts, but this approach remains more of a curiosity than a useful technology.

Heterogeneous Catalysts

Heterogeneous catalysts for hydrogenation are more common industrially. As in homogeneous catalysts, the activity is adjusted through changes in the environment around the metal, i.e. the coordination sphere. Different faces of a crystalline heterogeneous catalyst display distinct activities, for example. Similarly, heterogeneous catalysts are affected by their supports, i.e. the material upon with the heterogeneous catalyst is bound.

In many cases, highly empirical modifications involve selective "poisons". Thus, a carefully chosen catalyst can be used to hydrogenate some functional groups without affecting others, such as the hydrogenation of alkenes without touching aromatic rings, or the selective hydrogenation of alkynes to alkenes using Lindlar's catalyst. For example, when the catalyst palladium is placed on barium sulfate and then treated with quinoline, the resulting catalyst reduces alkynes only as far as alkenes. The Lindlar catalyst has been applied to the conversion of phenylacetylene to styrene.

Illustrative Hydrogenations

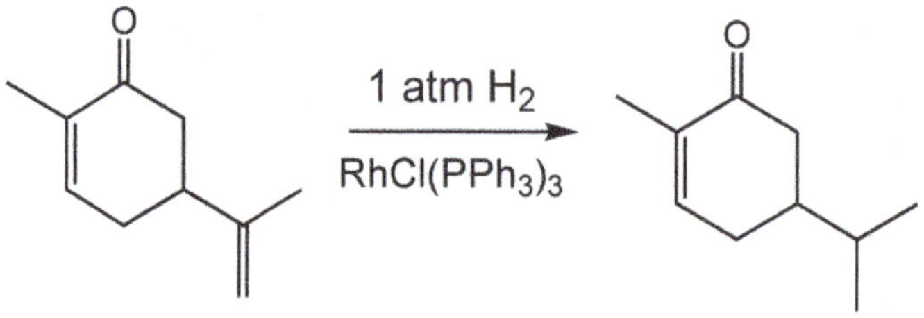

Selective hydrogenation of the less hindered alkene group in carvone using a homogeneous catalyst (Wilkinson's catalyst).

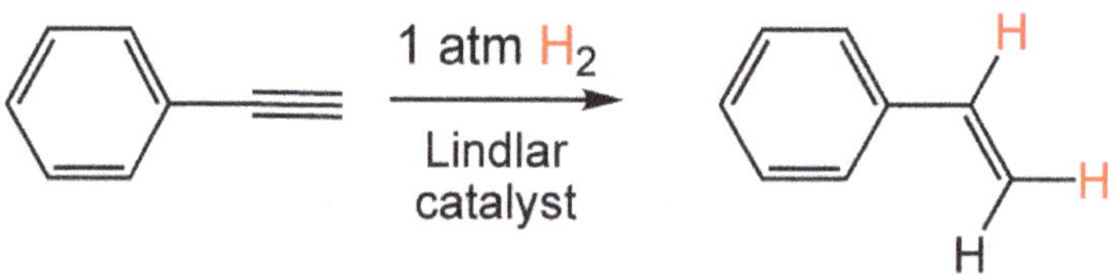

Partial hydrogenation of phenylacetylene using the Lindlar catalyst.

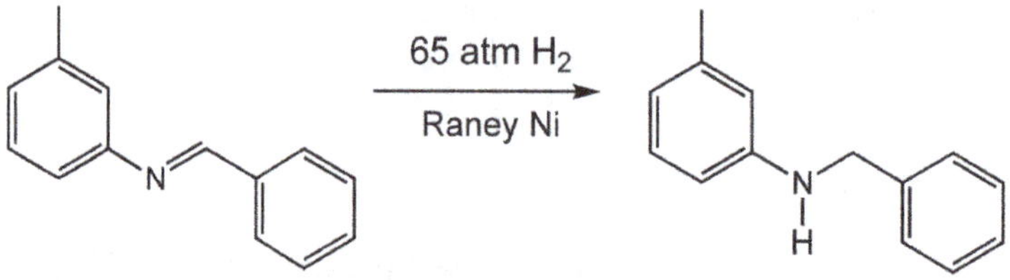

Hydrogenation of an imine using a Raney nickel catalyst, a popular heterogeneous catalyst.

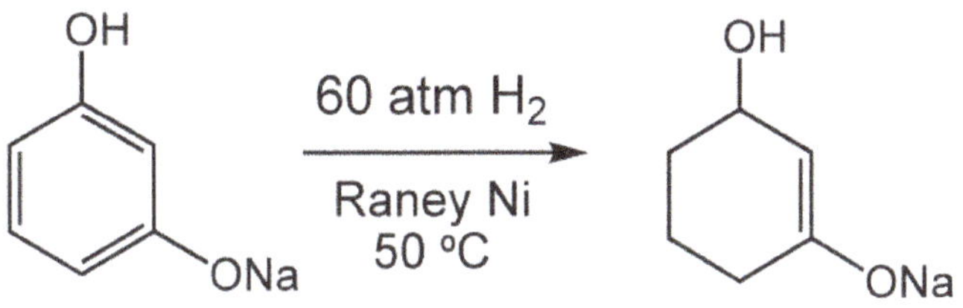

Partial hydrogenation of a resorcinol derivative using a Raney-Nickel catalyst.

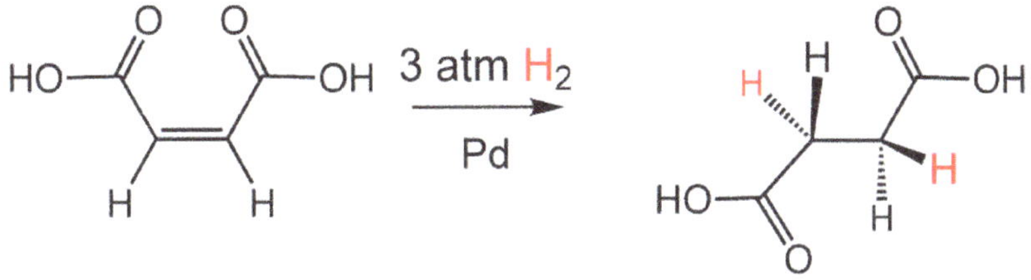

Hydrogenation of maleic acid to succinic acid.

Transfer Hydrogenation

Hydrogen also can be extracted ("transferred") from "hydrogen-donors" in place of H_2 gas. Hydrogen donors, which often serve as solvents include hydrazine, dihydronaphthalene, dihydroanthracene, isopropanol, and formic acid.

In organic synthesis, transfer hydrogenation is useful for the asymmetric reduction of polar unsaturated substrates, such as ketones, aldehydes, and imines. The hydrogenation of polar substrates such as ketones and aldehydes typically require transfer hydrogenation, at least in homogeneous catalysis. These catalysts are readily generated in chiral forms, which is the basis of asymmetric hydrogenation of ketones.

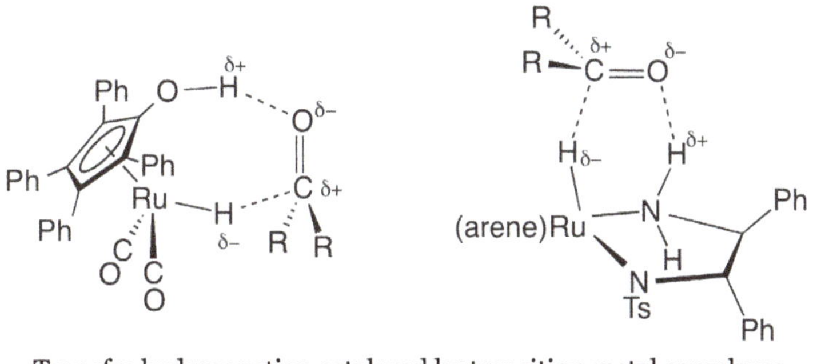

Transfer hydrogenation catalyzed by transition metal complexes proceeds by an "outer sphere mechanism."

Electrolytic Hydrogenation

Polar substrates such as nitriles can be hydrogenated electrochemically, using protic solvents and reducing equivalents as the source of hydrogen.

Thermodynamics and Mechanism

The addition of hydrogen to double or triple bonds in hydrocarbons is a type of redox reaction that can be thermodynamically favorable. For example, the addition of hydrogen to an alkene has a Gibbs free energy change of -101 kJ·mol^{-1}. However, the reaction rate for most hydrogenation reactions is negligible in the absence of catalysts. Hydrogenation is a strongly exothermic reaction. In the hydrogenation of vegetable oils and fatty acids, for example, the heat released is about 25 kcal per mole (105 kJ/mol), sufficient to raise the temperature of the oil by 1.6–1.7 °C per iodine number drop. The mechanism of metal-catalyzed hydrogenation of alkenes and alkynes has been extensively studied. First of all isotope labeling using deuterium confirms the regiochemistry of the addition:

$$RCH{=}CH_2 + D_2 \rightarrow RCHDCH_2D$$

Heterogeneous Catalysis

On solids, the accepted mechanism is the Horiuti-Polanyi mechanism:

1. Binding of the unsaturated bond, and hydrogen dissociation into atomic hydrogen onto the catalyst
2. Addition of one atom of hydrogen; this step is reversible
3. Addition of the second atom; effectively irreversible under hydrogenating conditions.

In the second step, the metallointermediate formed is a saturated compound that can rotate and then break down, again detaching the alkene from the catalyst. Consequently, contact with a hydrogenation catalyst necessarily causes *cis-trans*-isomerization, because the isomerization is thermodynamically favorable. This is a problem in partial hydrogenation, while in complete hydrogenation the produced *trans*-alkene is eventually hydrogenated.

For aromatic substrates, the first bond is hardest to hydrogenate because of the free energy penalty for breaking the aromatic system. The product of this is a cyclohexadiene, which is extremely active and cannot be isolated; in conditions reducing enough to break the aromatization, it is immediately reduced to a cyclohexene. The cyclohexene is ordinarily reduced immediately to a fully saturated cyclohexane, but special modifications to the catalysts (such as the use of the anti-solvent water on ruthenium) can preserve some of the cyclohexene, if that is a desired product.

Homogeneous Catalysis

In many homogeneous hydrogenation processes, the metal binds to both components to give an intermediate alkene-metal$(H)_2$ complex. The general sequence of reactions is assumed to be as follows or a related sequence of steps:

- binding of the hydrogen to give a dihydride complex via oxidative addition (preceding the oxidative addition of H_2 is the formation of a dihydrogen complex):

$$L_nM + H_2 \rightarrow L_nMH_2$$

- binding of alkene:

$$L_nM(\eta^2H_2) + CH_2{=}CHR \rightarrow L_{n-1}MH_2(CH_2{=}CHR) + L$$

- transfer of one hydrogen atom from the metal to carbon (migratory insertion)

$$L_{n-1}MH_2(CH_2{=}CHR) \rightarrow L_{n-1}M(H)(CH_2\text{-}CH_2R)$$

- transfer of the second hydrogen atom from the metal to the alkyl group with simultaneous dissociation of the alkane ("reductive elimination").

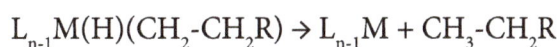

$$L_{n-1}M(H)(CH_2\text{-}CH_2R) \rightarrow L_{n-1}M + CH_3\text{-}CH_2R$$

Inorganic Substrates

The hydrogenation of nitrogen to give ammonia is conducted on a vast scale by the Haber-Bosch process, consuming an estimated 1% of the world's energy supply.

$$\overbrace{N \equiv N}^{\textit{nitrogen}} + \overbrace{3H_2}^{\substack{\textit{hydrogen}\\ (200\ atm)}} \xrightarrow[350-550^{\circ}\ C]{Fe\ catalyst} \overbrace{2NH_3}^{\textit{ammonia}}$$

Oxygen can be partially hydrogenated to give hydrogen peroxide, although this process has not been commercialized.

Industrial Applications

Catalytic hydrogenation has diverse industrial uses. Most frequently, industrial hydrogenation relies on heterogeneous catalysts.

Food Industry

The largest scale application of hydrogenation is for the processing of vegetable oils. Typical vegetable oils are derived from polyunsaturated fatty acids (containing more than one carbon-carbon double bond). Their partial hydrogenation reduces most, but not all, of these carbon-carbon double bonds. The degree of hydrogenation is controlled by restricting the amount of hydrogen, reaction temperature and time, and the catalyst.

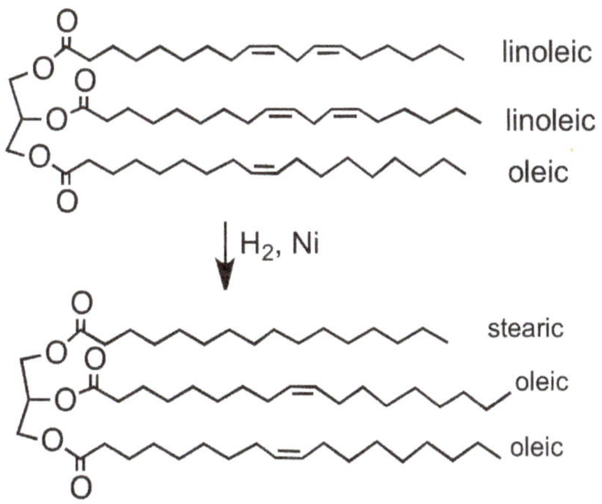

Partial hydrogenation of a typical plant oil to a typical component of margarine. Most of the C=C double bonds are removed in this process, which elevates the melting point of the product.

Hydrogenation converts liquid vegetable oils into solid or semi-solid fats, such as those present in margarine. Changing the degree of saturation of the fat changes some important physical properties, such as the melting range, which is why liquid oils become semi-solid. Solid or semi-solid fats are preferred for baking because the way the fat mixes with flour produces a more desirable texture in the baked product. Because partially hydrogenated vegetable oils are cheaper than animal fats, are available in a wide range of consistencies, and have other desirable characteristics (such as increased oxidative stability and longer shelf life), they are the predominant fats used as shortening in most commercial baked goods.

A side effect of incomplete hydrogenation having implications for human health is the isomerization of some of the remaining unsaturated carbon bonds, resulting in the trans isomers, which have been implicated in circulatory diseases including heart disease. The conversion from cis to trans bonds is favored because the trans configuration has lower energy than the natural cis one. At equilibrium, the trans/cis isomer ratio is about 2:1. Many countries and regions have introduced mandatory labeling of trans fats on food products and appealed to the industry for voluntary reductions.

Petrochemical Industry

In petrochemical processes, hydrogenation is used to convert alkenes and aromatics into saturated alkanes (paraffins) and cycloalkanes (naphthenes), which are less toxic and less reactive. Relevant to liquid fuels that are stored sometimes for long periods in air, saturated hydrocarbons exhibit superior storage properties. On the other hand, alkene tend to form hydroperoxides, which can form gums that interfere with fuel handing equipment. For example, mineral turpentine is usually hydrogenated. Hydrocracking of heavy residues into diesel is another application. In isomerization and catalytic reforming processes, some hydrogen pressure is maintained to hydrogenolyze coke formed on the catalyst and prevent its accumulation.

Organic Chemistry

Hydrogenation is a useful means for converting unsaturated compounds into saturated derivatives. Substrates include not only alkenes and alkynes, but also aldehydes, imines, and nitriles, which are converted into the corresponding saturated compounds, i.e. alcohols and amines. Thus, alkyl aldehydes, which can be synthesized with the oxo process from carbon monoxide and an alkene, can be converted to alcohols. E.g. 1-propanol is produced from propionaldehyde, produced from ethene and carbon monoxide. Xylitol, a polyol, is produced by hydrogenation of the sugar xylose, an aldehyde. Primary amines can be synthesized by hydrogenation of nitriles, while nitriles are readily synthesized from cyanide and a suitable electrophile. For example, isophorone diamine, a precursor to the polyurethane monomer isophorone diisocyanate, is produced from isophorone nitrile by a tandem nitrile hydrogenation/reductive amination by ammonia, wherein hydrogenation converts both the nitrile into an amine and the imine formed from the aldehyde and ammonia into another amine.

History

Heterogeneous Catalytic Hydrogenation

The earliest hydrogenation is that of platinum catalyzed addition of hydrogen to oxygen in the Döbereiner's lamp, a device commercialized as early as 1823. The French chemist Paul Sabatier is considered the father of the hydrogenation process. In 1897, building on the earlier work of James Boyce, an American chemist working in the manufacture of soap products, he discovered that traces of nickel catalyzed the addition of hydrogen to molecules of gaseous hydrocarbons in what is now known as the Sabatier process. For this work, Sabatier shared the 1912 Nobel Prize in Chemistry. Wilhelm Normann was awarded a patent in Germany in 1902 and in Britain in 1903 for the hydrogenation of liquid oils, which was the beginning of what is now a worldwide industry. The commercially important Haber–Bosch process, first described in 1905, involves hydrogenation of nitrogen. In the Fischer–Tropsch process, reported in 1922 carbon monoxide, which is easily derived from coal, is hydrogenated to liquid fuels.

In 1922, Voorhees and Adams described an apparatus for performing hydrogenation under pressures above one atmosphere. The Parr shaker, the first product to allow hydrogenation using elevated pressures and temperatures, was commercialized in 1926 based on Voorhees and Adams' research and remains in widespread use. In 1924 Murray Raney developed a finely powdered form of nickel, which is widely used to catalyze hydrogenation reactions such as conversion of nitriles to amines or the production of margarine.

Homogeneous Catalytic Hydrogenation

In the 1930s, Calvin discovered that copper(II) complexes oxidized H_2. The 1960s witnessed the development of well defined homogeneous catalysts using transition metal complexes, e.g., Wilkinson's catalyst ($RhCl(PPh_3)_3$). Soon thereafter cationic Rh and Ir were found catalyze the hydrogenation of alkenes and carbonyls. In the 1970s, asymmetric hydrogenation was demonstrated in the synthesis of L-DOPA, and the 1990s saw the invention of Noyori asymmetric hydrogenation. The development of homogeneous hydrogenation was influenced by work started in the 1930s and 1940s on the oxo process and Ziegler–Natta polymerization.

Metal-free Hydrogenation

For most practical purposes, hydrogenation requires a metal catalyst. Hydrogenation can, however, proceed from some hydrogen donors without catalysts, illustrative hydrogen donors being diimide and aluminium isopropoxide, the latter illustrated by the Meerwein–Ponndorf–Verley reduction. Some metal-free catalytic systems have been investigated in academic research. One such system for reduction of ketones consists of *tert*-butanol and potassium tert-butoxide and very high temperatures. The reaction depicted below describes the hydrogenation of benzophenone:

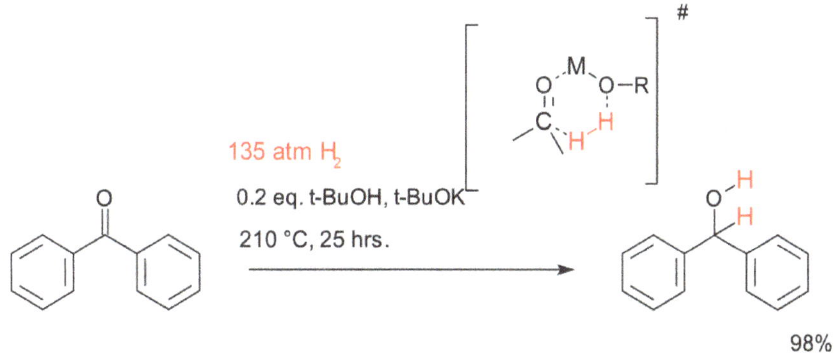

A chemical kinetics study found this reaction is first-order in all three reactants suggesting a cyclic 6-membered transition state.

Another system for metal-free hydrogenation is based on the phosphine-borane, compound 1, which has been called a *frustrated Lewis pair*. It reversibly accepts dihydrogen at relatively low temperatures to form the phosphonium borate 2 which can reduce simple hindered imines.

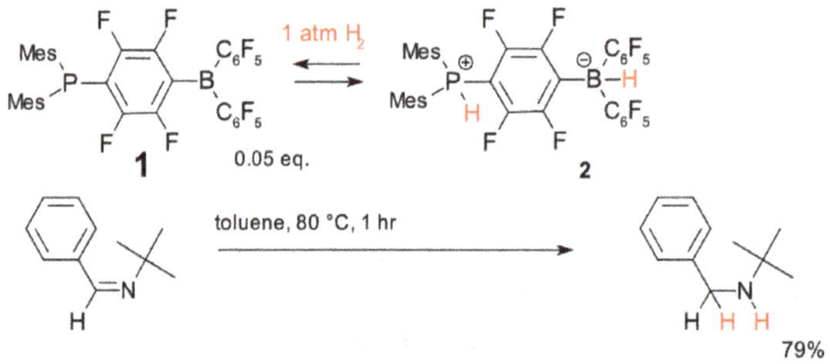

The reduction of nitrobenzene to aniline has been reported to be catalysed by fullerene, its mono-anion, atmospheric hydrogen and UV light.

Equipment Used for Hydrogenation

Today's bench chemist has three main choices of hydrogenation equipment:

- Batch hydrogenation under atmospheric conditions
- Batch hydrogenation at elevated temperature and/or pressure
- Flow hydrogenation

Batch Hydrogenation Under Atmospheric Conditions

The original and still a commonly practised form of hydrogenation in teaching laboratories, this process is usually effected by adding solid catalyst to a round bottom flask of dissolved reactant which has been evacuated using nitrogen or argon gas and sealing the mixture with a penetrable rubber seal. Hydrogen gas is then supplied from a H_2-filled balloon. The resulting three phase mixture is agitated to promote mixing. Hydrogen uptake can be monitored, which can be useful for monitoring progress of a hydrogenation. This is achieved by either using a graduated tube containing a coloured liquid, usually aqueous copper sulfate or with gauges for each reaction vessel.

Batch Hydrogenation at Elevated Temperature and/or Pressure

Since many hydrogenation reactions such as hydrogenolysis of protecting groups and the reduction of aromatic systems proceed extremely sluggishly at atmospheric temperature and pressure, pressurised systems are popular. In these cases, catalyst is added to a solution of reactant under an inert atmosphere in a pressure vessel. Hydrogen is added directly from a cylinder or built in laboratory hydrogen source, and the pressurized slurry is mechanically rocked to provide agitation, or a spinning basket is used. Heat may also be used, as the pressure compensates for the associated reduction in gas solubility.

Flow Hydrogenation

Flow hydrogenation has become a popular technique at the bench and increasingly the process scale. This technique involves continuously flowing a dilute stream of dissolved reactant over a fixed bed catalyst in the presence of hydrogen. Using established HPLC technology, this technique allows the application of pressures from atmospheric to 1,450 psi (100 bar). Elevated temperatures may also be used. At the bench scale, systems use a range of pre-packed catalysts which eliminates the need for weighing and filtering pyrophoric catalysts.

Industrial Reactors

Catalytic hydrogenation is done in a tubular plug-flow reactor (PFR) packed with a supported catalyst. The pressures and temperatures are typically high, although this depends on the catalyst. Catalyst loading is typically much lower than in laboratory batch hydrogenation, and various promoters are added to the metal, or mixed metals are used, to improve activity, selectivity and catalyst stability. The use of nickel is common despite its low activity, due to its low cost compared to precious metals.

Gas Liquid Induction Reactors (Hydrogenator) are also used for carrying out catalytic hydrogenation.

Asymmetric Hydrogenation

Asymmetric hydrogenation is a chemical reaction that adds two atoms of hydrogen preferentially to one of two faces of an unsaturated substrate molecule, such as an alkene or ketone. The selectivity derives from the manner that the substrate binds to the chiral catalysts. In jargon, this binding transmits spatial information (what chemists refer to as chirality) from the catalyst to the target, favoring the product as a single enantiomer. This enzyme-like selectivity is particularly applied to bioactive products such as pharmaceutical agents and agrochemicals.

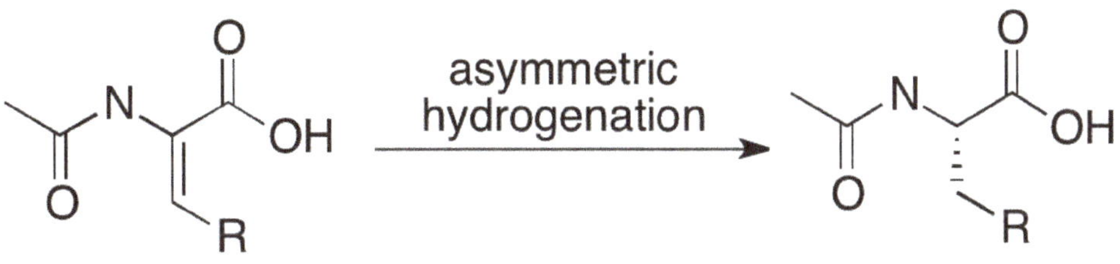

History

In 1956 a heterogeneous catalyst made of palladium deposited on silk was shown to effect asymmetric hydrogenation. Later, in 1968, the groups of William Knowles and Leopold Horner independently published the examples of asymmetric hydrogenation using a homogeneous catalysts. While exhibiting only modest enantiomeric excesses, these early reactions demonstrated feasibility. By 1972, enantiomeric excess of 90% was achieved, and the first industrial synthesis of the Parkinson's drug L-DOPA commenced using this technology.

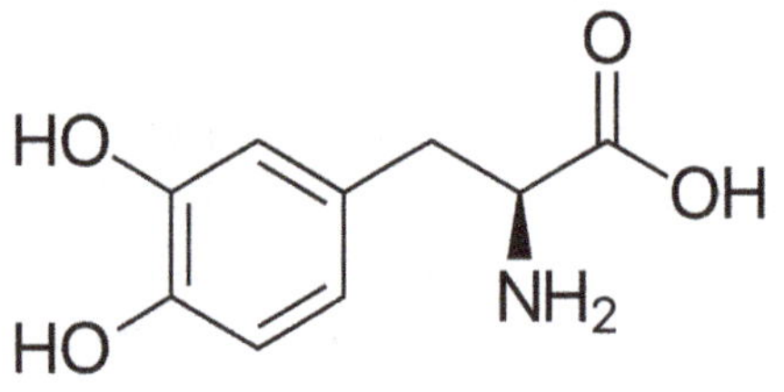

L-DOPA

The field of asymmetric hydrogenation continued to experience a number of notable advances. Henri Kagan developed DIOP, an easily prepared C_2-symmetric diphosphine that gave high ee's in certain reactions. Ryōji Noyori introduced the ruthenium-based catalysts for the asymmetric hydrogenated polar substrates, such as ketones and aldehydes. The introduction of P,N ligands then further expanded the scope of the C_2-symmetric ligands, although they are not fundamentally superior to chiral ligands lacking rotational symmetry. Today, asymmetric hydrogenation is a routine methodology in laboratory and industrial scale organic chemistry.

The importance of asymmetric hydrogenation was recognized by the 2001 Nobel Prize in Chemistry awarded to William Standish Knowles and Ryōji Noyori.

Mechanism

Two major mechanisms have been proposed for catalytic hydrogenation with rhodium complexes: the unsaturated mechanism and the dihydride mechanism. While distinguishing between the two mechanisms is difficult, the difference between the two for asymmetric hydrogenation is relatively unimportant since both converge to a common intermediate before any stereochemical information is transferred to the product molecule.

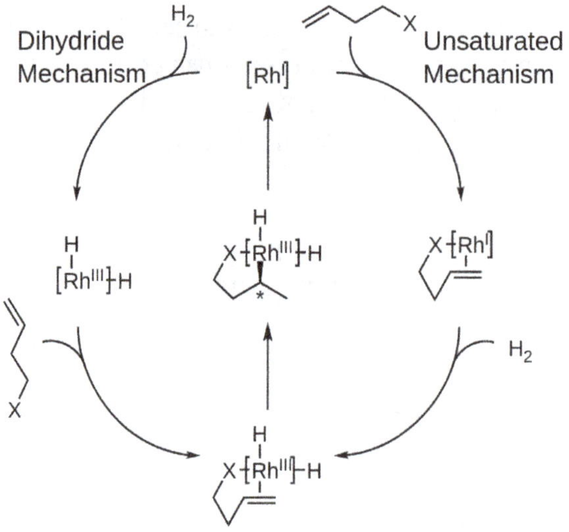

Proposed mechanisms for asymmetric hydrogenation

The preference for producing one enantiomer instead of another in these reactions is often explained in terms of steric interactions between the ligand and the prochiral substrate. Consideration of these interactions has led to the development of quadrant diagrams where "blocked" areas are denoted with a shaded box, while "open" areas are left unfilled. In the modeled reaction, large groups on an incoming olefin will tend to orient to fill the open areas of the diagram, while smaller groups will be directed to the blocked areas and hydrogen delivery will then occur to the back face of the olefin, fixing the stereochemistry. Note that only part of the chiral phosphine ligand is shown for the sake of clarity.

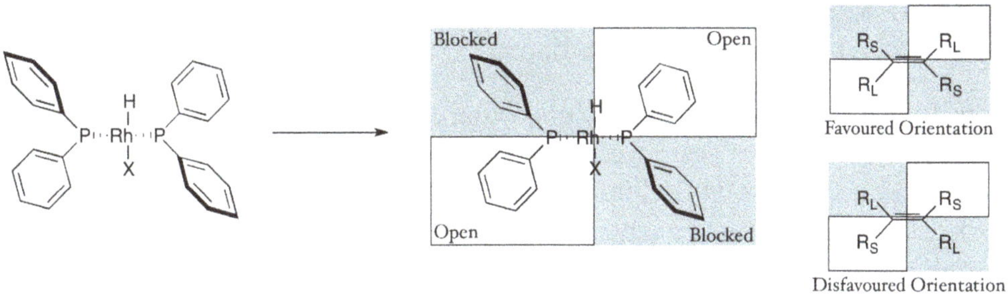

Quadrant model for asymmetric hydrogenation

Metals

Platinum-group Metals

Rhodium, the first metal to be used in a homogeneous asymmetric hydrogenation, continues to be widely used. Targets for asymmetric hydrogenation with rhodium generally require a coordinating group close to the olefin. While this requirement is a limitation, many classes of substrates possess such functionalization, e.g. unsaturated amides.

The Noyori asymmetric hydrogenation is based on ruthenium. Subsequent work has expanded upon Noyori's original catalyst template, leading to the inclusion of traditionally difficult substrates like *t*-butyl ketones and 1-tetralones as viable substrates for hydrogenation with ruthenium catalysts. Transfer hydrogenation based on the Ru and TsDPEN has also enjoyed commercial success.

Iridium catalysts are useful for a number of "non-traditional" substrates for which good catalysts had not been found with Ru and Rh. Unfunctionalized olefins are the archetypal case, but other examples including ketones exist. A common difficulty with iridium-based catalyst is their tendency to trimerize in solution. The use of a BAr_4^{F-} anions has proven to be the most widely applicable solution to the aggregation problem. Other strategies to enhance catalyst stability include the addition of an additional coordinating arm to the chiral ligand, increasing the steric bulk of the ligand, using a dendrimeric ligand, increasing the rigidity of the ligand, immobilizing the ligand, and using heterobimetallic systems (with iridium as one of the metals).

Base Metals

Iron is a popular research target for many catalytic processes, owing largely to its low cost and low toxicity relative to other transition metals. Asymmetric hydrogenation methods using iron have been realized, although in terms of rates and selectivity, they are inferior to catalysts based on precious metals. In some cases, structurally ill-defined nanoparticles have proven to be the active species *in situ* and the modest selectivity observed may result from their uncontrolled geometries.

Ligand Classes

Phosphine Ligands

Chiral phosphine ligands are the source of chirality in most asymmetric hydrogenation catalysts. Of these the BINAP ligand is perhaps the best-known, as a result of its Nobel Prize-winning application in the Noyori asymmetric hydrogenation.

Chiral phosphine ligands can be generally classified as mono- or bidentate. They can be further classified according to the location of the stereogenic centre – phosphorus vs the organic substituents. Ligands with a C_2 symmetry element have been particularly popular, in part because the presence of such an element reduces the possible binding conformations of a substrate to a metal-ligand complex dramatically (often resulting in exceptional enantioselectivity).

Chiral Monophosphine, Monodentate Ligands

Monophosphine-type ligands were among the first to appear in asymmetric hydrogenation, e.g., the ligand CAMP. Continued research into these types of ligands has explored both P-alkyl and P-heteroatom bonded ligands, with P-heteroatom ligands like the phosphites and phosphoramidites generally achieving more impressive results. Structural classes of ligands that have been successful include those based on the binapthyl structure of MonoPHOS or the spiro ring system of SiPHOS. Notably, these monodentate ligands can be used in combination with each other to achieve a synergistic improvement in enantioselectivity; something that is not possible with the diphosphine ligands.

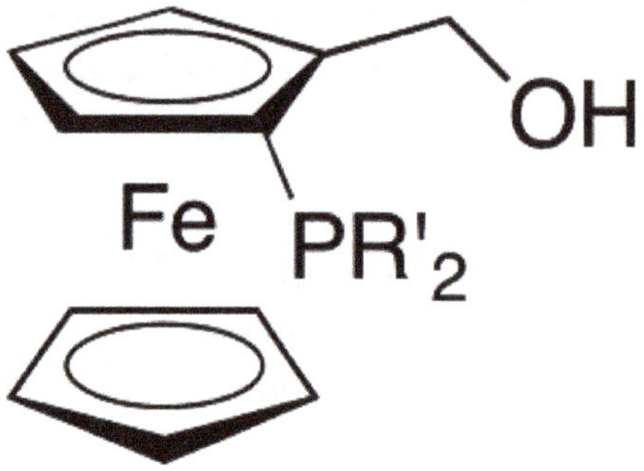

A ferrocene derivative

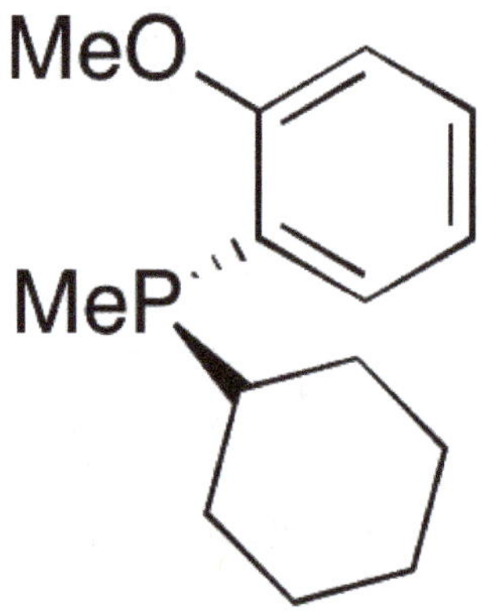

The CAMP ligand

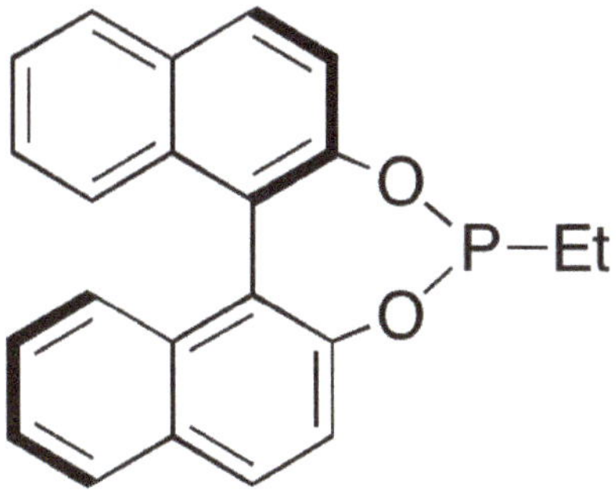

A BINOL derivative

Chiral Diphosphine Ligands

The diphosphine ligands have received considerably more attention than the monophosphines and, perhaps as a consequence, have a much longer list of achievement. This class includes the first ligand to achieve high selectivity (DIOP), the first ligand to be used in industrial asymmetric synthesis (DIPAMP) and what is likely the best known chiral ligand (BINAP). Chiral diphosphine ligands are now ubiquitous in asymmetric hydrogenation.

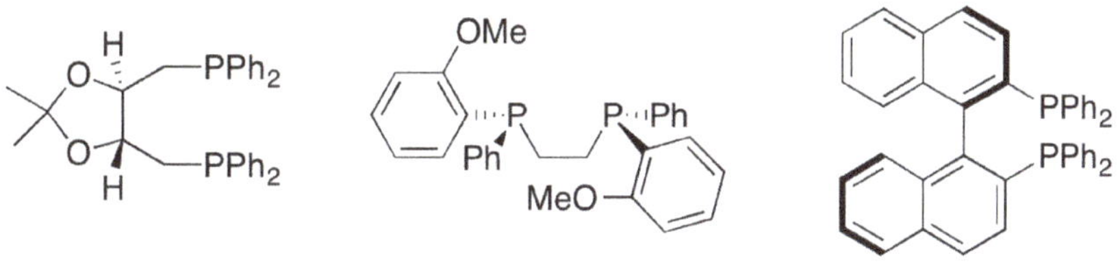

(*R,R*)-DIOP (*R,R*)-DIPAMP (*R*)-BINAP

Historically important diphosphine ligands

P,N and P,O Ligands

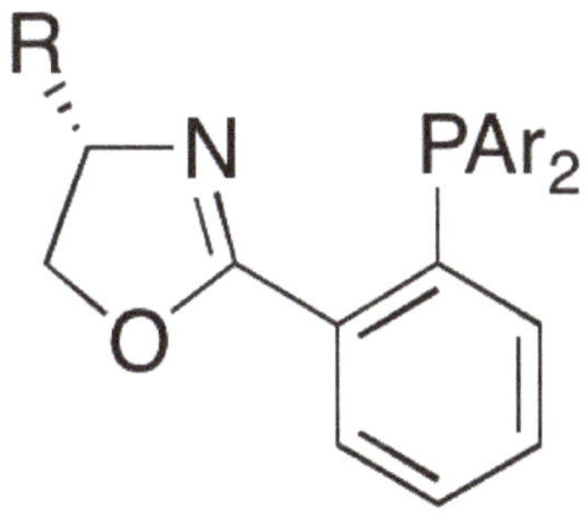

Generic PHOX ligand architecture

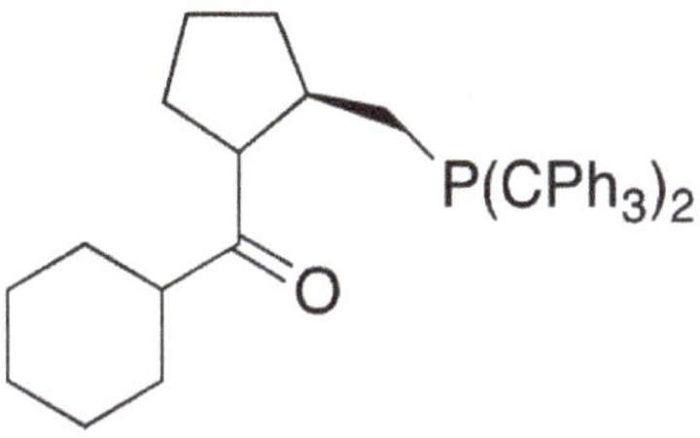

Effective ligand for various asymmetric-hydrogenation processes

The use of P,N ligands in asymmetric hydrogenation can be traced to the C_2 symmetric bisoxazoline ligand. However, these symmetric ligands were soon superseded by mono(oxazoline) ligands whose lack of C_2 symmetry has in no way limits their efficacy in asymmetric catalysis. Such ligands generally consist of an achiral nitrogen-containing heterocycle that is functionalized with a pendant phosphorus-containing arm, although both the exact nature of the heterocycle and the chemical environment phosphorus center has varied widely. No single structure has emerged as consistently effective with a broad range of substrates, although certain privileged structures (like the phosphine-oxazoline or PHOX architecture) have been established. Moreover, within a narrowly defined substrate class the performance of metallic complexes with chiral P,N ligands can closely approach perfect conversion and selectivity in systems otherwise very difficult to target. Certain complexes derived from chelating P-O ligands have shown promising results in the hydrogenation of α,β-unsaturated ketones and esters.

NHC Ligands

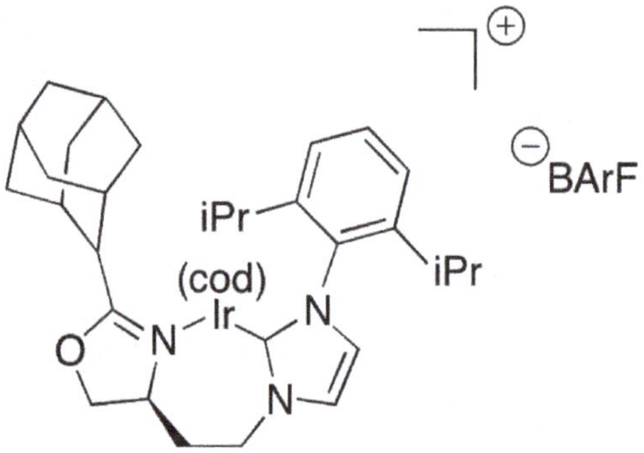

Catalyst developed by Burgess for asymmetric hydrogenation

Simple N-heterocyclic carbene (NHC)-based ligands have proven impractical for asymmetrical hydrogenation.

Some C,N ligands combine an NHC with a chiral oxazoline to give a chelating ligand. NHC-based ligands of the first type have been generated as large libraries from the reaction of smaller libraries of individual NHCs and oxazolines. NHC-based catalysts featuring a bulky seven-membered metallocycle on iridium have been applied to the catalytic hydrogenation of unfunctionalized olefins and vinyl ether alcohols with conversions and ee's in the high 80s or 90s. The same system has been applied to the synthesis of a number of aldol, vicinal dimethyl and deoxypolyketide motifs, and to the deoxypolyketides themselves.

C_2-symmetric NHCs have shown themselves to be highly useful ligands for the asymmetric hydrogenation.

Acyclic Substrates

Acyclic unsaturated substrates (olefins, ketones, enamines imines) represents the most common prochiral substrates. Substrates that are particularly amenable to asymmetric hydrogenation often feature a polar functional group adjacent to the site to be hydrogenates. In the absence of this functional) group, catalysis often results in low ee's. For unfunctionalized olefins, iridium with P,N-based ligands) have proven successful catalysts. Catalyst utility within this category is unusually narrow; consequently, many different categories of solved and unsolved catalytic problems have developed. 1,1-disubstituted, 1,2-diaryl trisubstituted, 1,1,2-trialkyl and tetrasubstituted olefins represent classes that have been investigated separately, and even within these classes variations may exist that make different solutions optimal.

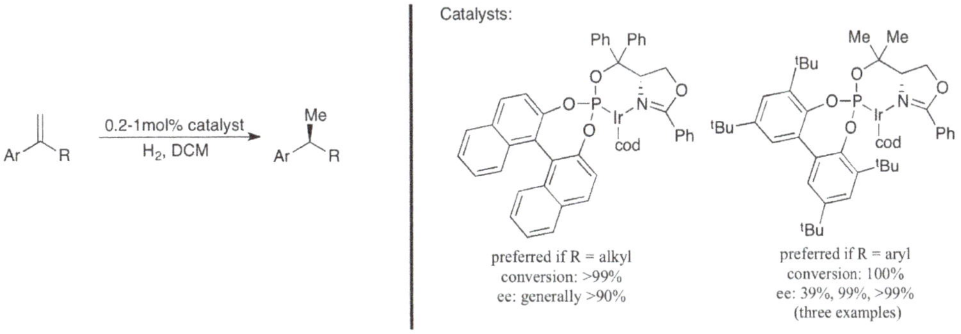

Example of asymmetric hydrogenation of unfunctionalized olefins

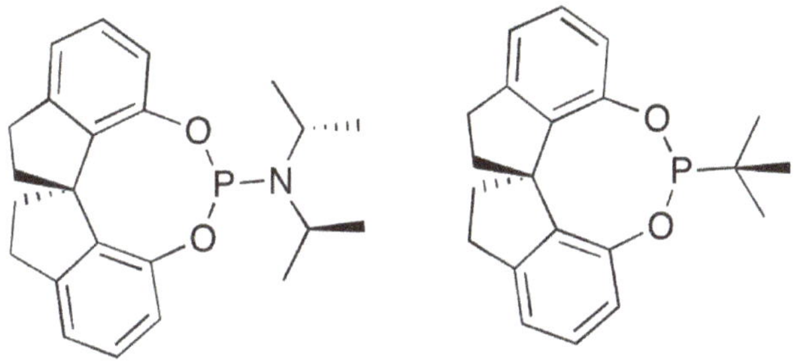

Chiral phosphoramidite and phosphonite ligands used in the asymmetric hydrogenation of enamines.

Conversely to the case of olefins, asymmetric hydrogenation of enamines has favoured diphosphine-type ligands; excellent results have been achieved with both iridium- and rhodium-based systems. However, even the best systems often suffer from low ee's and a lack of generality. Certain pyrrolidine-derived enamines of aromatic ketones are amenable to asymmetrically hydrogenation with cationic rhodium(I) phosphonite systems, and I_2 and acetic acid system with ee values usually above 90% and potentially as high as 99.9%. A similar system using iridium(I) and a very closely related phosphoramidite ligand is effective for the asymmetric hydrogenation of pyrrolidine-type enamines where the double bond was inside the ring: in other words, of dihydropyrroles. In both cases, the enantioselectivity dropped substantially when the ring size was increased from five to six.

Imines and Ketones

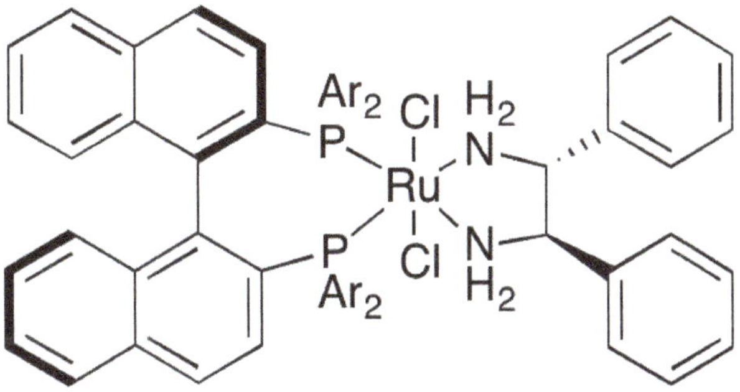

Archetype of Noyori's catalysts for asymmetric hydrogenation of ketones

Ketones and imines are related functional groups, and effective technologies for the asymmetric hydrogenation of each are also closely related. Of these, Noyori's ruthenium-chiral diphosphine-diamine system is perhaps one of the best known. It can be employed in conjunction with a wide range of phosphines and amines (where the amine may be, but need not be, chiral) and can be easily adjusted for an optimal match with the target substrate, generally achieving enantiomeric excesses (ee's) above 90%.

For carbonyl and imine substrates, end-on, η^1 coordination can compete with η^2 mode. For η^1-bound substrates, the hydrogen-accepting carbon is removed from the catalyst and resists hydrogenation.

Iridium/P,N ligand-based systems are also commonly used for the asymmetric hydrogenation of ketones and imines. For example, a consistent system for benzylic aryl imines uses the P,N ligand SIPHOX in conjunction with iridium(I) in a cationic complex to achieve asymmetric hydrogenation with ee >90%. One of the most efficient and effective catalysts ever developed for the asymmetric hydrogenation of ketones, with a turnover number (TON) up to 4,550,000 and ee up to 99.9%, uses another iridium(I) system with a closely related tridentate ligand.

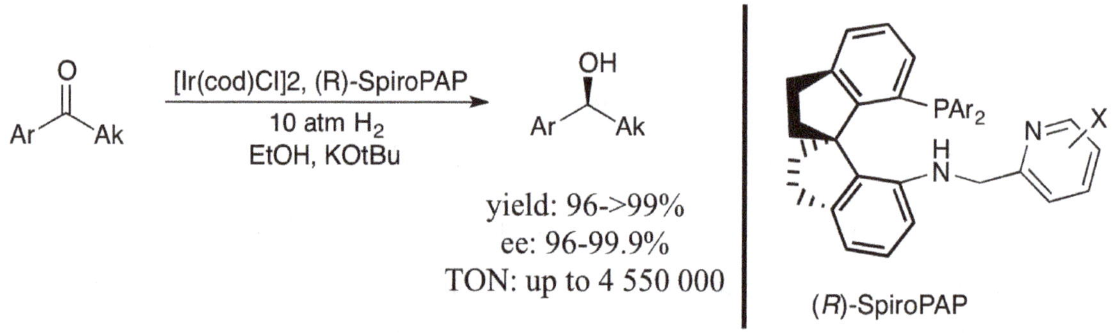

Highly effective system for the asymmetric hydrogenation of ketones

Despite their similarities, the two functional groups are not identical; there are many areas where they diverge significantly. One of these is in the asymmetric hydrogenation of N-unfunctionalized imines to give primary amines. Such species can be difficult to selectively reduce because they tend to exist in complex equilibria of imine and enamine tautomers, as well as (E) and (Z) isomers. One approach to this problem has been to use ketimines as their hydrochloride salt and rely on the steric properties of the adjacent alkyl or aryl groups to allow the catalyst to differentiate between the two enantiotopic faces of the ketimine.

Aromatic Substrates

The asymmetric hydrogenation of aromatic (especially heteroaromatic), substrates is a very active field of ongoing research. Catalysts in this field must contend with a number of complicating factors, including the tendency of highly stable aromatic compounds to resist hydrogenation, the potential coordinating (and therefore catalyst-poisoning) abilities of both substrate and product, and the great diversity in substitution patterns that may be present on any one aromatic ring. Of these substrates the most consistent success has been seen with nitrogen-containing heterocycles, where the aromatic ring is often activated either by protonation or by further functionalization of the nitrogen (generally with an electron-withdrawing protecting group). Such strategies are less applicable to oxygen- and sulfur-containing heterocycles, since they are both less basic and less nucleophilic; this additional difficulty may help to explain why few effective methods exist for their asymmetric hydrogenation.

Quinolines, Isoquinolines and Quinoxalines

Two systems exist for the asymmetric hydrogenation of 2-substituted quinolines with isolated yields generally greater than 80% and ee values generally greater than 90%. The first is an iridium(I)/chiral phosphine/I_2 system, first reported by Zhou *et al.* While the first chiral phosphine used in this system was MeOBiPhep, newer iterations have focused on improving the performance of this ligand. To this end, systems use phosphines (or related ligands) with improved air stability, recyclability, ease of preparation, lower catalyst loading and the potential role of achiral phosphine additives. As of October 2012 no mechanism appears to have been proposed, although both the necessity of I_2 or a halogen surrogate and the possible role of the heteroaromatic N in assisting reactivity have been documented.

The second is an organocatalytic transfer hydrogenation system based on Hantzsch esters and a chiral Brønsted acid. In this case, the authors envision a mechanism where the isoquinoline is alternately protonated in an activating step, then reduced by conjugate addition of hydride from the Hantzsch ester.

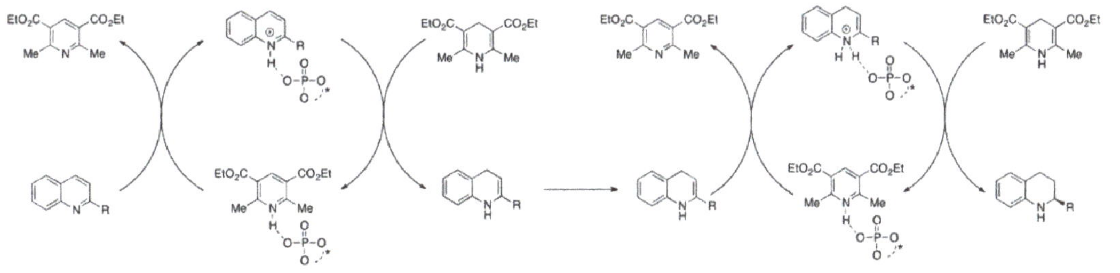

Proposed organocatalytic mechanism

Much of the asymmetric hydrogenation chemistry of quinoxalines is closely related to that of the structurally similar quinolines. Effective (and efficient) results can be obtained with an Ir(I)/phophinite/I_2 system and a Hantzsh ester-based organocatalytic system, both of which are similar to the systems discussed earlier with regards to quinolines.

Pyridines

Pyridines are highly variable substrates for asymmetric reduction (even compared to other heteroaromatics), in that five carbon centers are available for differential substitution on the initial ring. As of October 2012 no method seems to exist that can control all five, although at least one reasonably general method exists.

The most-general method of asymmetric pyridine hydrogenation is actually a heterogeneous method, where asymmetry is generated from a chiral oxazolidinone bound to the C2 position of the pyridine. Hydrogenating such functionalized pyridines over a number of different heterogeneous metal catalysts gave the corresponding piperidine with the substituents at C3, C4, and C5 positions in an all-*cis* geometry, in high yield and excellent enantioselectivity. The oxazolidinone auxiliary is also conveniently cleaved under the hydrogenation conditions.

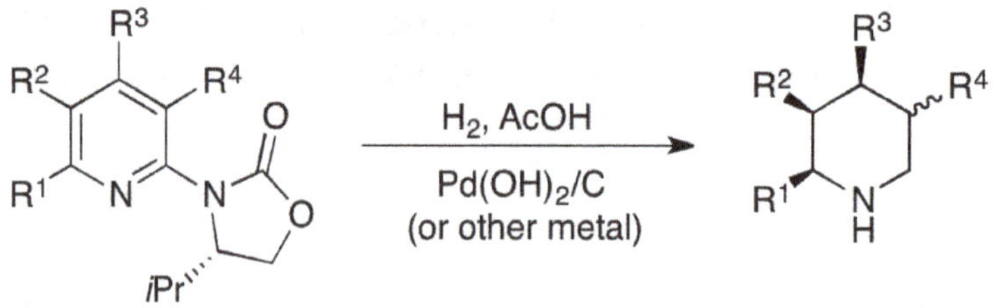

Asymmetric hydrogenation of pyridines with heterogeneous catalyst

Methods designed specifically for 2-substituted pyridine hydrogenation can involve asymmetric systems developed for related substrates like 2-substituted quinolines and quinoxalines. For example, an iridium(I)\chiral phosphine\I_2 system is effective in the asymmetric hydrogenation of activated (alkylated) 2-pyridiniums or certain cyclohexanone-fused pyridines. Similarly, chiral Brønsted acid catalysis with a Hantzsh ester as a hydride source is effective for some 2-alkyl pyridines with additional activating substitution.

Indoles

The asymmetric hydrogenation of indoles initially focused on N-protected indoles, where the protecting group could serve both to activate the heterocycle to hydrogenation and as a secondary coordination site for the metal. Later work allowed unprotected indoles to be targeted through Brønsted acid activation of the indole.

In the initial report on asymmetric indole hydrogenation, N-acetyl 2-substituted indoles could be protected with high yields and ee of 87-95%. 3-substituted indoles were less successful, with hydrolysis of the protecting group outcompeting the hydrogenation of the indole. Switching to an N-tosyl protecting group inhibited the hydrolysis reaction and allowed both 2- and 3-substituted indoles to be hydrogenated in high yield and ee. The problem with both methods, however, is that N-acetyl and N-tosyl groups require harsh cleavage conditions that might be incompatible with complex substrates. Using an easily cleaved N-Boc group would relieve this problem, and highly effective methods for the asymmetric hydrogenation of such indoles (both 2- and 3-substituted) were soon developed.

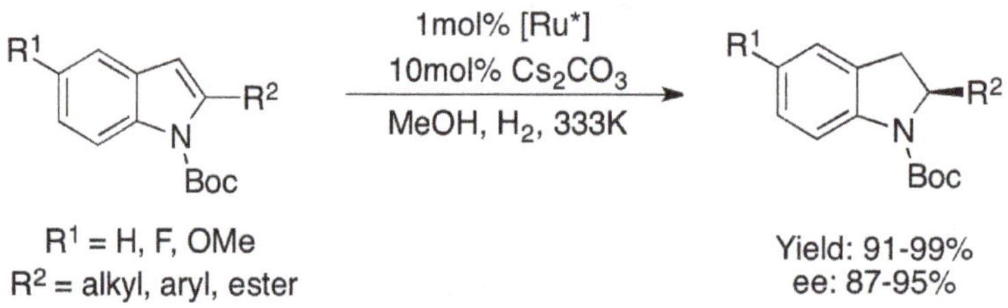

Method for asymmetric hydrogenation of boc-protected indoles

Despite these advances in the asymmetric hydrogenation of protected indoles, considerable operational simplicity can be gained by removing the protecting group altogether. This has been achieved with catalytic systems utilizing Brønsted acids to activate the indole. The initial system used a $Pd(TFA)_2$/H8-BINAP system to achieve the enantioselective *cis*-hydrogenation of 2,3- and 2-substituted indoles with high yield and excellent ee. A similar process, where sequential Friedel-Crafts alkylation and asymmetric hydrogenation occur in one pot, allow asymmetric 2,3-substituted indolines to be selectively prepared from 2-substituted indoles in similarly high yields and ee.

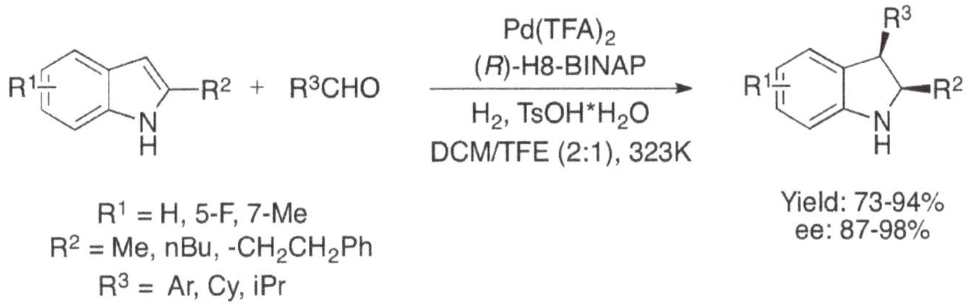

Sequential alkylation and asymmetric hydrogenation of 2-substituted indoles

A promising organocatalytic method for the asymmetric hydrogenation of 2,3-substituted indoles utilizing a chiral Lewis base also exists, although the observed ee's are not quite equivalent to those of the metal-based hydrogenations.

Pyrroles

Achieving complete conversion of pyrroles to pyrrolidines by asymmetric hydrogenation has so far proven difficult, with partial-hydrogenation products often being observed. Complete enantioselective reduction is possible, with the outcome depending on both the starting substrate and the method.

The asymmetric hydrogenation of 2,3,5-substituted pyrroles was achieved by the recognition that such substrates bear the same substitution pattern as 2-substituted indoles, and an asymmetric hydrogenation system that is effective for one of these substrates might be effective for both. Such an analysis led to the development of a ruthenium(I)/phosphine/amine base system for 2,3,5-substituted N-Boc pyrroles that can give either dihydro or tetrahydropyrroles (pyrrolidines), depending on the nature of the pyrrole substituents. An all-phenyl substitution pattern leads to dihydropyrroles in very high yield (>96%) and essentially perfect enantioselectivity. Access to the fully hydrogenated, all-*cis* dihydropyrrole may then be accessible through diastereoselective heterogeneous hydrogenation. Alkyl substitution may lead to either the dihydro or tetrahydropyrrole, although the yields (>70%) and enantioselectivities (often >90%) generally remain high. The regioselectivity in both cases appears to be governed by sterics, with the less-substituted double being preferentially hydrogenated.

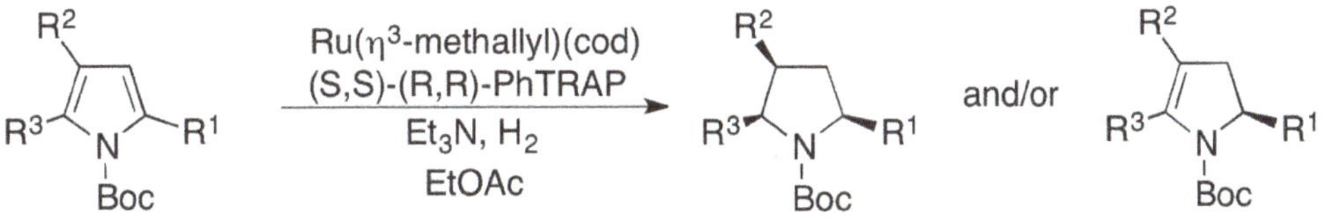

The asymmetric hydrogenation of 2,3,5-substituted N-Boc pyrroles

Unprotected 2,5-pyrroles may also be hydrogenated asymmetrically by a Brønsted acid/Pd(II)/chiral phosphine-catalyzed method, to give the corresponding 2,5-disubstituted 1-pyrrolines in roughly 70-80% yield and 80-90% ee.

Oxygen-containing Heterocycles

The asymmetric hydrogenation of furans and benzofurans has so far proven challenging. Some Ru-NHC complex catalyze asymmetric hydrogenations of benzofurans and furans. with high levels of enantioinduction.

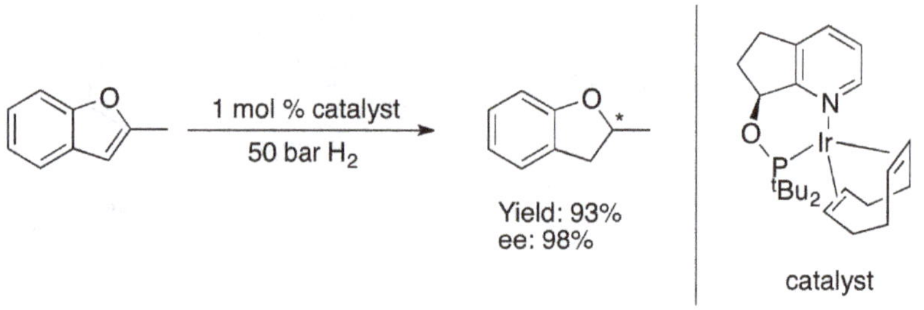

The asymmetric hydrogenation of furans and benzofurans

Sulfur-containing Heterocycles

As is the case with oxygen-containing heterocycles, the asymmetric hydrogenation of compounds where sulfur is part of the initial unsaturated pi-bonding system so far appears to be limited to thiophenes and benzothiophenes. The key approach to the asymmetric hydrogenation of these heterocycles involves a ruthenium(II) catalyst and chiral, C_2 symmetric N-heterocyclic carbene (NHC). Interestingly, this system appears to possess superb selectivity (ee > 90%) and perfect diastereoselectivity (all *cis*) if the substrate has a fused (or directly bound) phenyl ring but yields only racemic product in all other tested cases.

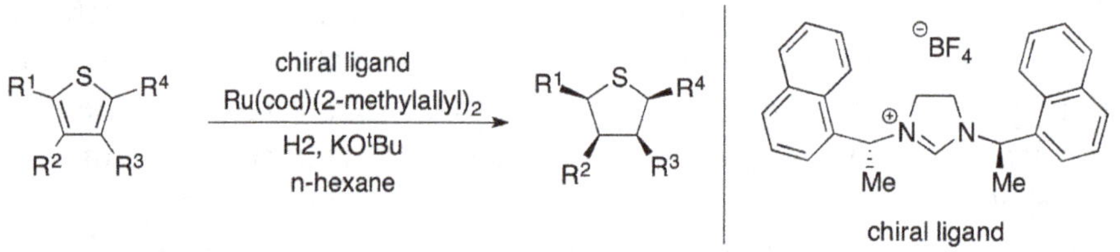

The asymmetric hydrogenation of thiophenes and benzothiophenes

Heterogeneous Catalysis

Research into asymmetric hydrogenation with heterogeneous catalysts has generally focused on three areas. The oldest, dating back to the first asymmetric hydrogenation with palladium deposited on a silk support, involves modifying a metal surface with a chiral molecule, usually one that can be harvested from nature. Alternatively, researchers have used various techniques to attempt to immobilize what would otherwise be homogeneous catalysts on heterogeneous supports or have used synthetic organic ligands and metal sources to build chiral metal-organic frameworks (MOFs).

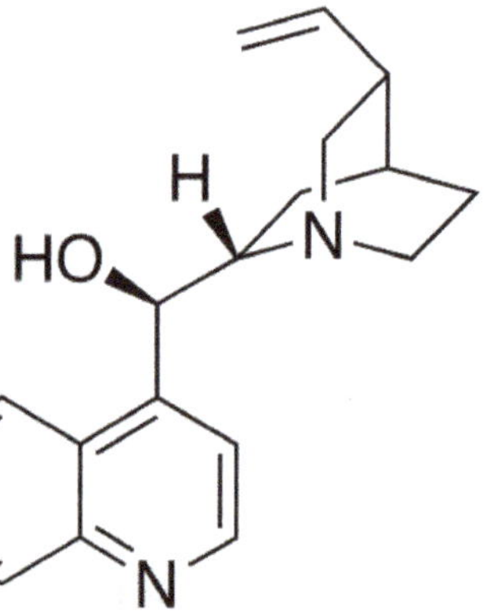

Cinchonidine, one of the cinchona alkaloids

The greatest successes in chiral modification of metal surfaces have come from the use of cinchona alkaloids, though numerous other classes of natural products have been evaluated. These alkaloids have been shown to enhance the rate of substrate hydrogenation by 10–100 times, such that less than one molecule of cinchona alkaloid is needed for every reactive site on the metal and, in fact, the presence of too much of the chiral modifier can cause a decrease in the enantioselectivity of the reaction.

An alternative technique and one that allows more control over the structural and electronic properties of active catalytic sites is the immobilization of catalysts that have been developed for homogeneous catalyis on a heterogeneous support. Covalent bonding of the catalyst to a polymer or other solid support is perhaps most common, though immobilization of the catalyst may also be achieved by adsorption onto a surface, ion exchange, or even physical encapsulation. One drawback of this approach is the potential for the proximity of the support to change the behaviour of the catalyst, lowering the enantioselectivity of the reaction. To avoid this, the catalyst is often bound to the support by a long linker though cases are known where the proximity of the support can actually enhance the performance of the catalyst.

The final approach involves the construction of MOFs that incorporate chiral reaction sites from a number of different components, potentially including chiral and achiral organic ligands, structural metal ions, catalytically active metal ions, and/or preassembled catalytically active organometallic cores. This field is relatively new, and few examples exist of chiral asymmetric hydrogenation using these frameworks. One of these was reported in 2003, when a heterogeneous catalyst was reported that included structural zirconium, catalytically active ruthenium, and a BINAP-derived phosphonate as both chiral ligand and structural linker. As little as 0.005 mol% of this catalyst proved sufficient to achieve the asymmetric hydrogenation of aryl ketones, though the usual conditions featured 0.1 mol % of catalyst and resulted in an enantiomeric excess of 90.6–99.2%.

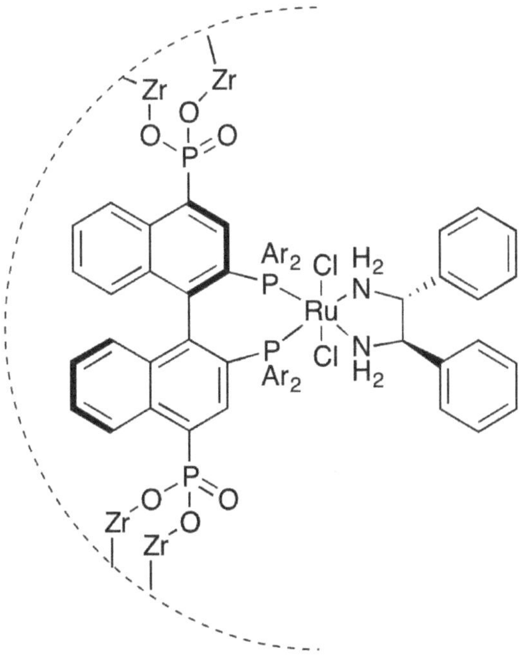

The active site of a heterogeneous zirconium phosphonate catalyst for asymmetric hydrogenation

Industrial Applications

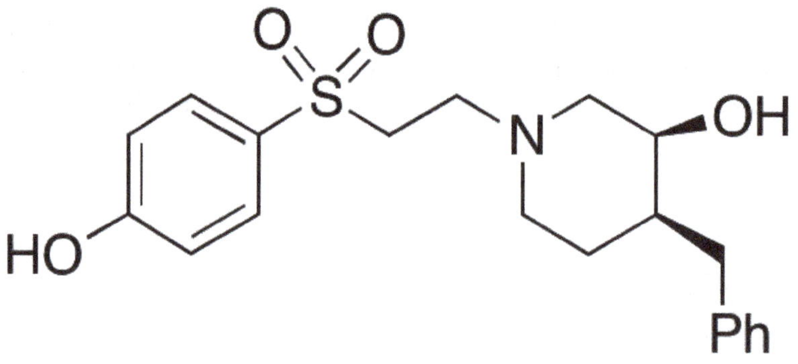

(*S,S*)-Ro 67-8867

Knowles' research into asymmetric hydrogenation and its application to the production scale synthesis of L-Dopa gave asymmetric hydrogenation a strong start in the industrial world. More recently, a 2001 review indicated that asymmetric hydrogenation accounted for 50% of production scale, 90% of pilot scale, and 74% of bench scale catalytic, enantioselective processes in industry, with the caveat that asymmetric catalytic methods in general were not yet widely used.

The success of asymmetric hydrogenation in industry can be seen in a number of specific cases where the replacement of kinetic resolution based methods has resulted in substantial improvements in the process's efficiency. For example, Roche's Catalysis Group was able to achieve the synthesis of (S,S)-Ro 67-8867 in 53% overall yield, a dramatic increase above the 3.5% that was achieved in the resolution based synthesis. Roche's synthesis of mibefradil was likewise improved by replacing resolution with asymmetric hydrogenation, reducing the step count by three and increasing the yield of a key intermediate to 80% from the original 70%.

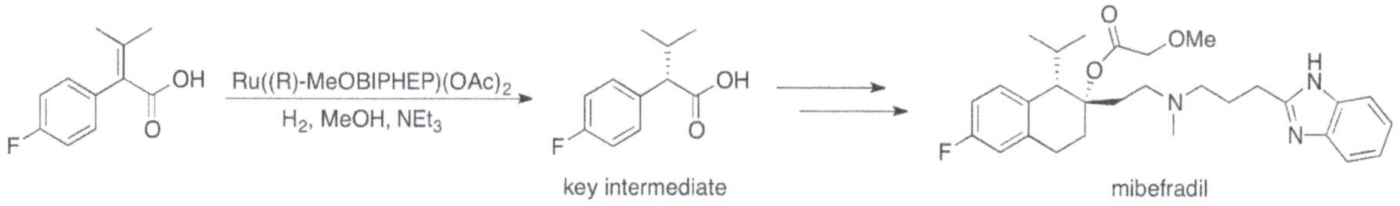

Asymmetric hydrogenation in the industrial synthesis of mibefradil

Ketone

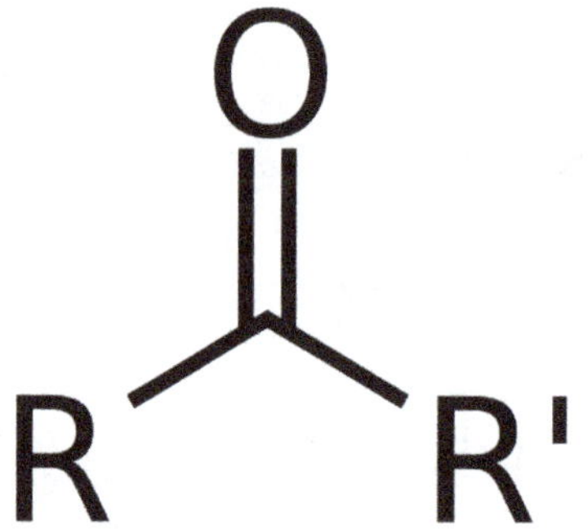

Ketone group

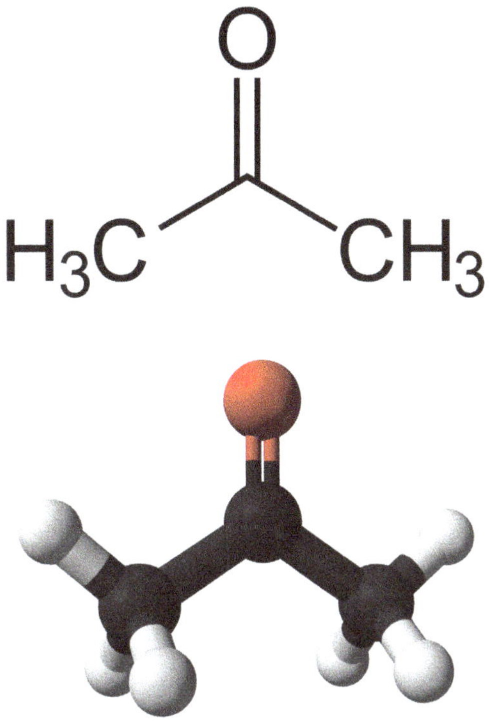

Acetone

In chemistry, a ketone (alkanone) is an organic compound with the structure RC(=O)R', where R and R' can be a variety of carbon-containing substituents. Ketones and aldehydes are simple compounds that contain a carbonyl group (a carbon-oxygen double bond). They are considered "simple" because they do not have reactive groups like –OH or –Cl attached directly to the carbon atom in the carbonyl group, as in carboxylic acids containing –COOH. Many ketones are known and many are of great importance in industry and in biology. Examples include many sugars (ketoses) and the industrial solvent acetone, which is the smallest ketone.

Nomenclature and Etymology

The word *ketone* is derived from *Aketon*, an old German word for acetone.

According to the rules of IUPAC nomenclature, ketones are named by changing the suffix *-ane* of the parent alkane to *-anone*. The position of the carbonyl group is usually denoted by a number. For the most important ketones, however, traditional nonsystematic names are still generally used, for example acetone and benzophenone. These nonsystematic names are considered retained IUPAC names, although some introductory chemistry textbooks use systematic names such as "2-propanone" or "propan-2-one" for the simplest ketone (CH_3–CO–CH_3) instead of "acetone".

The common names of ketones are obtained by writing separately the names of the two alkyl groups attached to the carbonyl group, followed by "ketone" as a separate word. The names of the alkyl groups are written alphabetically. When the two alkyl groups are the same, the prefix di- is added before the name of alkyl group. The positions of other groups are indicated by Greek letters, the α-carbon being the atom adjacent to carbonyl group. If both alkyl groups in a ketone are the same then the ketone is said to be symmetrical, otherwise unsymmetrical.

Although used infrequently, *oxo* is the IUPAC nomenclature for a ketone functional group. Other prefixes, however, are also used. For some common chemicals (mainly in biochemistry), *keto* or *oxo* refer to the ketone functional group. The term *oxo* is used widely through chemistry. For example, it also refers to an oxygen atom bonded to a transition metal (a metal oxo).

Structure and Properties

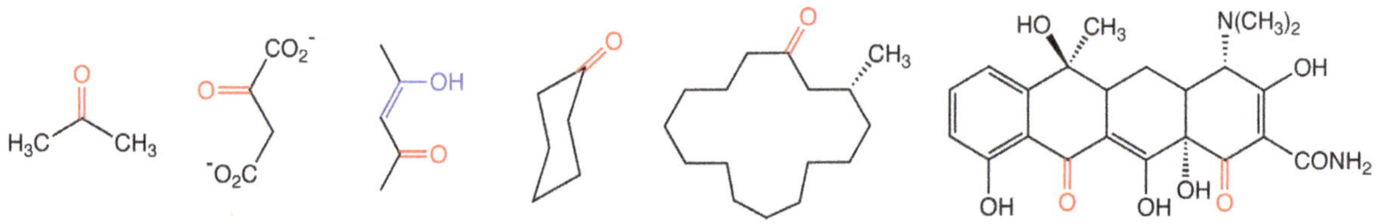

Representative ketones, from the left: acetone, a common solvent; oxaloacetate, an intermediate in the metabolism of sugars; acetylacetone in its (mono) enol form (the enol highlighted in blue); cyclohexanone, precursor to nylon; muscone, an animal scent; and tetracycline, an antibiotic.

The ketone carbon is often described as "sp^2 hybridized", a description that includes both their electronic and molecular structure. Ketones are trigonal planar around the ketonic carbon, with C–C–O and C–C–C bond angles of approximately 120°. Ketones differ from aldehydes in that the carbonyl group (CO) is bonded to two carbons within a carbon skeleton. In aldehydes, the carbonyl is bonded to one carbon and one hydrogen and are located at the ends of carbon chains. Ketones are also distinct from other carbonyl-containing functional groups, such as carboxylic acids, esters and amides.

The carbonyl group is polar because the electronegativity of the oxygen is greater than that for carbon. Thus, ketones are nucleophilic at oxygen and electrophilic at carbon. Because the carbonyl group interacts with water by hydrogen bonding, ketones are typically more soluble in water than the related methylene compounds. Ketones are hydrogen-bond acceptors. Ketones are not usually hydrogen-bond donors and cannot hydrogen-bond to themselves. Because of their inability to serve both as hydrogen-bond donors and acceptors, ketones tend not to "self-associate" and are more volatile than alcohols and carboxylic acids of comparable molecular weights. These factors relate to the pervasiveness of ketones in perfumery and as solvents.

Classes of Ketones

Ketones are classified on the basis of their substituents. One broad classification subdivides ketones into symmetrical and asymmetrical derivatives, depending on the equivalency of the two organic substituents attached to the carbonyl center. Acetone and benzophenone ($C_6H_5C(O)C_6H_5$) are symmetrical ketones. Acetophenone ($C_6H_5C(O)CH_3$) is an asymmetrical ketone. In the area of stereochemistry, asymmetrical ketones are known for being prochiral.

Diketones

Many kinds of diketones are known, some with unusual properties. The simplest is diacetyl ($CH_3C(O)C(O)CH_3$), once used as butter-flavoring in popcorn. Acetylacetone (pentane-2,4-dione) is virtually a misnomer (inappropriate name) because this species exists mainly as the monoenol $CH_3C(O)CH{=}C(OH)CH_3$. Its enolate is a common ligand in coordination chemistry.

Unsaturated Ketones

Ketones containing alkene and alkyne units are often called unsaturated ketones. The most widely used member of this class of compounds is methyl vinyl ketone, $CH_3C(O)CH=CH_2$, which is useful in the Robinson annulation reaction. Lest there be confusion, a ketone itself is a site of unsaturation; that is, it can be hydrogenated.

Cyclic Ketones

Many ketones are cyclic. The simplest class have the formula $(CH_2)_nCO$, where n varies from 2 for cyclopropanone to the teens. Larger derivatives exist. Cyclohexanone, a symmetrical cyclic ketone, is an important intermediate in the production of nylon. Isophorone, derived from acetone, is an unsaturated, asymmetrical ketone that is the precursor to other polymers. Muscone, 3-methylpentadecanone, is an animal pheromone. Another cyclic ketone is cyclobutanone, having the formula C_4H_6O.

Keto-enol Tautomerization

Keto-enol tautomerism. 1 is the keto form; 2 is the enol.

Ketones that have at least one alpha-hydrogen, undergo keto-enol tautomerization; the tautomer is an enol. Tautomerization is catalyzed by both acids and bases. Usually, the keto form is more stable than the enol. This equilibrium allows ketones to be prepared via the hydration of alkynes.

Acidity of Ketones

Ketones are far more acidic ($pK_a \approx 20$) than a regular alkane ($pK_a \approx 50$). This difference reflects resonance stabilization of the enolate ion that is formed upon deprotonation. The relative acidity of the α-hydrogen is important in the enolization reactions of ketones and other carbonyl compounds. The acidity of the α-hydrogen also allows ketones and other carbonyl compounds to undergo nucleophilic reactions at that position, with either stoichiometric and catalytic base.

Characterization

An aldehyde differs from a ketone because of its hydrogen atom attached to its carbonyl group, making aldehydes easier to oxidize. Ketones don't have a hydrogen atom bonded to the carbonyl group, and are more resistant to oxidation. They are only oxidized by powerful oxidizing agents which have the ability to cleave carbon-carbon bonds.

Spectroscopy

Ketones and aldehydes absorb strongly in the infra-red spectrum near 1700 cm^{-1}. The exact position of the peak depends on the substituents.

Whereas H NMR spectroscopy is generally not useful for establishing the presence of a ketone, C NMR spectra exhibit signals somewhat downfield of 200 ppm depending on structure. Such signals are typically weak due to the absence of nuclear Overhauser effects. Since aldehydes resonate at similar chemical shifts, multiple resonance experiments are employed to definitively distinguish aldehydes and ketones.

Qualitative Organic Tests

Ketones give positive results in Brady's test, the reaction with 2,4-dinitrophenylhydrazine to give the corresponding hydrazone. Ketones may be distinguished from aldehydes by giving a negative result with Tollens' reagent or with Fehling's solution. Methyl ketones give positive results for the iodoform test.

Synthesis

Many methods exist for the preparation of ketones in industrial scale and academic laboratories. Ketones are also produced in various ways by organisms.

In industry, the most important method probably involves oxidation of hydrocarbons, often with air. For example, a billion kilograms of cyclohexanone are produced annually by aerobic oxidation of cyclohexane. Acetone is prepared by air-oxidation of cumene.

For specialized or small scale organic synthetic applications, ketones are often prepared by oxidation of secondary alcohols:

$$R_2CH(OH) + O \rightarrow R_2C{=}O + H_2O$$

Typical strong oxidants (source of "O" in the above reaction) include potassium permanganate or a Cr(VI) compound. Milder conditions make use of the Dess–Martin periodinane or the Moffatt–Swern methods.

Many other methods have been developed, examples include:

- By geminal halide hydrolysis.
- By hydration of alkynes. Such processes occur via enols and require the presence of an acid and $HgSO_4$. Subsequent enol–keto tautomerization gives a ketone. This reaction always produces a ketone, even with a terminal alkyne.
- From Weinreb Amides using stoichiometric organometallic reagents.
- Aromatic ketones can be prepared in the Friedel–Crafts acylation, the related Houben–Hoesch reaction, and the Fries rearrangement.
- Ozonolysis, and related dihydroxylation/oxidative sequences, cleave alkenes to give aldehydes and/or ketones, depending on alkene substitution pattern.
- In the Kornblum–DeLaMare rearrangement ketones are prepared from peroxides and base.

- In the Ruzicka cyclization, cyclic ketones are prepared from dicarboxylic acids.
- In the Nef reaction, ketones form by hydrolysis of salts of secondary nitro compounds.
- In the Fukuyama coupling, ketones form from a thioester and an organozinc compound.
- By the reaction of an acid chloride with organocadmium compounds or organocopper compounds.
- The Dakin–West reaction provides an efficient method for preparation of certain methyl ketones from carboxylic acids.
- Ketones can also be prepared by the reaction of Grignard reagents with nitriles, followed by hydrolysis.
- By decarboxylation of carboxylic anhydride.
- Ketones can be prepared from haloketones in reductive dehalogenation of halo ketones.
- In ketonic decarboxylation symmetrical ketones are prepared from carboxylic acids.
- Oxidation of amines with iron(III) chloride.
- Hydrolysis of unsaturated secondary amides, β-Keto acid esters, or β-diketones.
- Acid-catalysed rearrangement of 1,2-diols.

Reactions

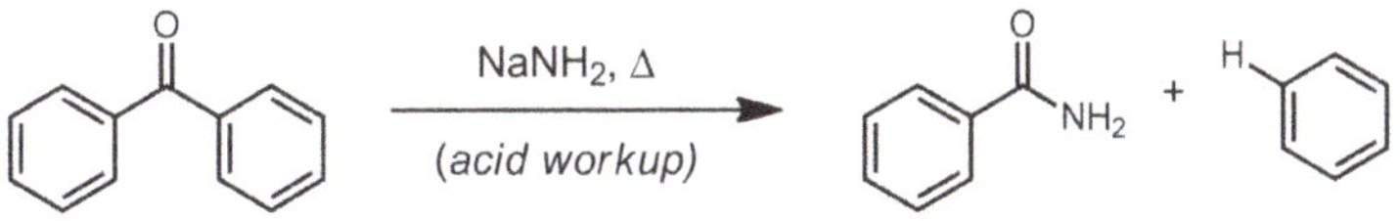

The Haller-Bauer reaction occurs between a *non-enolizable* ketone and a strong amide base. In this prototypical example involving benzophenone, the tetrahedral intermediate expels phenyl anion to give benzamide and benzene as the organic products

Ketones engage in many organic reactions. The most important reactions follow from the susceptibility of the carbonyl carbon toward nucleophilic addition and the tendency for the enolates to add to electrophiles. Nucleophilic additions include in approximate order of their generality:

- With water (hydration) gives geminal diols, which are usually not formed in appreciable (or observable) amounts
- With an acetylide to give the α-hydroxyalkyne
- With ammonia or a primary amine gives an imine
- With secondary amine gives an enamine
- With Grignard and organolithium reagents to give, after aqueous workup, a tertiary alcohol
- With an alcohols or alkoxides to gives the hemiketal or its conjugate base. With a diol to the ketal. This reaction is employed to protect ketones.

- With sodium amide resulting in C–C bond cleavage with formation of the amide $RCONH_2$ and the alkane R'H, a reaction called the Haller–Bauer reaction.
- With strong oxidizing agents to give carboxylic acids.
- Electrophilic addition, reaction with an electrophile gives a resonance stabilized cation.
- With phosphonium ylides in the Wittig reaction to give the alkenes.
- With thiols to give the thioacetal.
- With hydrazine or 1-disubstituted derivatives of hydrazine to give hydrazones.
- With a metal hydride gives a metal alkoxide salt, hydrolysis of which gives the alcohol, an example of ketone reduction.
- With halogens to form an α-haloketone, a reaction that proceeds via an enol.
- With heavy water to give an α-deuterated ketone.
- Fragmentation in photochemical Norrish reaction.
- Reaction of 1,4-aminodiketones to oxazoles by dehydration in the Robinson–Gabriel synthesis.
- In the case of aryl–alkyl ketones, with sulfur and an amine give amides in the Willgerodt reaction.
- With hydroxylamine to produce oximes.
- With reducing agents to form secondary alcohols.
- With peroxy acids to form esters in the Baeyer–Villiger oxidation.

Biochemistry

Ketones are pervasive in nature. The formation of organic compounds in photosynthesis occurs via the ketone ribulose-1,5-bisphosphate. Many sugars are ketones, known collectively as ketoses. The best known ketose is fructose, which exists as a cyclic hemiketal, which masks the ketone functional group. Fatty acid synthesis proceeds via ketones. Acetoacetate is an intermediate in the Krebs cycle which releases energy from sugars and carbohydrates.

In medicine, acetone, acetoacetate, and beta-hydroxybutyrate are collectively called ketone bodies, generated from carbohydrates, fatty acids, and amino acids in most vertebrates, including humans. Ketone bodies are elevated in the blood (ketosis) after fasting, including a night of sleep; in both blood and urine in starvation; in hypoglycemia, due to causes other than hyperinsulinism; in various inborn errors of metabolism, and intentionally induced via a ketogenic diet, and in ketoacidosis (usually due to diabetes mellitus). Although ketoacidosis is characteristic of decompensated or untreated type 1 diabetes, ketosis or even ketoacidosis can occur in type 2 diabetes in some circumstances as well.

Applications

Ketones are produced on massive scales in industry as solvents, polymer precursors, and pharmaceuticals. In terms of scale, the most important ketones are acetone, methylethyl ketone, and cyclohexanone. They are also common in biochemistry, but less so than in organic chemistry in general. The combustion of hydrocarbons is an uncontrolled oxidation process that gives ketones as well as many other types of compounds.

Toxicity

Although it is difficult to generalize on the toxicity of such a broad class of compounds, simple ketones are, in general, not highly toxic. This characteristic is one reason for their popularity as solvents. Exceptions to this rule are the unsaturated ketones such as methyl vinyl ketone with LD_{50} of 7 mg/kg (oral).

Reactions of Ketones

Enantioselective reduction of C=O double bond in organic synthesis has important application in synthesis of many natural products as well as pharmaceutical products. The section covers the representative examples of metal catalyzed reactions. The frequently used chiral ligands for the metal catalyzed enantioselective reduction reactions of ketones are listed in figure.

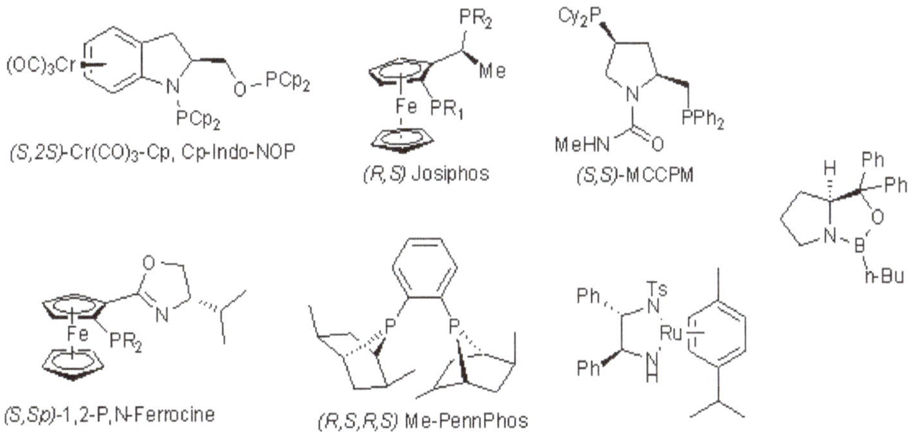

Reactions of α-Keto Amides

Asymmetric hydrogenation of α -keto esters has been studied with several rhodium catalysts. Neutral rhodium catalysts with chiral ligands such as $Cr(CO)_3$-Cp,Cp-Indo-NOP demonstrate excellent enantioselectivity and reactivity in the hydrogenation of amides.

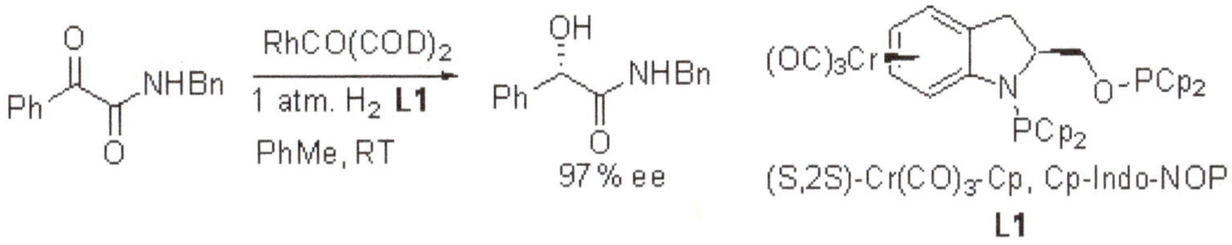

Enantioselective Hydrogenation of α -Keto Amide

Reactions of β - Keto Esters

Asymmetric hydrogenation of β -keto esters has been extensively studied using chiral ruthenium catalysts. However, only handful of examples analogous to rhodium-catalyzed reaction are explored. The Rh-(R,S)-Josiphos complex provides an effective catalyst for the asymmetric hydrogenation of ethyl 3-oxobutanoate affording the corresponding β -hydroxy ester in 97% ee. The above ligands Josiphos family such as chiral Walphos, Joshiphos, BPPFOH, TRAP and PIGIPHOS ligands could be easily synthesized from commercially available Ugi amine.

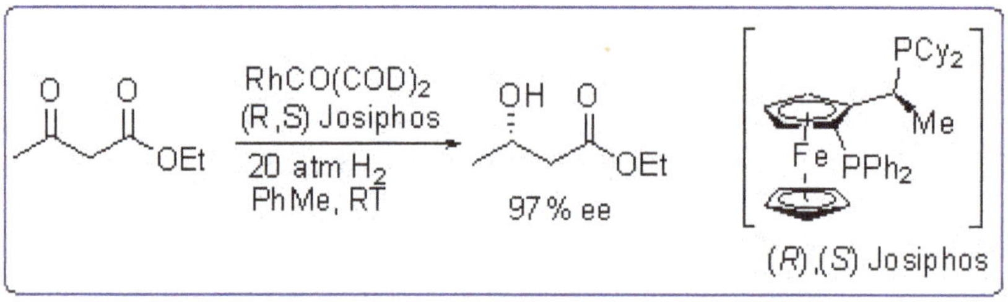

Enantioselective Hydrogenation of β -Keto ester

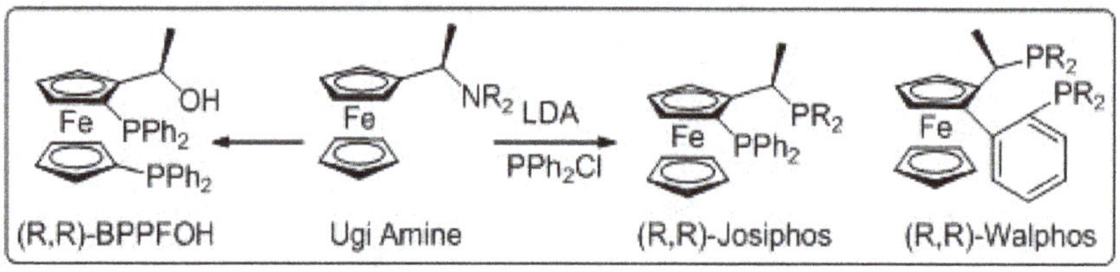

Synthesis of Josiphos Type Ligands

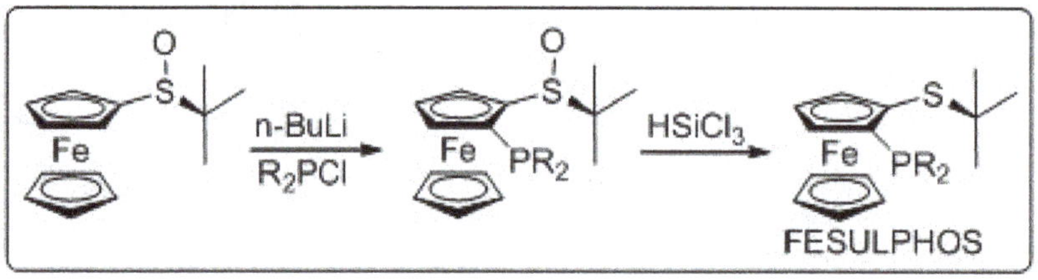

Synthesis of Feluphos

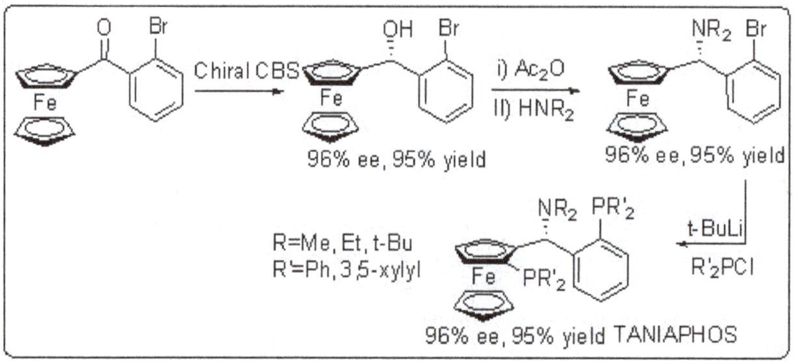

Synthesis of Taniaphos

Iridium/spiro PAP has been used as effective catalyst for the asymmetric hydrogenation of β-aryl β-ketoesters. The reaction provides a readily accessible method for the synthesis of β-hydroxy esters in high enantioselectivity up to 99.8% ee and high TONs up to 1230000.

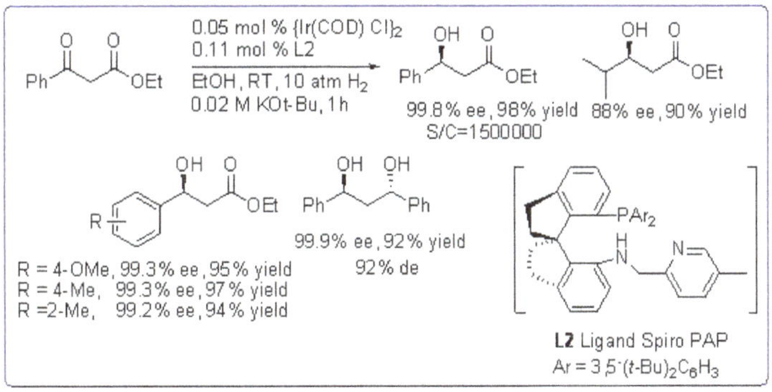

Enantioselective hydrogenation of β- ketoesters

Reactions of Aromatic Ketones

Amino ketones and their hydrochloride salts can be effectively hydrogenated with chiral rhodium catalysts. The rhodium precatalysts, combined with chiral phosphorous ligands *(S,S)*- MCCPM provide excellent enantioselectivity and reactivity for the asymmetric hydrogenation of α, β, and γ -alkyl amino ketone hydrochloride salts with S/C=100000.

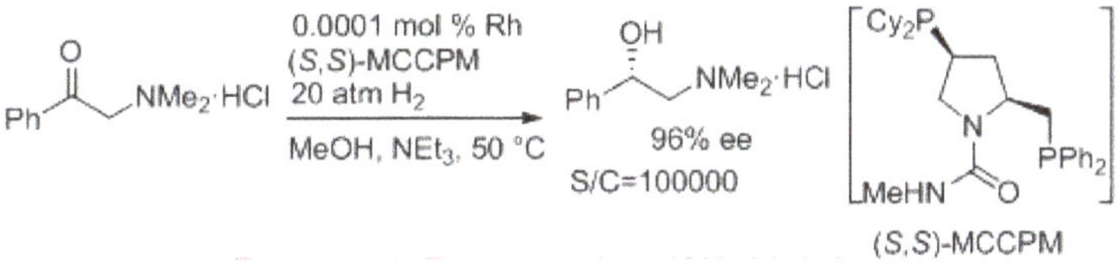

Enantioselective Hydrogenation of α -Aryl Amino Ketone

The enantioselective hydrogenation of 3,5-bistrifluoromethyl acetophenone (BTMA) can be carried out using a Ru/phosphine-oxazoline complex . The reaction is compatible with 140-kg scale at 20 bar and 25°C with S/C ratios of 20,000. The synthesis of the ligand is shown in Figure.

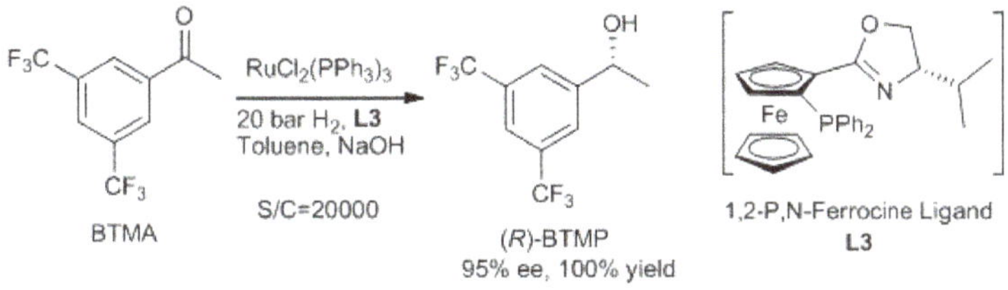

Hydrogenation of α -Aryl Ketone

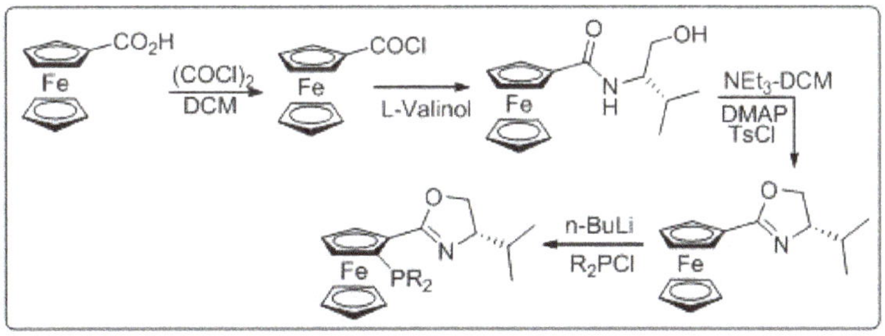

Synthesis of (S,Sp)- 1,2-P,N-Ferrocine

The enantioselective hydrogenation of amino ketones has been applied extensively to the synthesis of chiral drugs and pharmaceuticals. For example, direct enantioselective hydrogenation of

3-aryloxy-2-oxo-1-propylamine leads to 1-amino-3-aryloxy-2-propanol using 0.01 mol % of the neutral Rh-(S, S)-MCCPM complex. The chiral product 1-amino-3-aryloxy-2-propanol serves as β-adrenergic blocking agents. (S)-Propranolol is obtained in 90.8% ee from the corresponding α-amino ketone.

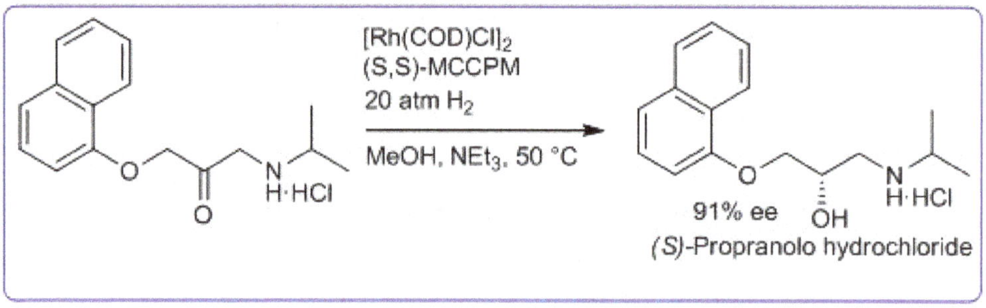

Key step for the Direct Synthesis of (S)- Propranolol

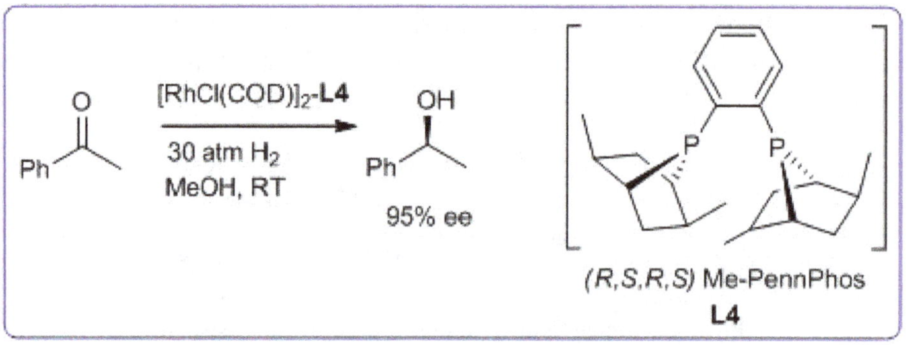

Asymmetric Reduction of Acetophenone

Reactions of Aliphatic Ketones

The asymmetric hydrogenation of simple aliphatic ketones remains still a challenging problem. This is due to the difficulty to design the appropriate chiral catalyst that will easily differentiate between the two-alkyl substituents of the ketone. Promising results have been obtained in asymmetric hydrogenation of aliphatic ketones using the (R,S,R,S)-PennPhos- Rh complex in combination with 2,6-lutidine and KBr. For example, the reaction of tert -butyl methyl ketone takes place with 94% ee . Similarly, isopropyl-, n -butyl- and cyclohexyl methyl ketones can be reduced with 85% ee , 75% ee and 92% ee, respectively.

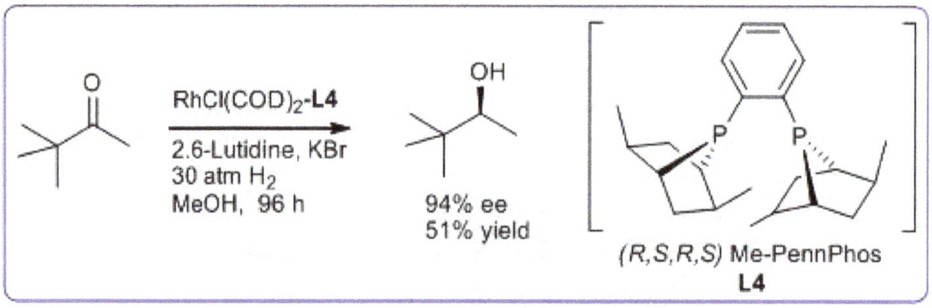

The chiral Ru-diphosphine/diamine derived from chiral BINAP, DPEN (diphenylethylene diamine) and indanol effect enantioselective hydrogenation of certain amino or amido ketones via a non-chelate mechanism without interaction between Ru and nitrogen or oxygen. The diamine catalyst can be synthesized from chiral 1,2- diphenylethylene diamine.

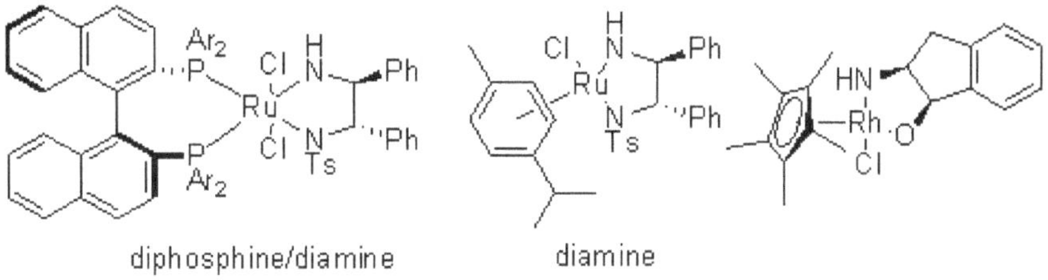

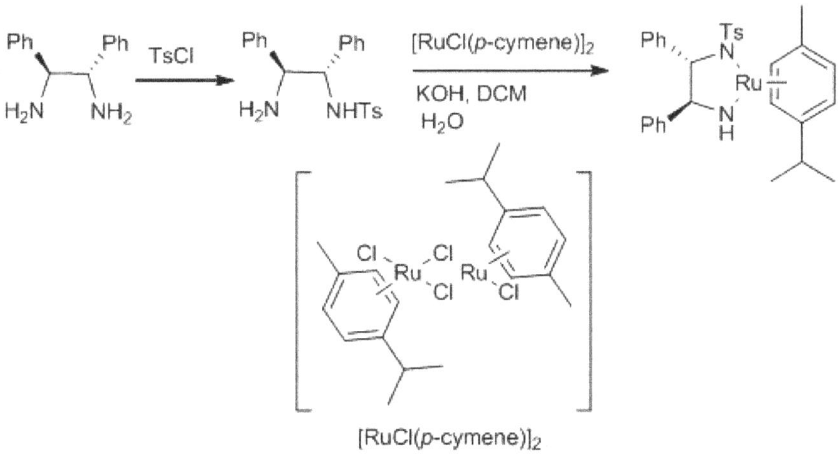

These catalysts have been employed for the asymmetric synthesis of various important pharmaceuticals, including (R)-denopamine, a β 1-receptor agonist, the anti -depressant (R)-fluoxetine, the anti -psychotic BMS 181100 and (S)-duloxetine.

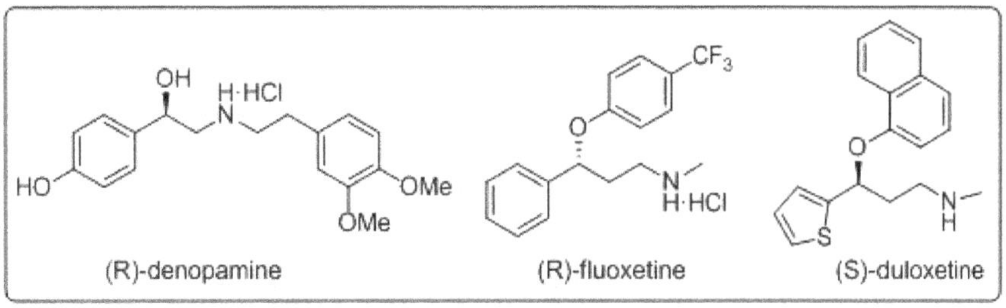

Unsymmetric benzophenones could also be hydrogenated with high S/C ratio of up to 20000 without over-reduction. Enantioselective hydrogenation of certain ortho -substituted benzophenones leads to the unsymmetrically substituted benzhydrols, allowing convenient synthesis of the anticholinergic and anti -histaminic (S)-orphenadrine and antihistaminic (R)-neobenodine.

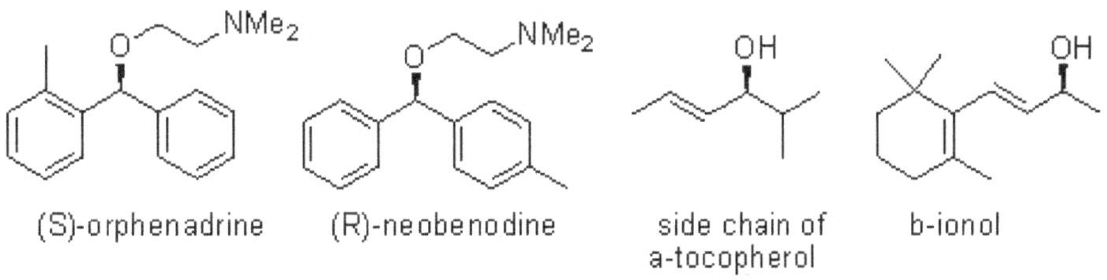

Asymmetric Synthesis of Some of the Important Pharmaceuticals

The asymmetric hydrogenation of simple ketone is generally achieved by the combined use of an (S)-BINAP and an (S)-1,2-diphenylethylenediamine. However, the reaction of 2,4,4-trimethyl-2-cyclohexenone can be effectively done with racemic $RuCl_2$ [-tol-BINAP]- and chiral DPEN with up to >95% ee.

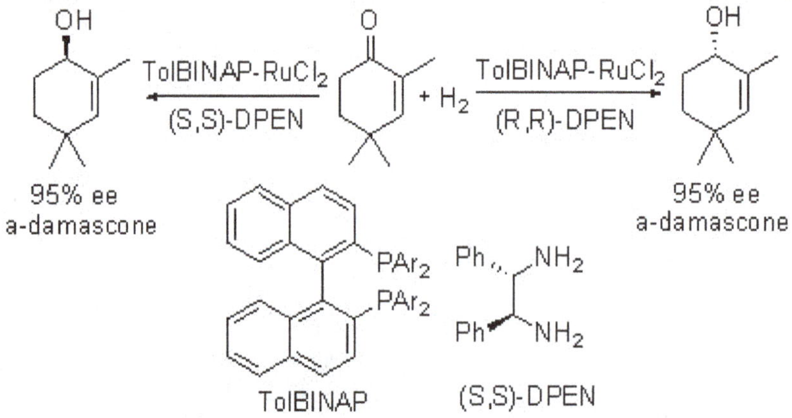

Hydrogenation of Carbon–nitrogen Double Bonds

In chemistry, the hydrogenation of carbon–nitrogen double bonds is the addition of the elements of dihydrogen (H_2) across a carbon–nitrogen double bond, forming amines or amine derivatives. Although a variety of general methods have been developed for the enantioselective hydrogenation of ketones, methods for the hydrogenation of carbon–nitrogen double bonds are less general. Hydrogenation of imines is complicated by both *syn/anti* isomerization and tautomerization to enamines, which may be hydrogenated with low enantioselectivity in the presence of a chiral catalyst. Additionally, the substituent attached to nitrogen affects both the reactivity and spatial properties of the imine, complicating the development of a general catalyst system for imine hydrogenation. Despite these challenges, methods have been developed that address particular substrate classes, such as *N*-aryl, *N*-alkyl, and endocyclic imines.

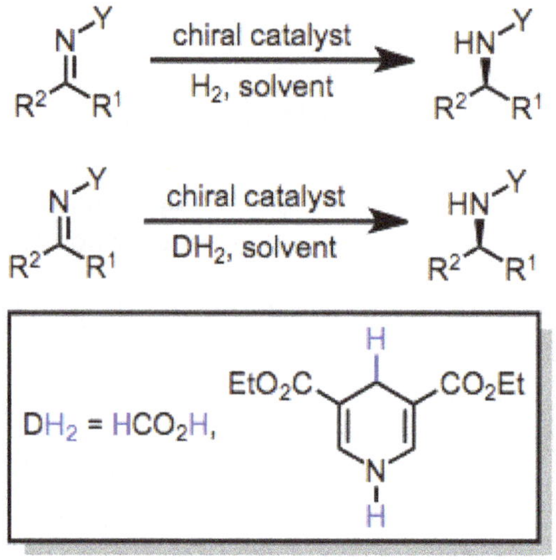

As in hydrogenation reactions of other functional groups, the reductant in C=N hydrogenations is either hydrogen gas or a transfer hydrogenation reductant such as formic acid. The process is usually catalyzed by a transition metal complex. If the complex is chiral and non-racemic and the substrate is prochiral, an excess of a single enantiomer of a chiral product can result.

Mechanism and Stereochemistry

Inner Sphere Mechanisms

The mechanism of imine hydrogenation depends on the nature of the catalyst .Catalysis by some rhodium(I) complexes proceeds through the dihydride species I. The mechanism is proposed to involve both η¹ (σ-type) and η² (π-type) coordination of the imine followed by transfer of hydrogen to generate the amine complex. Dissociation of the amine product and oxidative addition of H_2 then occur, preparing the catalyst to bind another imine molecule. In some iridium-catalyzed hydrogenations, the mechanism is believed to proceed via a monohydride species. The oxidation state of iridium is always +3.

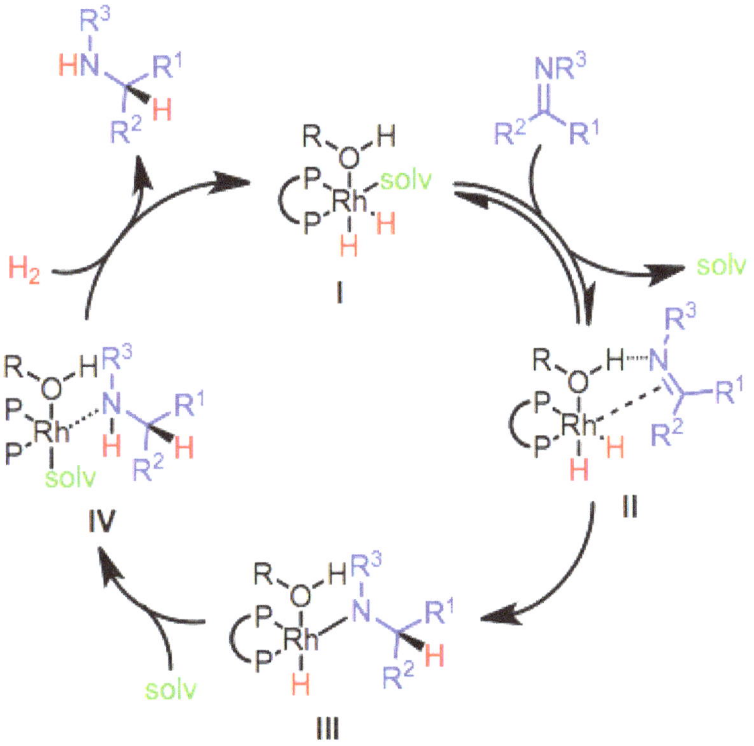

Outer Sphere Mechanisms

Ruthenium(II) catalysts incorporating chiral diphosphine ligands operate according to the inner-sphere mechanisms described above, in which the imine must coordinate to the catalyst and insert into the metal-hydrogen bond. Notably many ruthenium(II) catalyst operate through an "outer-sphere mechanism," during which the imine never interacts with the metal center indirectly. Instead, it receives the elements of H_2 from Ru-H and N-H in a concerted, polarized fashion. This process is utilized by the Shvo catalyst and many ruthenium amine complexes.

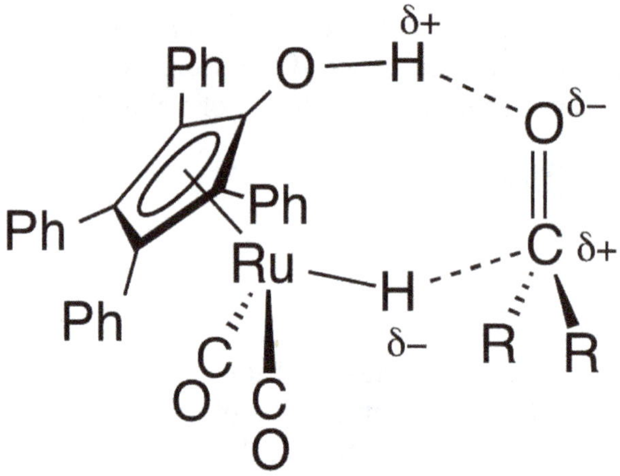

Structure of proposed intermediate in transfer hydrogenation of a ketone (iminium ions bind similarly) by Shvo's catalyst.

Scope and Limitations

Because the substituents attached to the imine nitrogen exert a profound influence on reactivity, few general catalyst systems exist for the enantioselective hydrogenation of imines and imine derivatives. However, catalyst systems have been developed that catalyze hydrogenation of particular classes of imines with high enantioselectivity and yield. This topic describes some of these systems and is organized by the substitution pattern of the imine.

Imines

N-Aryl imines are efficiently hydrogenated using a preformed iridium(I) catalyst with low pressures of hydrogen. Although enantioselectivities vary, conversion is very high across a variety of iridium-based catalyst systems.

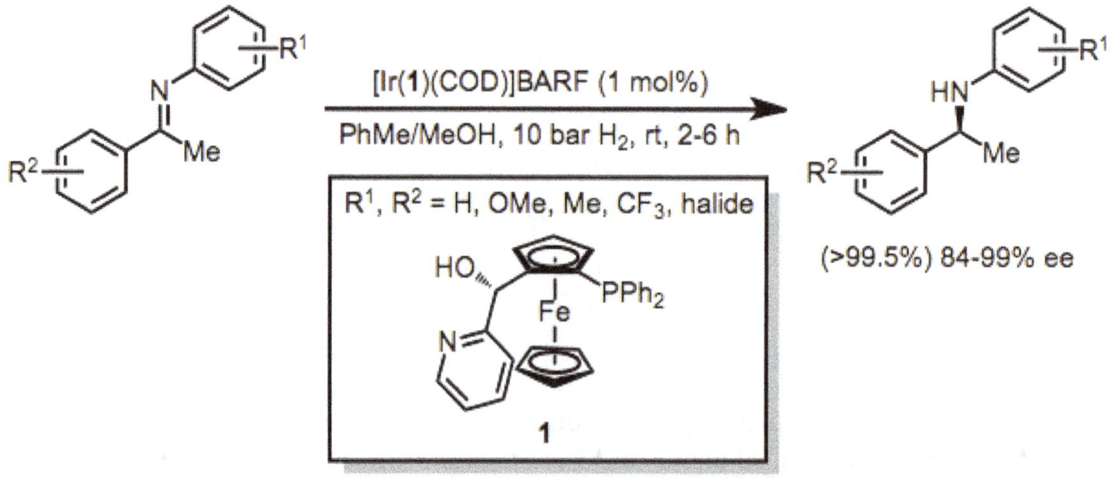

N-Alkyl imines are generally more difficult to reduce with high enantioselectivity than *N*-aryl imines. Most catalyst systems that reduce *N*-alkyl imines enantioselectively are based on rhodium(I) or ruthenium(II).

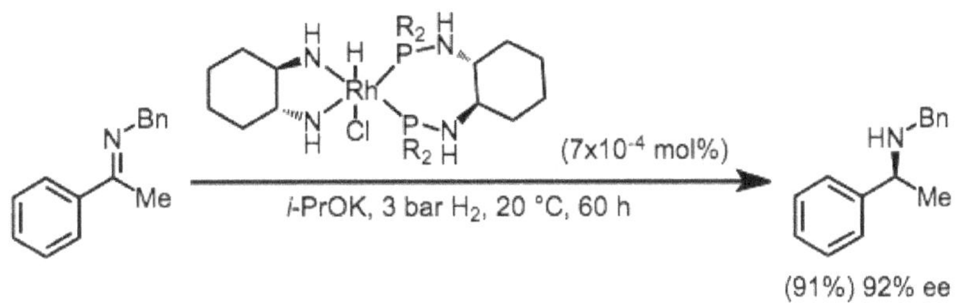

Endocyclic imines do not suffer from complications due to *syn/anti* isomerization; however, this has not proven to be a significant advantage in enantioselective hydrogenations. 3-Hydroindoles may be reduced with hydrogen gas and an iridium catalyst.

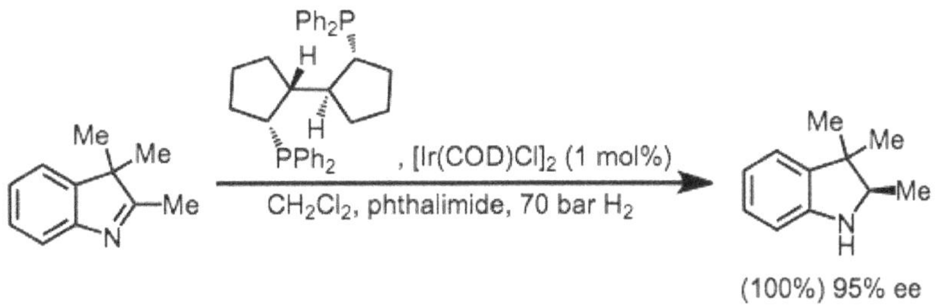

Alternatively, and illustrating the outer sphere mechanism, transfer hydrogenation from formic acid/triethylamine is catalyzed by a ruthenium(II)/diamine complex.

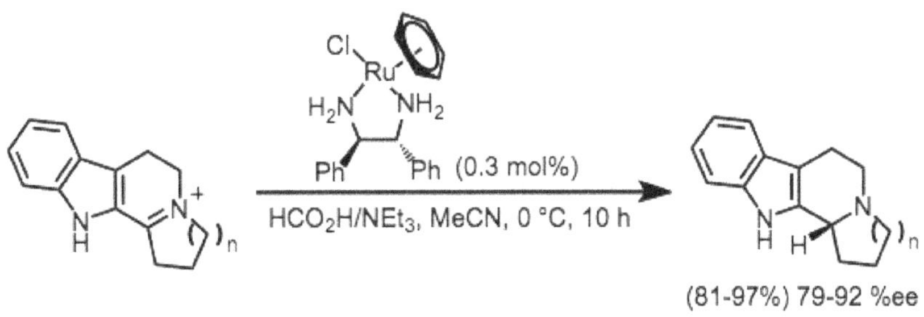

Heteroaromatic substrates may be reduced by hydrogen in the presence of a transition-metal complex. Although catalyst systems tend to be substrate specific, and enantioselectivities are often low, quinolines may be reduced to tetrahydroquinolies enantioselectively using several catalyst systems.

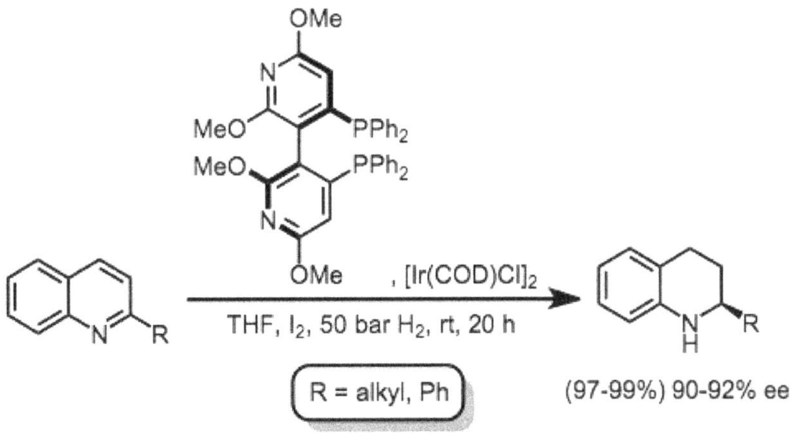

Imine Derivatives

Electronically polarized *N*-tosyl imines are reduced with high enantioselectivity and yield by palladium(II) and ruthenium(II) catalyst systems, although catalyst loadings tend to be relatively high under both direct and transfer hydrogenative conditions.

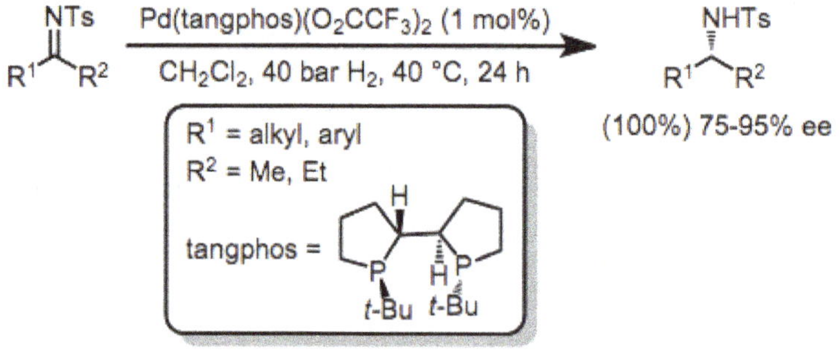

The presence of a second coordinating group on the substrate enhances enantioselectivity in a number of cases. *C*-Acyl hydrazones are hydrogenated with much higher enantioselectivity than the corresponding *C*-alkyl hydrazones, for instance.

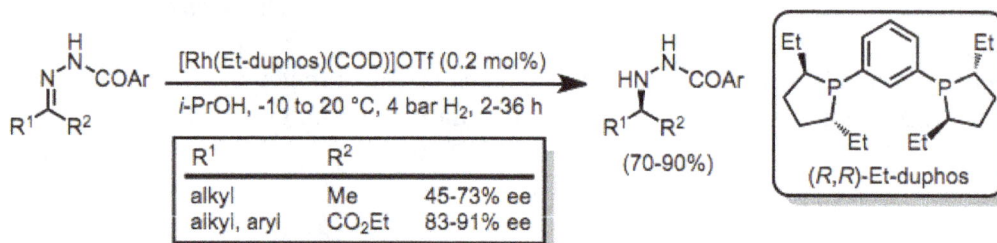

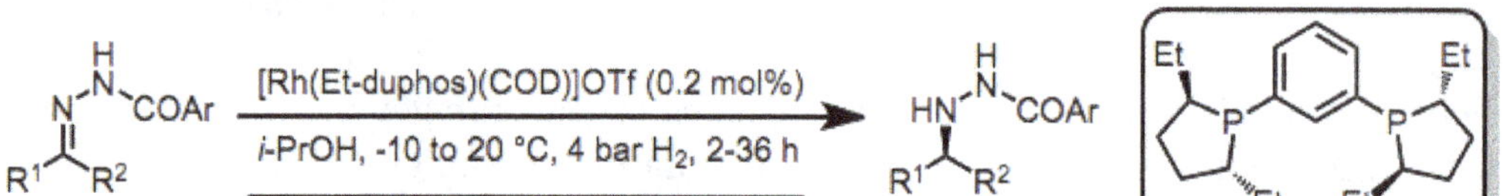

R¹	R²	
alkyl	Me	45-73% ee
alkyl, aryl	CO_2Et	83-91% ee

α-Carboxy imines are attractive precursors for α-amino acids. Organocatalytic reduction of these substrates is possible using a Hantzsch ester and a chiral phosphoric acid catalyst.

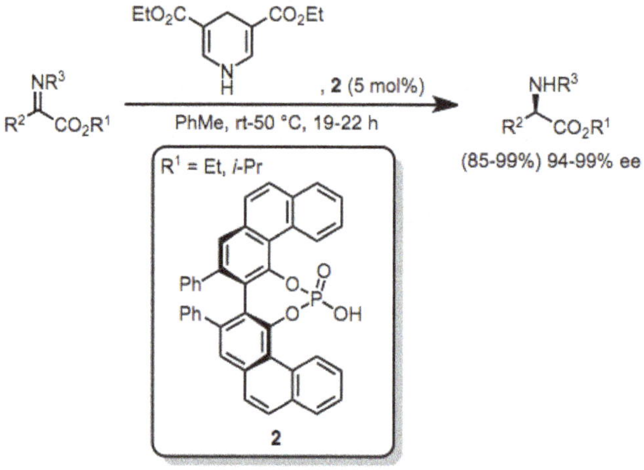

Applications

Metolachlor is the active ingredient in the widely used herbicide Dual Magnum®. A key step in its industrial production involves the enantioselective reduction of an *N*-aryl imine. This reduction is achieved with extremely high turnover number (albeit moderate enantioselectivity) through the use of a specialized catalyst system consisting of $[Ir(COD)Cl]_2$, modified Josiphos ligand 3, and acid and iodide additives.

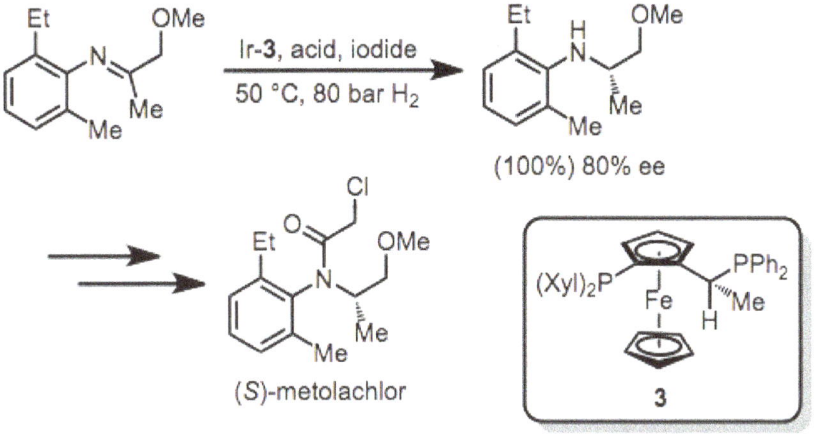

Transfer hydrogenation of endocyclic imines has been applied to the synthesis of tetrahydroisoquinoline alkaloids, such as cryspine A.

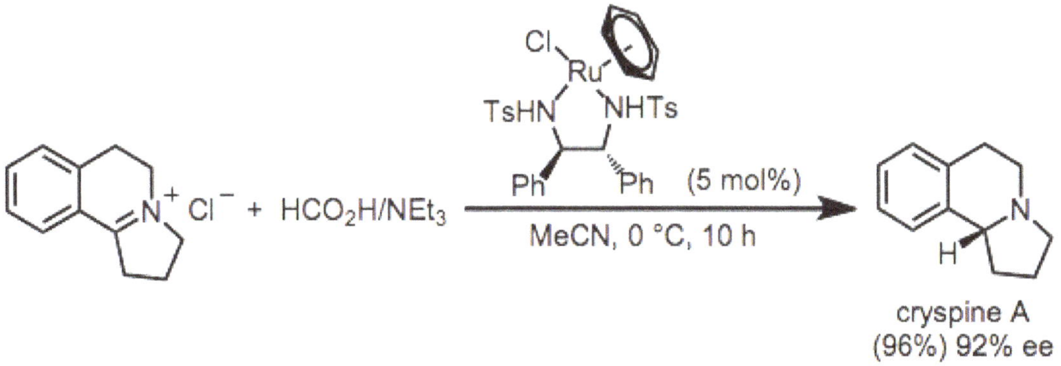

Comparison with Other Methods

Imines may be reduced enantioselectively using stoichiometric amounts of chiral metal hydrides. Such methods have the advantage that they are easy to implement. Reduction with hydrosilanes is a second alternative to transition-metal catalyzed hydrogenation.

Aminotransferase enzymes can be used to synthesize chiral amines in non-racemic form either through kinetic resolution of a racemate or reductive amination of a prochiral ketone. Enantioselectivities are often extremely high; however, optimization of the reaction conditions can be laborious.

Experimental Conditions

The ideal conditions for imine hydrogenation depend on the nature of the substrate and catalyst system. Generally, higher pressures of hydrogen gas are needed in order to achieve reasonable reaction times; however, hydrogen pressure does not appreciably affect enantioselectivity. If the active catalyst is generated *in situ*, Schlenck techniques or a glove box should be employed. Parr shakers may be used for reaction pressures up to 4 bar, although precise temperature control is difficult to achieve in the reaction vessel. For higher hydrogen pressures, specialized autoclave equipment is required.

Hydrogen forms flammable mixtures with air, and any apparatus used for hydrogenation should be checked for leaks prior to the introduction of substrates and catalysts.

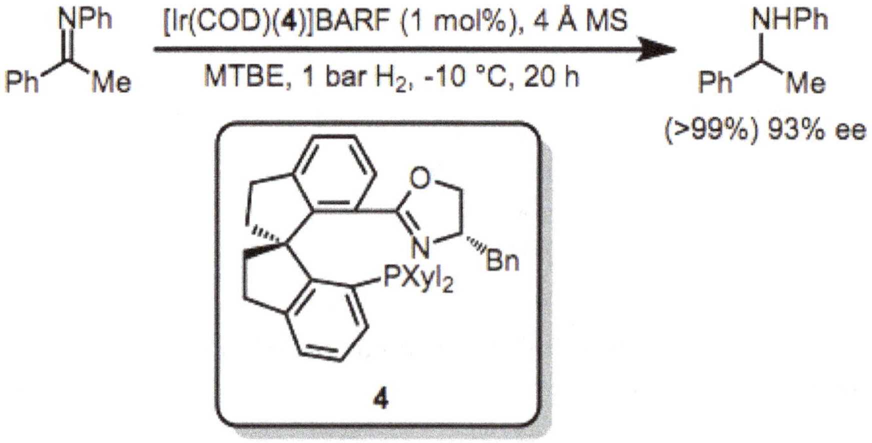

Imine

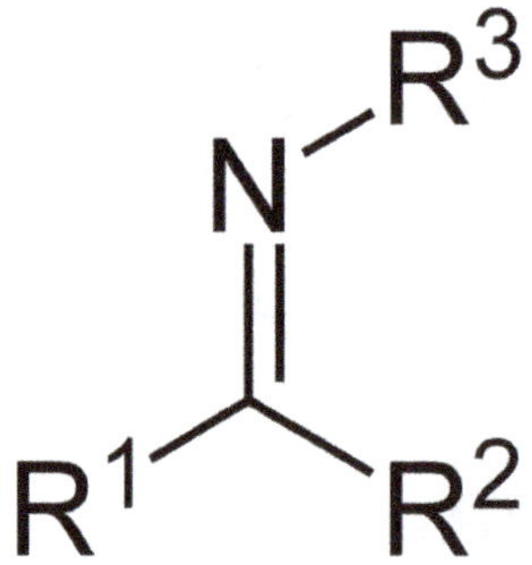

The general structure of an imine

An imine is a functional group or chemical compound containing a carbon–nitrogen double bond. The Nitrogen atom can be attached to a hydrogen (H) or an organic group (R). If this group is *not* a hydrogen atom, then the compound can sometimes be referred to as a Schiff base. The carbon atom has two additional single bonds.

Nomenclature and Classification

Usually imines refer to compounds with the connectivity $R_2C=NR$, as discussed below. In the older literature, imine refers to the aza analogue of an epoxide. Thus, ethyleneimine is the three-membered ring species C_2H_4NH.

Imines are related to ketones and aldehydes by replacement of the oxygen with an NR group. When R = H, the compound is a primary imine, when R is hydrocarbyl, the compound is a secondary imine. Imines exhibit diverse reactivity and are commonly encountered throughout chemistry. When R^3 is OH, the imine is called an oxime, and when R^3 is NH_2 the imine is called a hydrazone.

A primary imine in which C is attached to both a hydrocarbyl and a H is called a primary aldimine; a secondary imine with such groups is called a secondary aldimine. A primary imine in which C is attached to two hydrocarbyls is called a primary ketimine; a secondary imine with such groups is called a secondary ketimine .

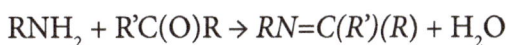

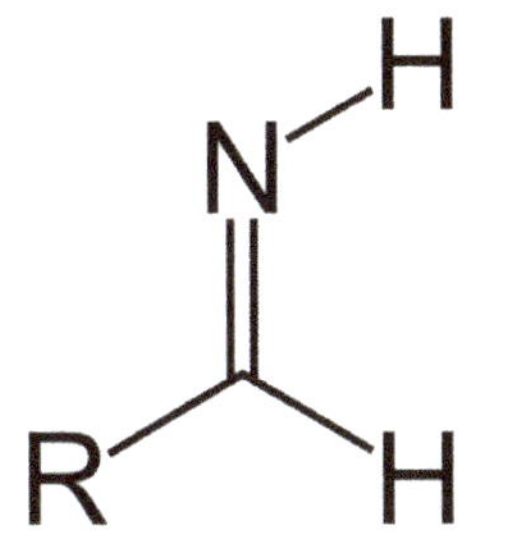	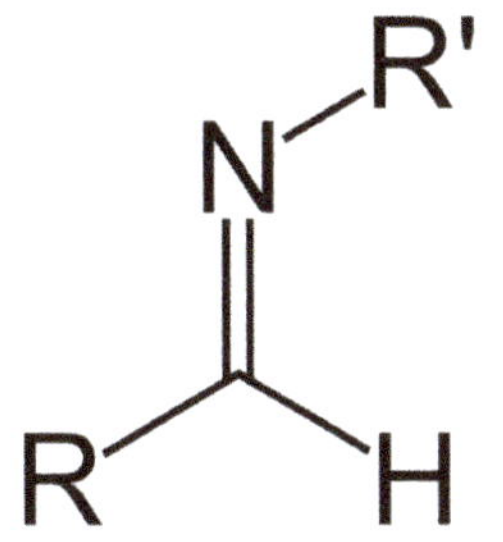	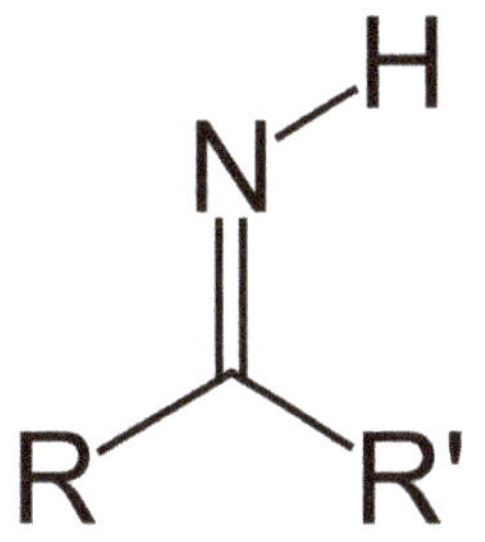
Primary aldimine	Secondary aldimine	Primary ketimine

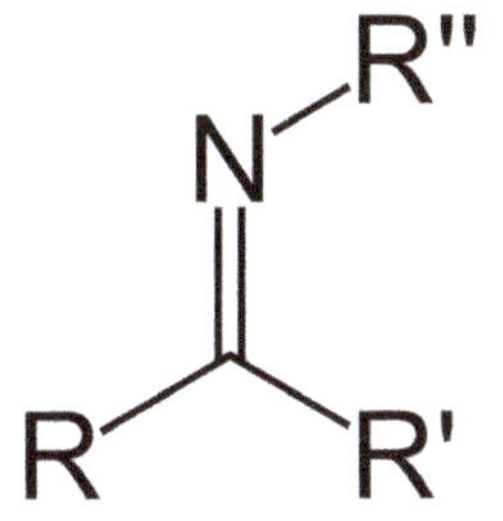	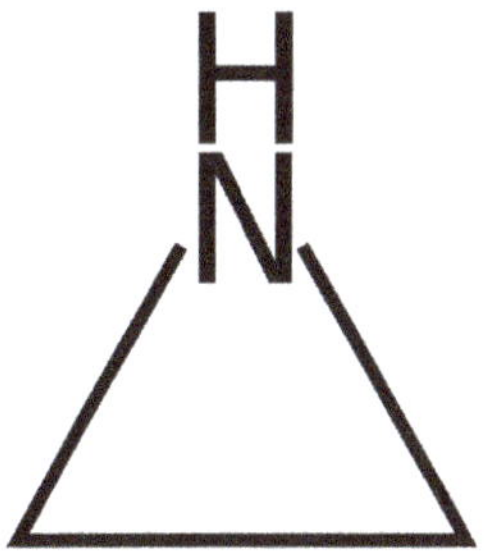
Secondary ketimine	Aziridine and its derivatives are sometimes referred to as imines.

One way of naming aldimines is to take the name of the radical, remove final "e", and add "-imine", for example ethanimine. Alternately, an imine is named as a derivative of a carbonyl, adding the word "imine" to the name of a carbonyl compound whose oxo group is replaced by an imino group, for example sydnone imine and acetone imine (an intermediate in the synthesis of acetone azine).

N-Sulfinyl imines are a special class of imines having a sulfinyl group attached to the nitrogen atom.

Synthesis of Imines

Imines are typically prepared by the condensation of primary amines and aldehydes and less commonly ketones:

$$RNH_2 + R'C(O)R \rightarrow \mathit{RN{=}C(R')(R)} + H_2O$$

In terms of mechanism, such reactions proceed via the nucleophilic addition giving a hemiaminal -C(OH)(NHR)- intermediate, followed by an elimination of water to yield the imine. The equilibrium in this reaction usually favors the carbonyl compound and amine, so that azeotropic distillation or use of a dehydrating agent, such as molecular sieves or magnesium sulfate, is required to push the reaction in favor of imine formation. In recent years, several reagents such as Tris(2,2,2-trifluoroethyl) borate [$B(OCH_2CF_3)_3$], pyrrolidine or titanium ethoxide [$Ti(OEt)_4$] have been shown to catalyse imine formation.

More Specialized Methods

Several other methods exist for the synthesis of imines.

- Reaction of organic azides with metal carbenoids (produced from diazocarbonyl compounds).
- Condensation of carbon acids with nitroso compounds.
- The rearrangement of trityl N-haloamines in the Stieglitz rearrangement.
- Dehydration of hemiaminals.
- By reaction of alkenes with hydrazoic acid in the Schmidt reaction.
- By reaction of a nitrile, hydrochloric acid, and an arene in the Hoesch reaction.
- Multicomponent synthesis of 3-thiazolines in the Asinger reaction.
- Primary ketimines can be synthesized via a Grignard reaction with a nitrile.

Imine Reactions

The most important reactions of imines are their hydrolysis to the corresponding amine and carbonyl compound. Otherwise this functional group participates in many other reactions, many of which are analogous to the reactions of aldehydes and ketones.

- An imine is reduced in reductive amination.
- An imine reacts with an amine to an aminal.
- An imine reacts with dienes in the Aza Diels-Alder reaction to a tetrahydropyridine.
- An imine can be oxidized with meta-chloroperoxybenzoic acid (mCPBA) to give an oxaziridine
- An aromatic imine reacts with an enol ether to a quinoline in the Povarov reaction.
- A tosylimine reacts with an α,β-unsaturated carbonyl compound to an allylic amine in the Aza-Baylis–Hillman reaction.
- Imines are intermediates in the alkylation of amines with formic acid in the Eschweiler-Clarke reaction.
- A rearrangement in carbohydrate chemistry involving an imine is the Amadori rearrangement.

- A methylene transfer reaction of an imine by an unstabilised sulphonium ylide can give an aziridine system.
- Imines react, thermally, with ketenes in [2+2] cycloadditions to form β-lactams in the Staudinger synthesis.

Acid-base Reactions

Somewhat like the parent amines, imines are mildly basic and reversibly protonate to give iminium salts. Iminium derivatives are particularly susceptible to reduction to the amines using transfer hydrogenation or by the stoichiometric action of sodium cyanoborohydride. Since imines derived from unsymmetrical ketones are prochiral, their reduction is a useful method for the synthesis of chiral amines.

As Ligands

Imines are common ligands in coordination chemistry. The condensation of salicylaldehyde and ethylenediamine give families of imine-containing chelating agents such as salen.

Imine Reductions

An imine can be reduced to an amine via hydrogenation for example in a synthesis of *m*-tolylbenzylamine:

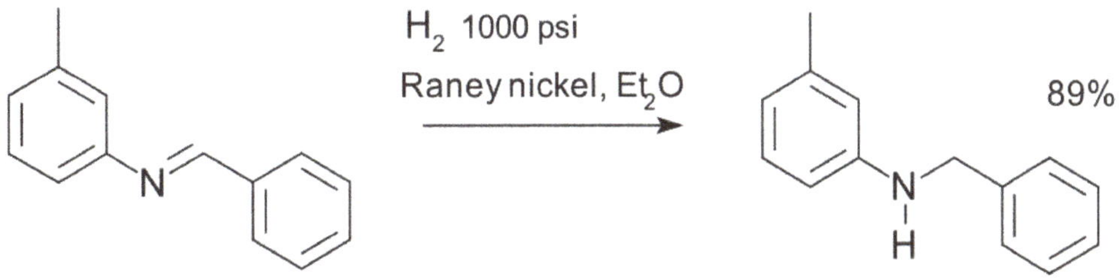

Other reducing agents are lithium aluminium hydride and sodium borohydride.

The first asymmetric imine reduction was reported in 1973 by Kagan using Ph(Me)C=NBn and $PhSiH_2$ in a hydrosilylation with chiral ligand DIOP and rhodium catalyst ($RhCl(CH_2CH_2)_2)_2$. Many systems have since been investigated.

Biological Role

Imines are common in nature. Vitamin B6 promotes the deamination of amino acids via the formation of imines, for example.

Reactions of Imines (C=N)

An important field of investigation for new industrial catalysts is the development of improved catalysts for the reduction of imines to obtain the corresponding chiral amines. These chiral amines are used as key components in many active pharmaceutical intermediates.

Synthesis of (S)-metolachlor (widely used as an herbicide) has been achieved by enantioselective hydrogenation of imine in presence of a catalyst generated in situ from [Ir(COD)Cl]2 and (R, S)-PPF–P(3,5-Xyl)$_2$ (xyliphos). This catalyst shows a high catalytic activity with TOF=396 h-1 and enantioselectivity of 79% ee.

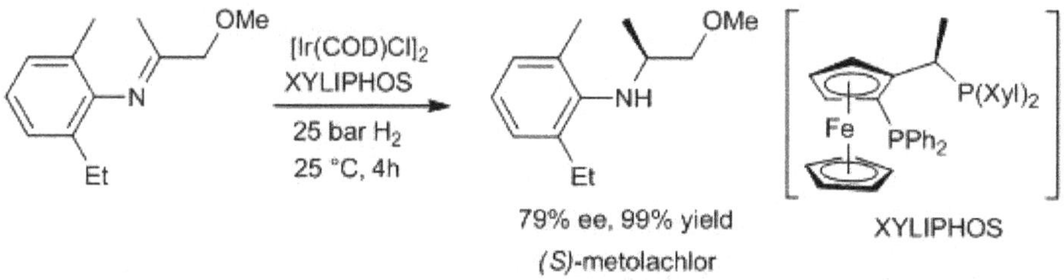

Preparation of (S)- Metolachlor by Enantioselective Hydrogenation

Subsequently, an air- and moisture-tolerant enantioselective reduction of N -phosphinyl imines has been performed with (CNbox)Re(O)Cl$_2$ (OPPh$_3$). A wide range of aromatic imines, including cyclic, acyclic and heteroaromatic, α -iminoesters, and α,β -unsaturated imines undergo reaction with good to excellent enantioselectivity.

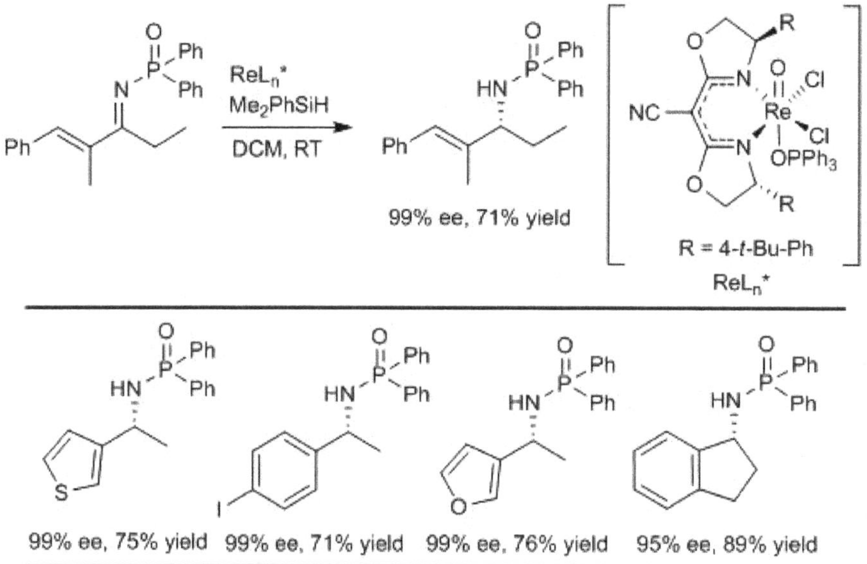

Enantioselective Reduction of Imines Catalyzed by Rhenium(V)-oxo Complex

The use of modified CBS-type catalysts has been extended to the reduction of oximes into chiral amines. The BINOL-proline-borate complex reduces acetophenone oxime into chiral 1-phenyl-ethylamine with 98% ee, but the ee drops when the borate complex is used catalytically.

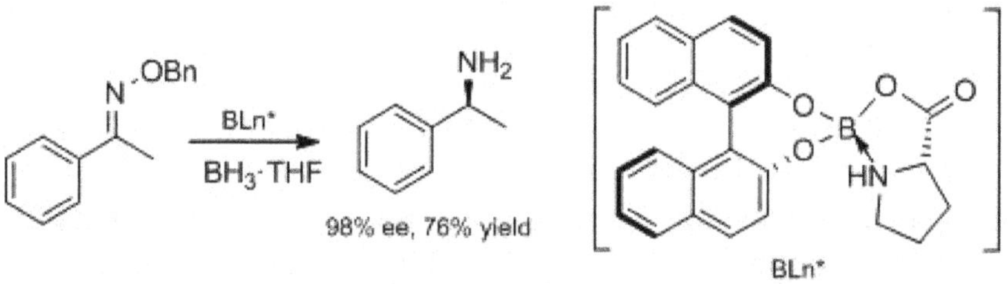

Modified CBS catalyst for Enantioselective Reduction of Imines

A new method for the reduction of α -imino esters using Hantzsch ester is reported with chiral phosphoric acid. A series of α -imino esters could be reduced to the corresponding α-amino esters in excellent yield with up to 94% ee.

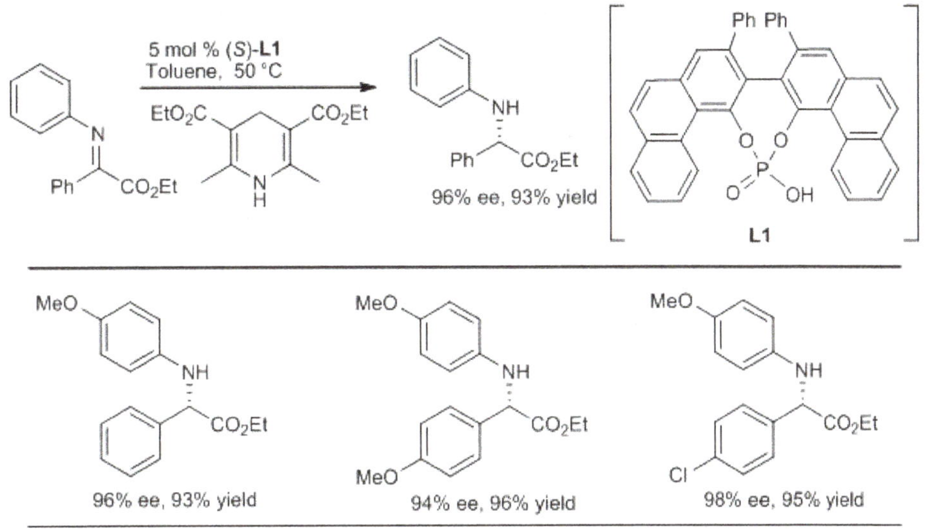

Chiral Biaryl Phosphoric Acid-Catalyzed Reduction of α-Imino Esters

An efficient metal/brønsted acid relay catalysis has been shown for the highly enantioselective hydrogenation of quinoxalines through convergent disproportionation of dihydroquinoxalines with up to 94%.

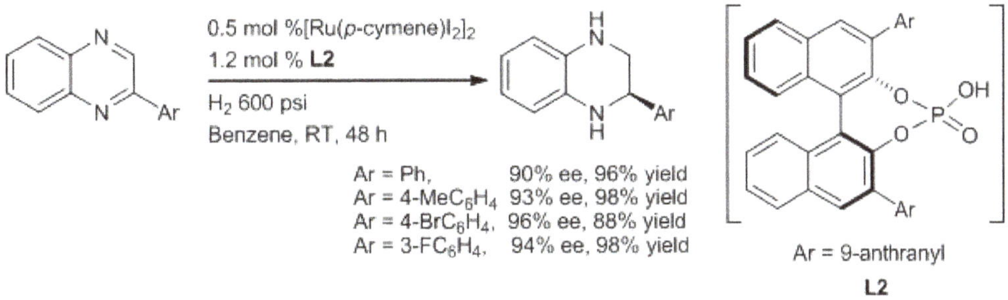

Metal/Brønsted Acid Catalysis for Enantioselective Reduction of Quinoxalines

Employing hydrogen gas as the reductant makes this convergent disproportionation an ideal atom-economical process. A dramatic reversal of enantioselectivity is observed for the hydrogenation relative to the transfer hydrogenation of quinoxalines promoted by chiral phosphoric acids L2 .

Transfer Hydrogenation

Transfer hydrogenation is the addition of hydrogen (H_2; dihydrogen in inorganic and organometallic chemistry) to a molecule from a source other than gaseous H_2. It is applied in industry and in organic synthesis, in part because of the inconvenience and expense of using gaseous H_2. One large scale application of transfer hydrogenation is coal liquefaction using "donor solvents" such as tetralin.

Organometallic Catalysts

In the area of organic synthesis, a useful family of hydrogen-transfer catalysts have been developed based on ruthenium and rhodium complexes, often with diamine and phosphine ligands. A representative catalyst precursor is derived from (cymene)ruthenium dichloride dimer and the tosylated diphenylethylenediamine. These catalysts are mainly employed for the reduction of ketones and imines to alcohols and amines, respectively. The hydrogen-donor (transfer agent) is typically isopropanol, which converts to acetone upon donation of hydrogen. Transfer hydrogenations can proceed with high enantioselectivities when the starting material is prochiral:

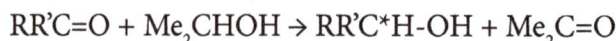

$$RR'C{=}O + Me_2CHOH \rightarrow RR'C^{*}H\text{-}OH + Me_2C{=}O$$

where RR'C*H-OH is a chiral product. A typical catalyst is (cymene)*R,R*-HNCHPhCHPhNTs, where Ts = $SO_2C_6H_4Me$ and *R,R* refers to the absolute configuration of the two chiral carbon centers. This work was recognized with the 2001 Nobel Prize in Chemistry to Ryōji Noyori.

Another family of hydrogen-transfer agents are those based on aluminium alkoxides, such as aluminium isopropoxide in the MPV reduction; however their activities are relatively low by comparison with the transition metal-based systems.

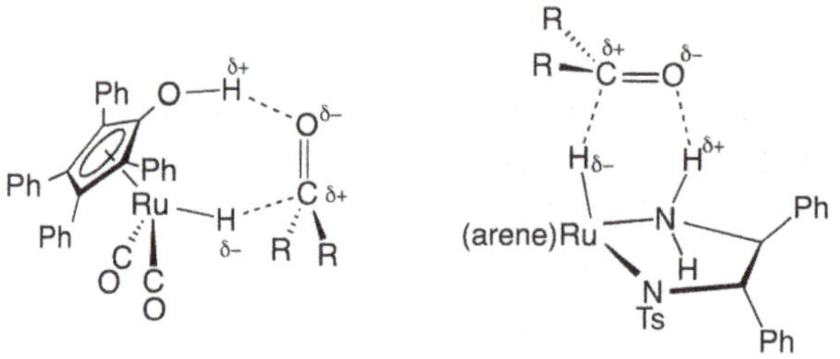

Transfer hydrogenation catalyzed by transition metal complexes proceeds by an "outer sphere mechanism."

Metal-free Routes

Prior to the development of catalytic hydrogenation, many methods were developed for the hydrogenation of unsaturated substrates. Many of these methods are only of historical and pedagogical interest. One prominent transfer hydrogenation agent is diimide or $(NH)_2$, also called diazene. This becomes oxidized to the very stable N_2:

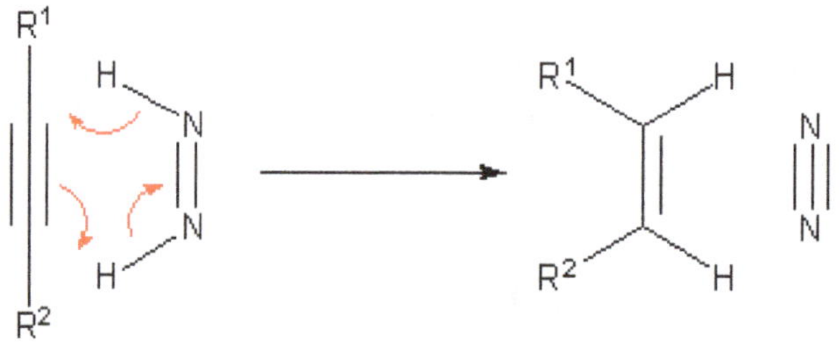

The diimide is generated from hydrazine. Two hydrocarbons that can serve as hydrogen donors are cyclohexene or cyclohexadiene. In this case, an alkane is formed, along with a benzene. The gain of aromatic stabilization energy when the benzene is formed is the driving force of the reaction. Pd can be used as a catalyst and a temperature of 100 °C is employed. More exotic transfer hydrogenations have been reported, including this intramolecular one:

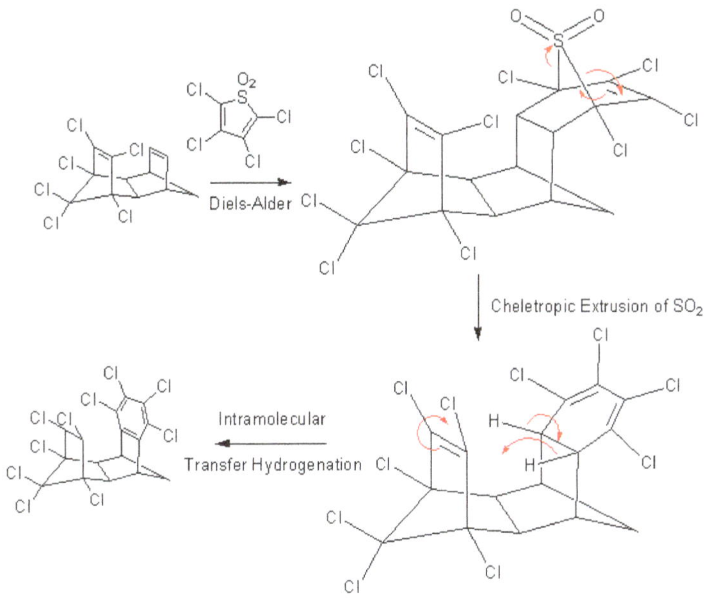

Many reactions exist with alcohol or amines as the proton donors, and alkali metals as electron donors. Of continuing value is the sodium metal-mediated Birch reduction of *arenes* (another name for aromatic hydrocarbons). Less important presently is the Bouveault–Blanc reduction of esters. The combination of magnesium and methanol is used in alkene reductions, e.g. the synthesis of asenapine:

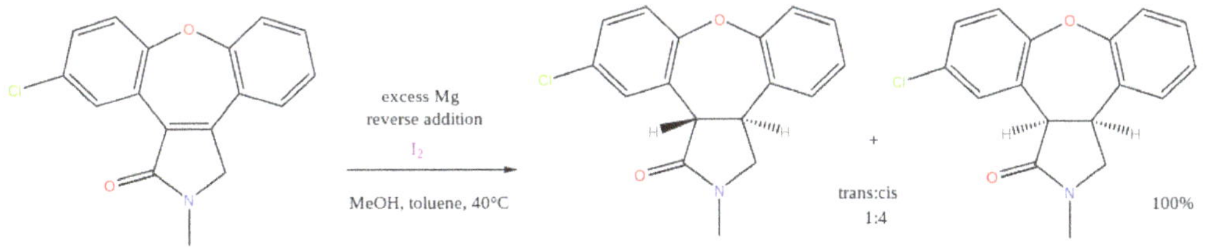

Organocatalytic Transfer Hydrogenation

Organocatalytic transfer hydrogenation has been described by the group of List in 2004 in a system with a Hantzsch ester as hydride donor and an amine catalyst:

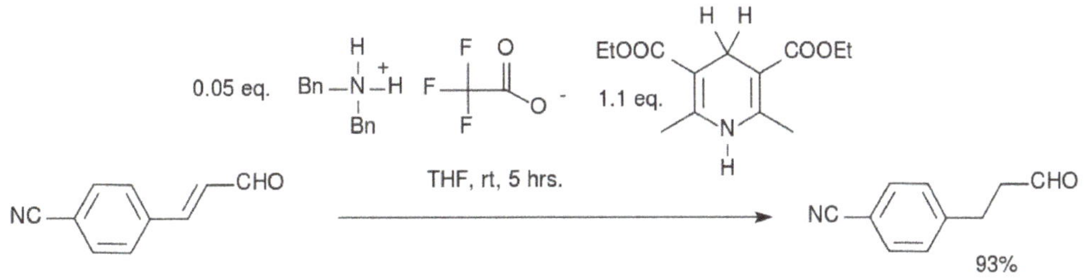

In this particular reaction the substrate is an α,β-unsaturated carbonyl compound. The proton donor is oxidized to the pyridine form and resembles the biochemically relevant coenzyme NADH. In the catalytic cycle for this reaction the amine and the aldehyde first form an iminium ion, then proton transfer is followed by hydrolysis of the iminium bond regenerating the catalyst. By adopting a chiral imidazolidinone MacMillan organocatalyst an enantioselectivity of 81% ee was obtained:

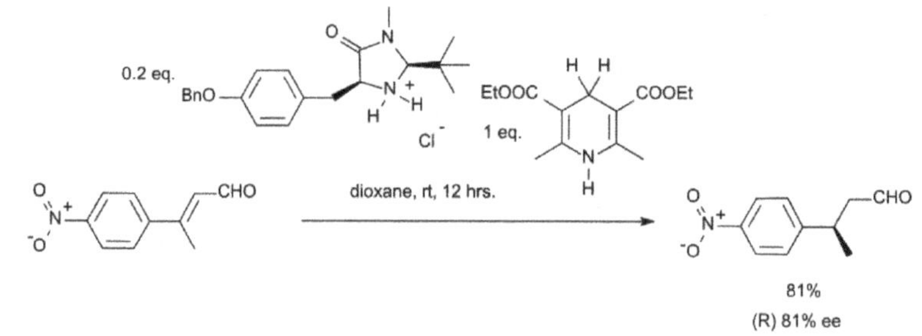

The group of MacMillan independently published a very similar asymmetric reaction in 2005:

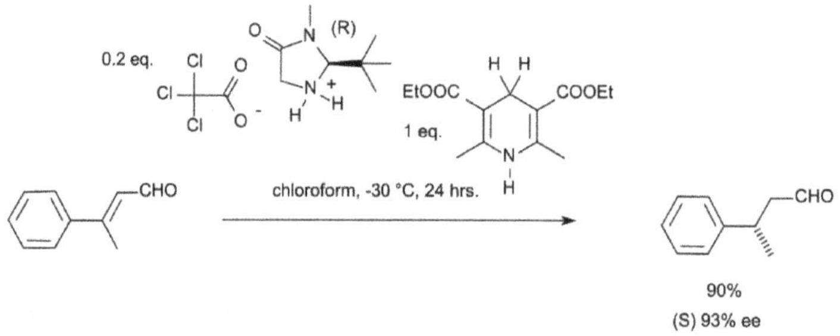

In an interesting case of stereoconvergence, both the E-isomer and the Z-isomer in this reaction yield the (S)-enantiomer.

Extending the scope of this reaction towards ketones or rather enones requires fine tuning of the catalyst (add a benzyl group and replace the t-butyl group by a furan) and of the Hantzsch ester (add more bulky t-butyl groups):

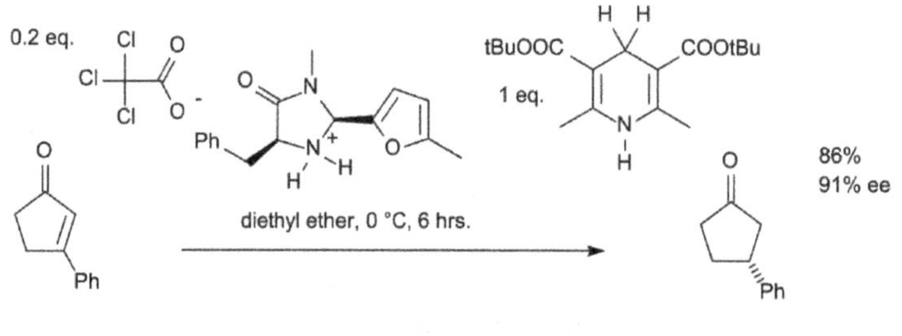

With a different organocatalyst altogether, hydrogenation can also be accomplished for imines. In one particular reaction the catalysts is a BINOL based phosphoric acid, the substrate a quinoline and the product a chiral tetradehydroquinoline in a 1,4-addition, isomerization and 1,2-addition cascade reaction:

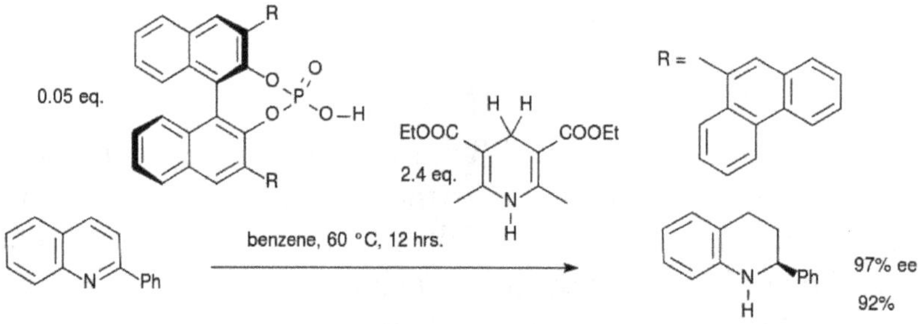

The first step in this reaction is protonation of the quinoline nitrogen atom by the phosphoric acid forming a transient chiral iminium ion. It is noted that with most traditional metal based catalysts, hydrogenation of aromatic or heteroaromatic substrates tend to fail.

Asymmetric Transfer Hydrogenation Reactions (ATHRs)

Another field where asymmetric transfer hydrogenation (ATH) catalysts have made an industrial impact is in the area of chiral amine synthesis by stereo controlled reduction of imines. The reduction of cyclic imines to yield chiral amines is proved to be a highly versatile and successful strategy for the synthesis of chiral tetrahydroisoquinolines and related compounds.

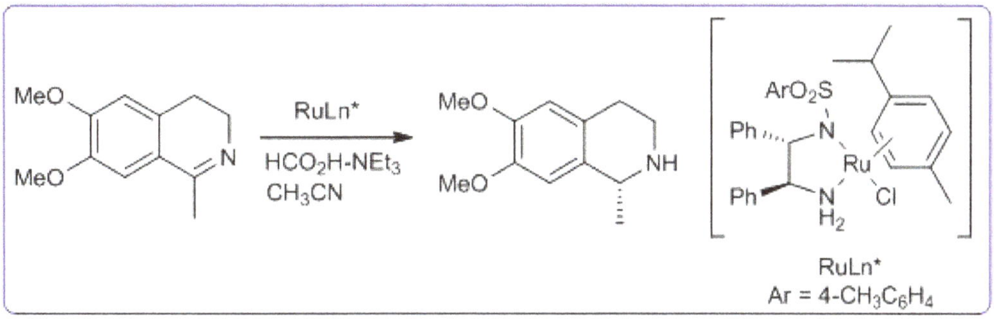

Catalytic enantioselective conjugate reduction of imines

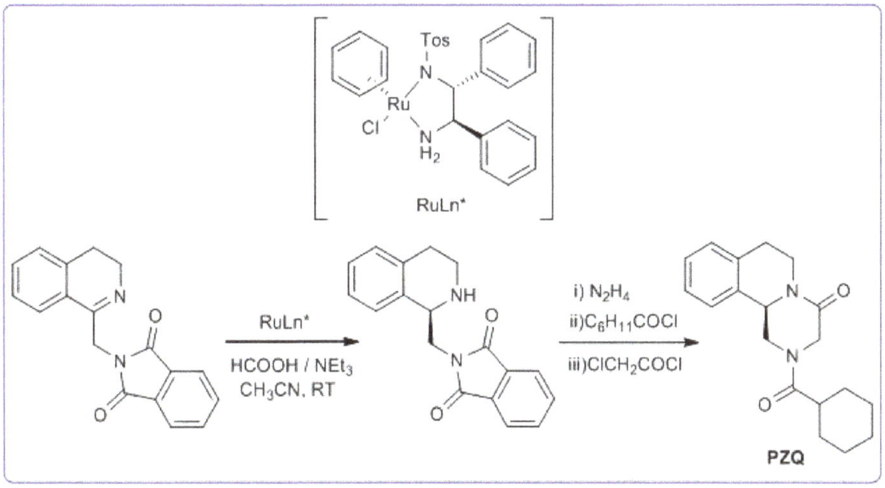

Enantioselective Synthesis of (R) - Praziquantel (PZQ)

The enantioselective preparation of Praziquantel (PZQ) a pharmaceutical for the treatment of schistosomiasis and soil-transmitted helminthiasis has been accomplished. The synthesis is completed from staring chiral reduction of imine which could be synthesized from readily available phenyl ethyl amine, phthalic anhydride and glycine.

In parallel to metal catalysis, organo catalyst like chiral thiourea and chiral imidazoilidines have been used for the asymmetric hydrogen transfer (ATS) reaction in presence of Hantzsch ester. For example, enantioselective Hantzsch ester mediated conjugate transfer hydrogenation of α,β -disubstituted nitro-alkenes has been shown using chiral thiourea. A broad range of substrates including β,β -unsaturated aldehydes and ketones, ketimines and aldimines, α -keto esters, and now nitro alkenes are successfully employed for hydrogenation.

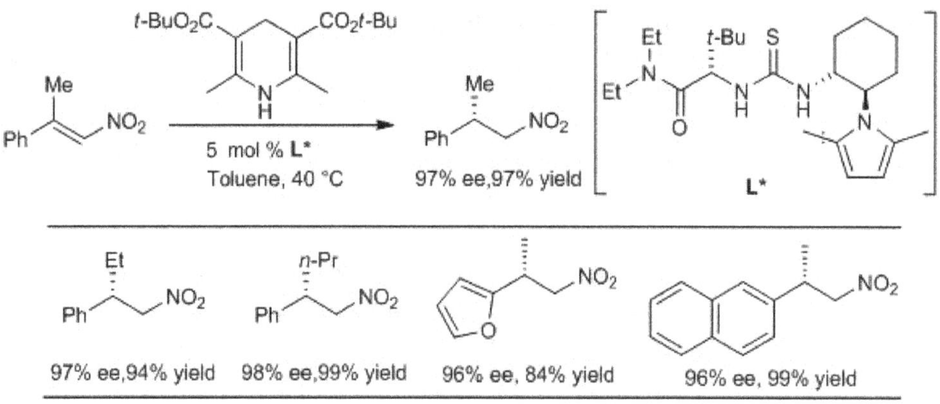

Transfer Hydrogenation of Nitro Styrene by Chiral Thiourea Catalyst

The above catalyst is also used for enantioselective Hantzsch ester mediated conjugate reduction of β -nitroacrylates. After subsequent reduction with Pd-H_2-MeOH, chiral β -amino acids can be synthesized with high yield and ee. This provides a key step in a new route to optically active $β_2$-amino acids.

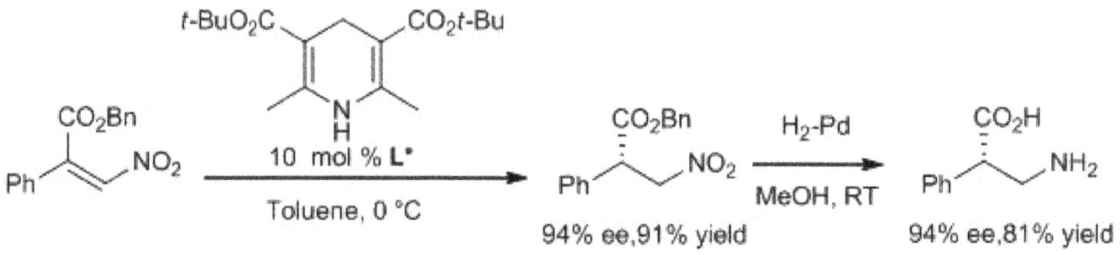

Transfer Hydrogenation of Nitro Styrene by Chiral Thiourea Catalyst

In parallel to the chiral thiourea catalyst, the use of iminium catalysis for the enantioselective reduction of β, β -substituted α, β -unsaturated aldehydes to generate β -stereogenic aldehydes has been shown. The capacity of the catalyst to accelerate (E)-(Z) isomerization prior to selective (E) -alkene reduction allows the implementation of geometrically impure enals in this operationally simple protocol.

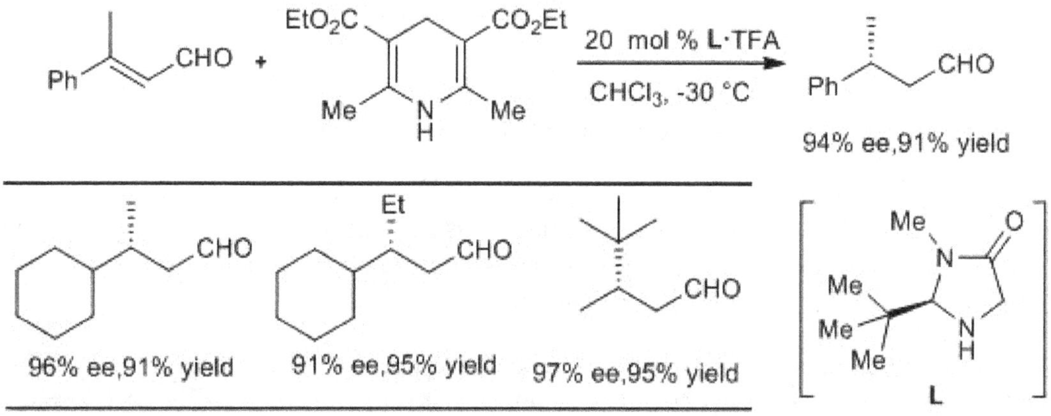

Transfer Hydrogenation of α , β -Unsaturated Aldehydes by Chiral Imidazolidinone

The above catalytic system is used for transfer hydrogenation of cyclic enones. Cycloalkenones with 5-, 6-, and 7-membered ring systems undergo reaction with high stereoselectivity.

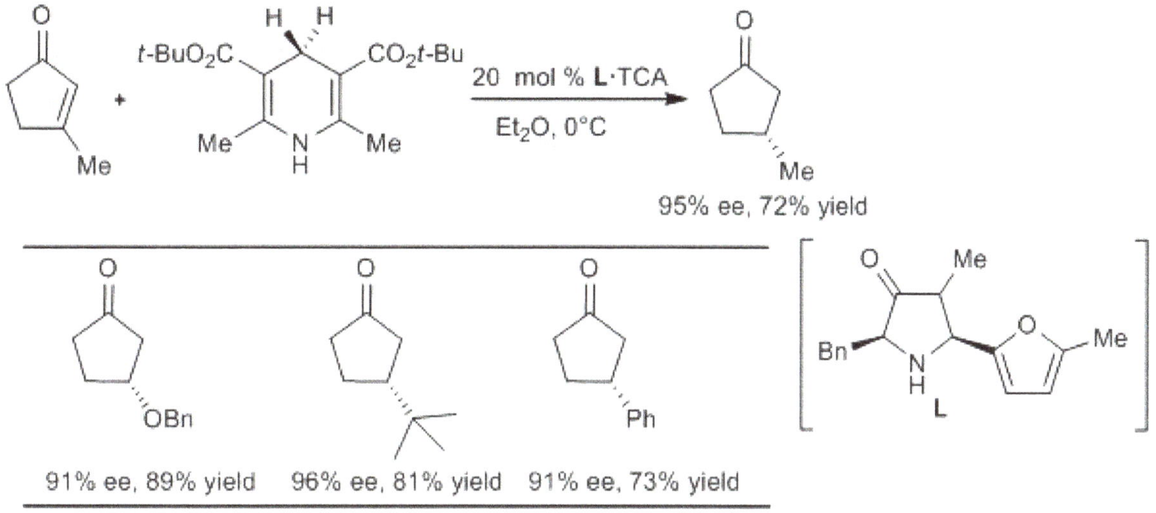

Transfer Hydrogenation of Cyclic Enones by Imidazolidinone

Reactions Carbon-Carbon Double Bonds

Enantioselective reduction of C=C double bond has important application in the synthesis of many natural products and pharmaceutically important compounds. Summarizes some of the common successful phosphine based chiral ligands developed for the catalytic asymmetric hydrogenation of alkenes.

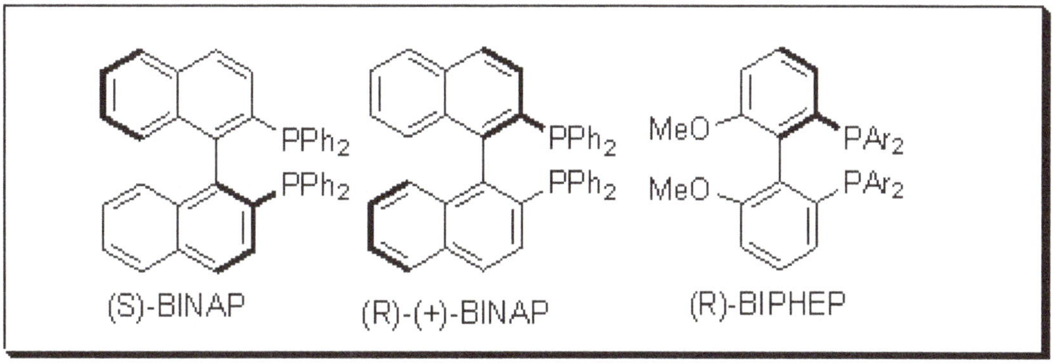

BINAP based ligands play an important role for asymmetric hydrogenation of alkenes. Both (S)-BINAP and (R)- BINAP could be synthesized by resolution methods using (1S,2S)- tartaric acid as well as (8R,9S)- N - benzylcinchonidinium chloride as the chiral sources. Synthesis of (S)- BINAP could be performed from racemic 2,2'-dibromo BINAP. Resolution of the corresponding phosphine oxide with (1S,2S)- tartaric acid and subsequent reduction with $HSiCl_3$ can afford (S)- BINAP in gram scale.

Alternatively, (S)- BINAP and (R)- BINAP can be synthesized by resolution of racemic BINOL using (8R,9S)- N - benzylcinchonidinium chloride. Converting them into triflate derivative and subsequent cross-coupling with Ph_2PH using $NiCl_2$ to afford (S)- BINAP and (R) - BINAP in gram scale. (S)- BINAP ; light brown solid, mp 205°C , 99 % ee, $[\alpha]^{21}_D = -29.4°$ (THF , c =1). (R)- BINAP; white crystalline solid, mp 207°C , 99% ee, $[\alpha]^{21}_D = 26.2 - 30.9°$ (THF , c 1).

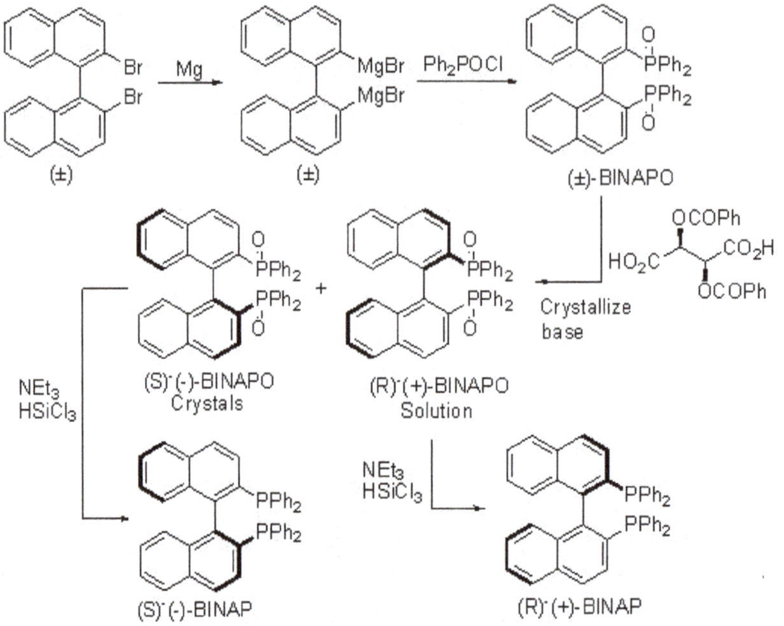

Gram Scale Synthesis of (S)- BINAP and (R)- BINAP

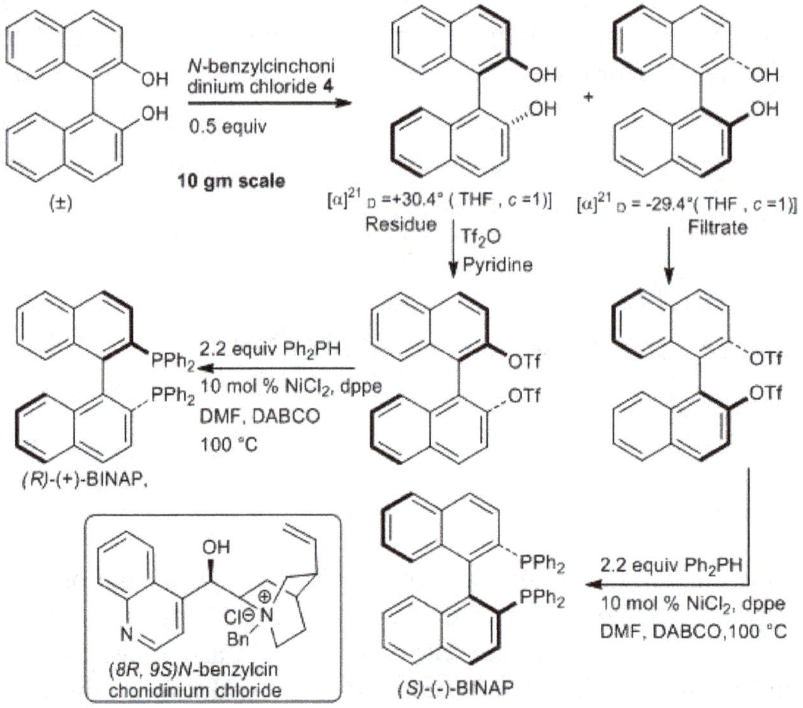

Alternative Synthesis of Chiral (S)- BINAP and (R)- BINAP

Reduction of α,β -Unsaturated Carboxylic Acids

Chiral Ru(II)-BINAP catalyzes the hydrogenation of α,β- unsaturated carboxylic acids. For example, the hydrogenation of naphthacrylic acid can be performed using a Ru-(S)-BINAP with 134 atm H_2 pressure. The reaction affords chiral (S)-naproxen with 98% ee, which is a nonsteroidal anti-inflammatory drug .

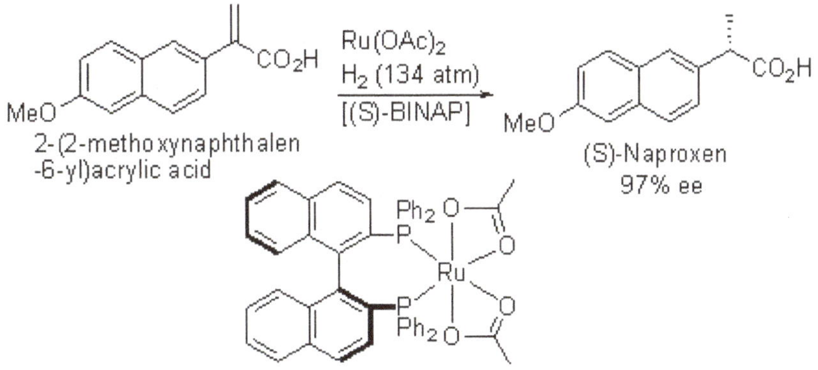

Synthesis of (S)- Naproxen by Chiral Reduction of α,β- Unsaturated Carboxylic Acids

Hydrogenation has been explored for the synthesis of intermediate of (S)- mibefradil. For this reaction chiral Ru-complex bearing (R)-MeO-BIPHEP is found to be effective affording the target intermediate with 92% ee.

Synthesis of Intermediate for (S)- Mibefradil

Reduction of Allylic Alcohol

Allylic alcohols can be reduced with high selectivity using chiral Ru-(S)-BINAP as a catalyst. For example, the reduction of geraniol can be accomplished with 94% ee. The reduced product is used for the large scale synthesis of L-(+)-menthol. Under these conditions, nerol undergoes reduction to give (S)-citronellol in 99% ee. Chiral iridium-based catalytic systems have also been subsequently explored for the asymmetric reduction of allylic alcohols. For example, the complex bearing chiral phosphanodihydrooxazole L_1 catalyzes asymmetric reduction of an allyl alcohol, which is used as a key step in the synthesis of lillial. Figure illustrates the synthesis of chiral phosphanodihydrooxazole L_1.

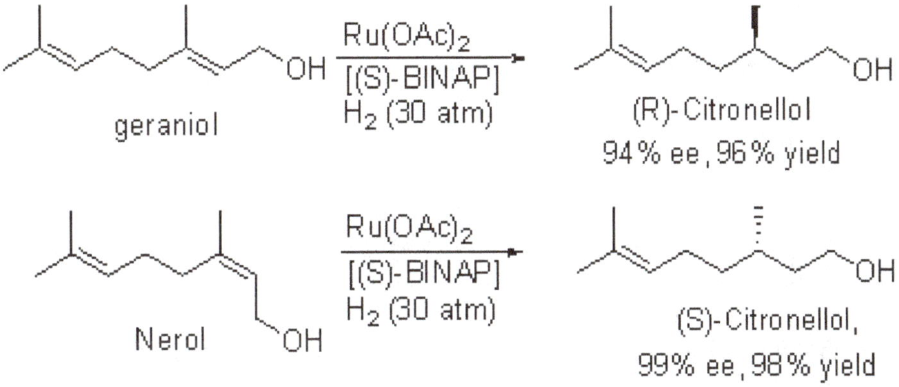

Synthesis of (S) and (R)- Citronellol by Chiral Reduction of Geraniol and Nerol

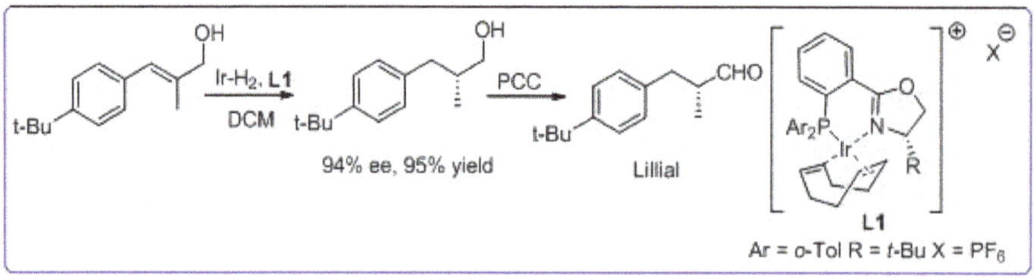

Asymmetric Synthesis of Lillial.

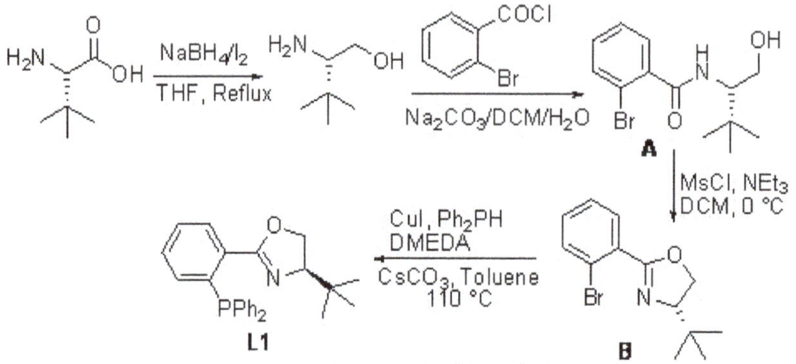

Synthesis of Phosphanodihydrooxazole L_1

Reduction of Allylic Amines

In parallel to the reduction of allylic alcohol, Rh-(S)-BINAP system has been used for the reduction of allylic amine. For example, the synthesis of (R)- citronellal can be accomplished via reduction of allylic amine. The key step is the isomerization of geranyl diethylamine forming (R)-citronellal enamine . The Rh-complex performs the rearrangement of this allylic amine to the enamine creating a new chiral centre with >98% ee, which upon hydrolysis gives (R)-citronellal in 96–99% ee. The latter serves as substrate precursor for the synthesis of L-(+)-menthol via intramolecular ene reaction followed by hydrogenation.

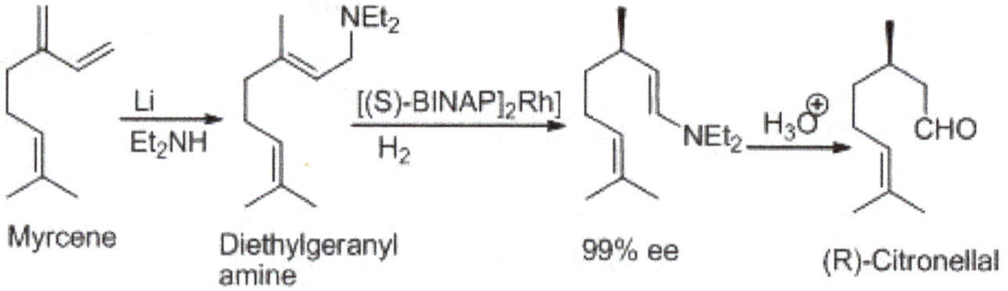

Chiral Reduction of Allylic Amine to Synthesize (R)-Citronellal

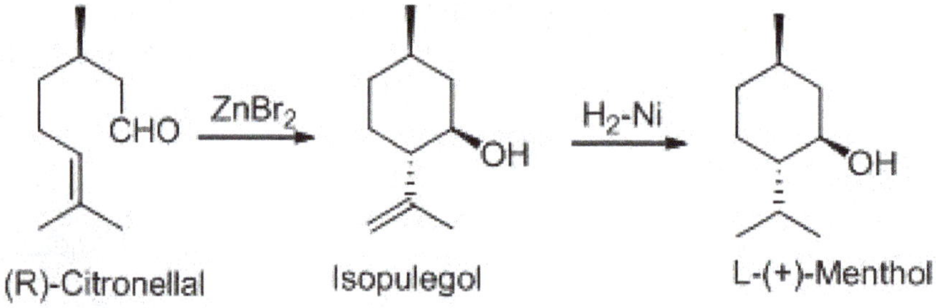

Industrial preparation of L-(+)-Menthol by Chiral Reduction of Allylic Amine

Reduction of α, β-Unsaturated Aldehydes

Asymmetric reduction of α, β -unsaturated aldehydes with transition metal catalysts has not yet proven ready for wide spread industrial application. In comparison to CBS catalyst, the Baker's yeast is most useful, since the precursor (R)-proline used to synthesize CBS is expensive. The chiral reduction of enals to chiral alcohols using Baker's yeast has been known for over 30 years. Figure summarizes some of the examples for the Baker yeast catalyzed reduction of C=C of α, β -unsaturated aldehydes.

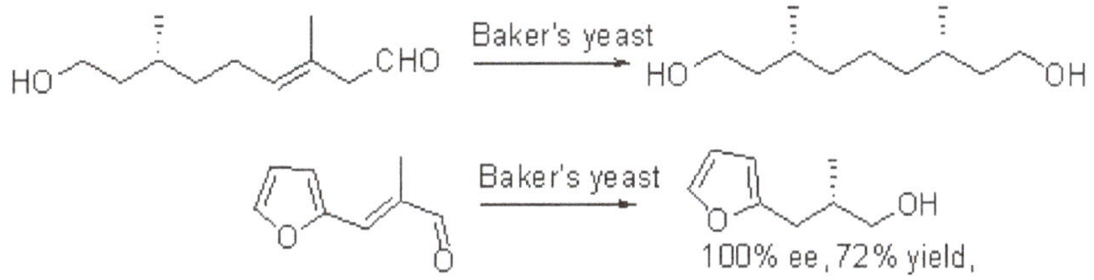

Baker's yeast cell for Reduction of α, β -Unsaturated Aldehydes

Subsequently, organocatalysis has been found be effective for the asymmetric reduction. A recent interesting development is the organocatalytic hydride transfer reductions of α, β -unsaturated aldehydes to chiral aldehyde. Hantzsch ester acts as a good NADH mimic in the hydride transfer to an iminium ion, formed when the α,β -unsaturated aldehyde reacts with the amine of the organocatalyst.

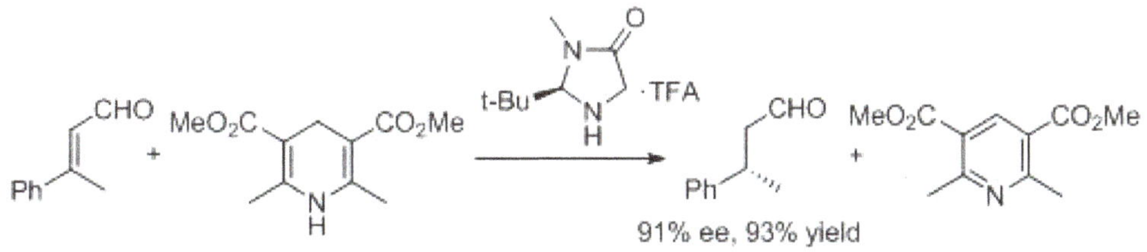

Organocatalytic Reduction of an Unsaturated Aldehyde

Similarly, chiral phosphoric acid L2 catalyses the reduction of C=C of α, β -unsaturated aldehyde with 90% ee and 98% yield in the presence of Hantzsch ester.

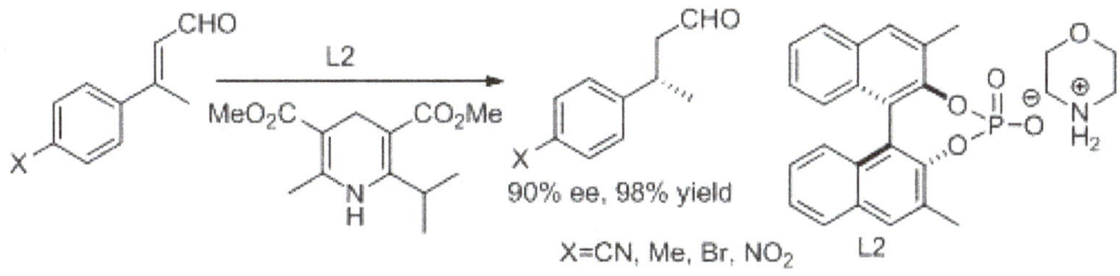

Organocatalytic Reduction of an α, β -Unsaturated Aldehyde

Reduction of α, β-Unsaturated α-Amino Acid

Asymmetric reduction of α, β -unsaturated α-amino acid has wide application in organic synthesis. Chiral biphosphines in combination with Rh acts as the best combination for the reduction α, β-unsaturated α -amino acids. Figure summarizes some of the successful chiral phosphines for the Rh-catalyzed reactions.

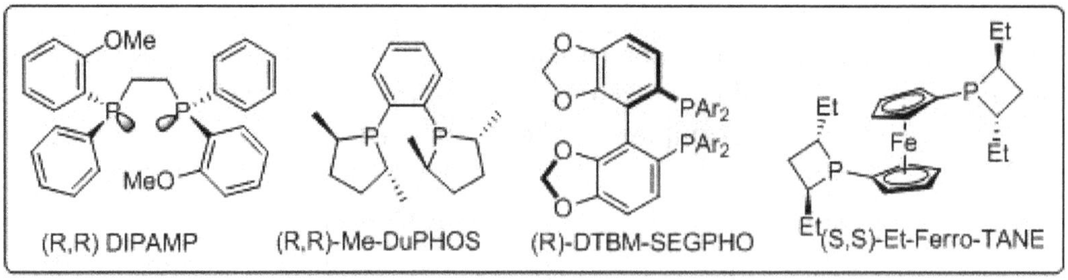

Ligands used for Chiral Reduction of α, β-Unsaturated α -Amino acid

Rh-DIPAMP has been explored for the reduction of α, β-unsaturated α-amino acids. For example, L-DOPA, a chiral drug for treating Parkinson's disease, is synthesized using Rh-(R,R)-DIPAMP catalyzed reduction of α, β-unsaturated α -amino acid as a key step.

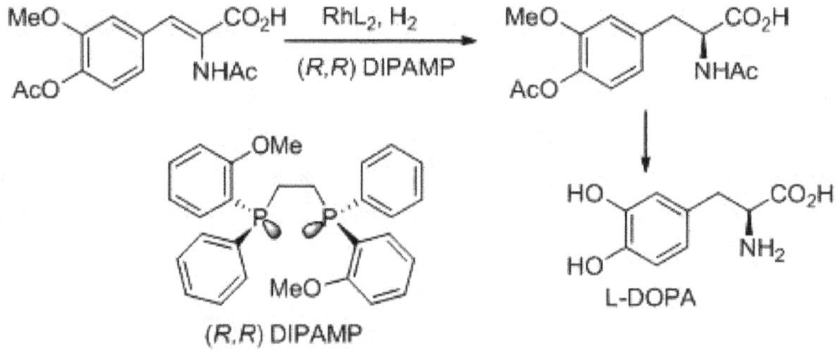

Key Step for Industrial Synthesis of L-DOPA

Rh -(R,R)- DuPHOS can be used for the reduction of α, β-unsaturated α-amino acid to give chiral amino acid. Using this procedure many of the unnatural α-amino acids can be obtained directly with enantioselectivity approaching 100% ee and S/C ratio 10000-50000. The rhodium-catalyzed hydrogenation of the E- and Z -isomers, with BINAP in THF, affords products with opposite absolute configurations. Remarkably, the (R,R)- DuPHOS system provides excellent enantioselectivity for both isomeric substrates with the same absolute configuration, irrespective of the E/Z -geometry. This result is particularly important for the construction of alkyl dehydroamino acid derivatives, which are difficult to prepare in enantiomerically pure form.

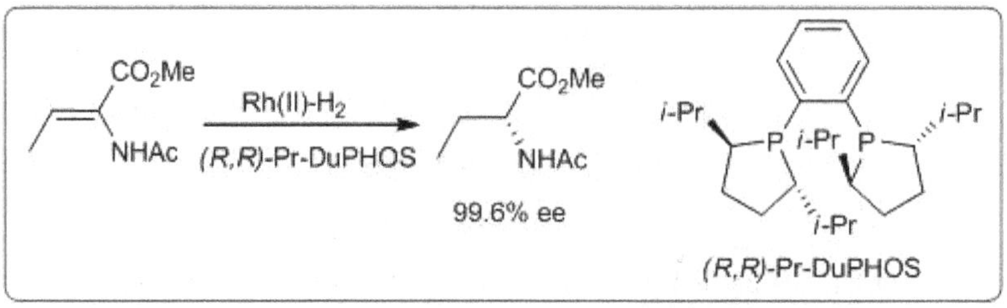

Reduction of α, β-Unsaturated α-Amino Ester

The hydrogenation of the (E)- or (Z)- isomer of β-(acetylamino)- β-methyl- α-dehydroamino acids with Rh(I)-Me-DuPHOS provides either diastereomers of the N, N -protected 2,3-diaminobutanoic acid derivatives with 98% ee.

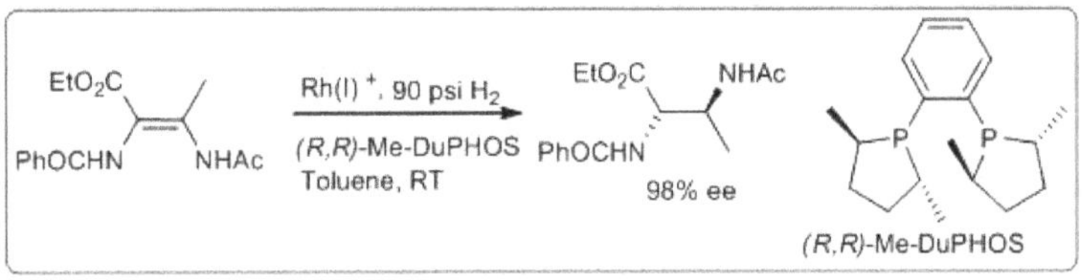

Reduction of α, β-Unsaturated α-Amino β-ester

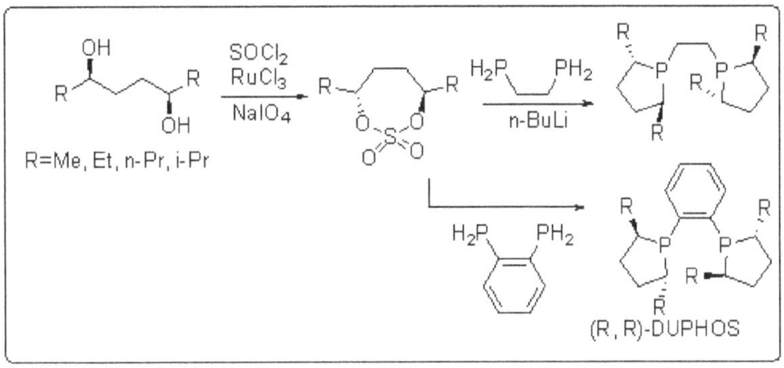

Synthesis of 1,2-Bis(phospholano) (DuPHOS) Ligands

(S)- SEGPHOS and its analogous provide superior results in Ru-catalyzed hydrogenation of four and five-membered cyclic lactones or carbonates bearing an exocyclic methylene group. For example, the reduction of the four membered lactone can be achieved with excellent enantioselectivity using S/C=12270.

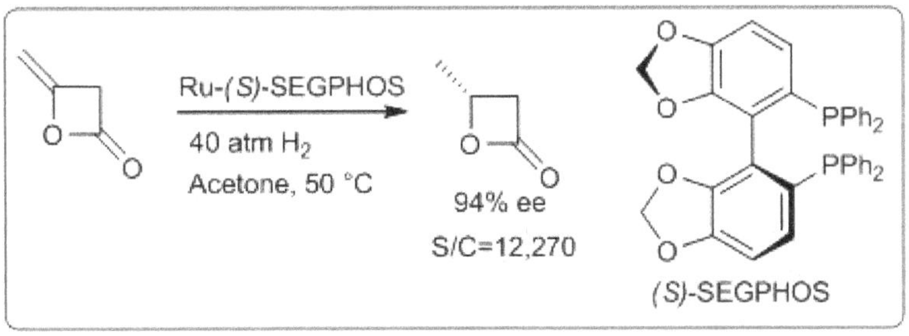

Reduction of α,β -Unsaturated Lactone using (S)- SEGPHOS

Figure describes the synthesis of SEGPHOS. The key step is the resolution of racemic phosphine oxide with (S,S)- DBTA (di-benzoyl-tartaric acid) to provide chiral phosphine oxide. Subsequent reduction with $HSiCl_3$ affords the target SEGPHOS in good yield.

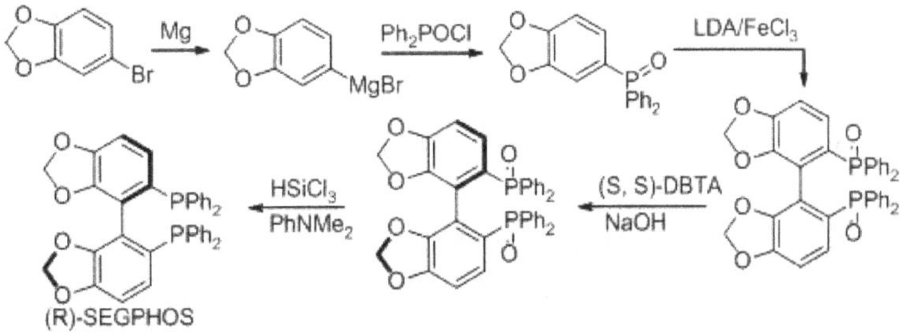

Synthesis of (R)- SEGPHOS Ligands

Moreover, chiral 1,10-diphosphetanylferrocene Et-FerroTANE serves as an effective ligand for the rhodium-catalyzed hydrogenation of β -aryl- and β -alkyl-substituted monoamido itaconate.

For example, Et-DuPHOS–Rh is utilized for the asymmetric hydrogenation of the trisubstituted alkene to afford the reduced product, which is used for synthesis of intermediate of the drug candoxatril in 99% ee . Candoxatril is the orally active prodrug of candoxatril (UK-73967) human neutral endopeptidase (Neprilysin).

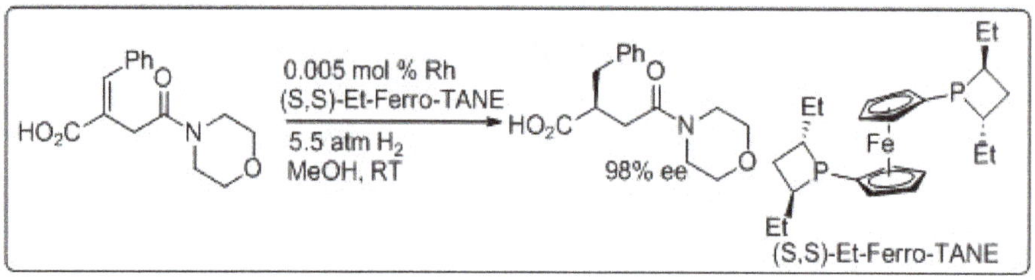

Reduction of α,β-Unsaturated Carboxylic using Et-Ferro TANE

The above described alkyl/aryl-ferro-TANE family ligands could be synthesized from optically active diols. Cyclization with SO_2Cl_2 in presence of $RuCl_3$ and $NaIO_4$ affords chiral cyclized sulfonate, which reacts with ferro-phosphine in the presence of n-BuLi to give the target chiral alkyl/aryl-Ferro-TANE family in good yield.

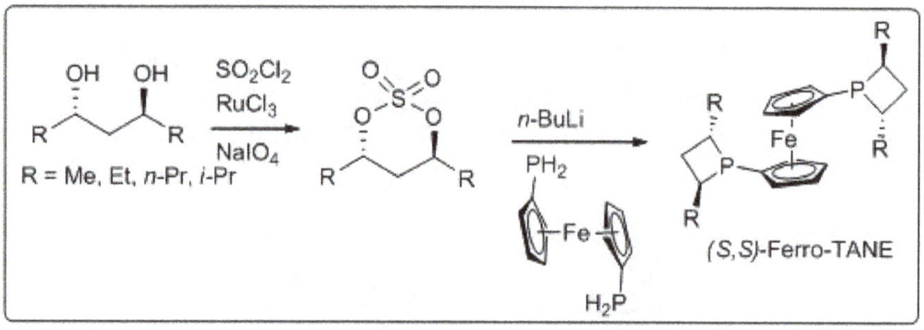

Synthesis of Chiral Et-Ferro TANE Ligands

Similarly, the reduction of α,α -disubstituted α, β-unsaturated ester can be carried out using chiral Ru-Et-Ferro TANE. The reaction is compatible with different electron donating and withdrawing groups attached to benzene ring.

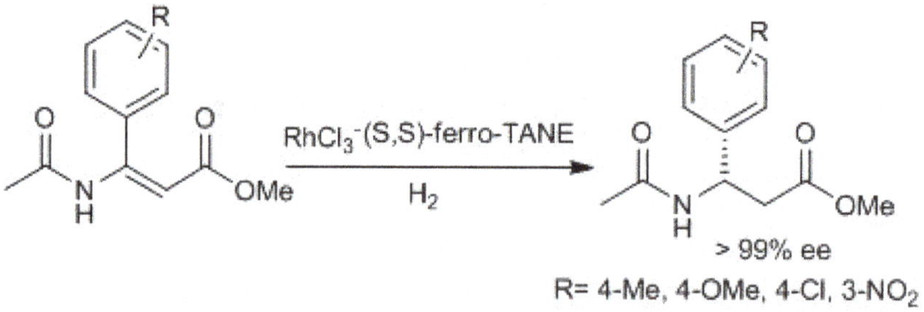

Chiral Reduction of α,α -Disubstituted α,β-Unsaturated Ester.

Reduction of α -Alkyl Substituted Acids

Another important chiral acid is the α -alkyl substituted acid which is used in the synthesis of aliskiren (the active ingredient of Tekturna1). The key step for the synthesis requires the hydrogenation of cinnamic acid derivative in the presence of Rh-phosphoramidite . The reduction also affords 97% ee using Rh-WALPHOS.

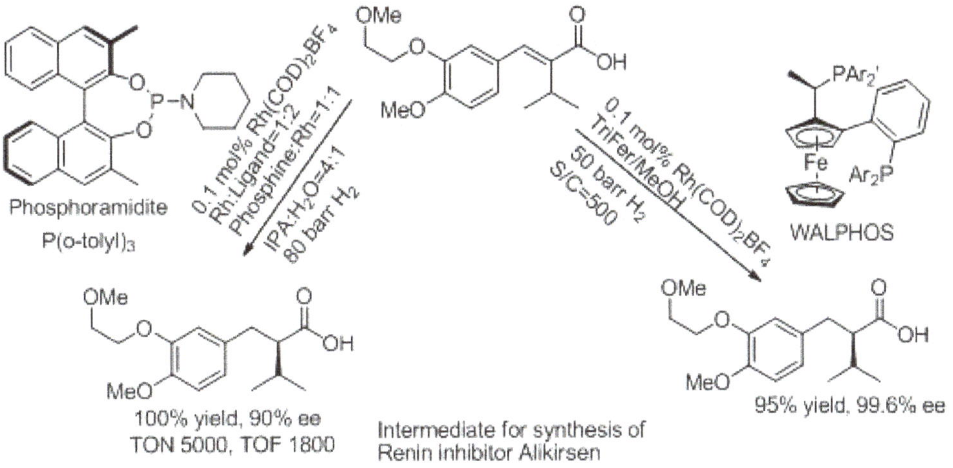

Key Step for Synthesis of Renin Inhibitors Aliskiren

Reduction of α, β-Unsaturated Nitriles

The asymmetric reduction of unsaturated nitriles is a very useful process for the synthesis of many pharmaceutical intermediates. An important application of this strategy involves the further reduction of the nitrile group to yield chiral amines. For example, chiral Rh-phosphine catalyzes the asymmetric hydrogenation of an unsaturated nitrile. The reduced product is used for the synthesis of the Pregabalin .

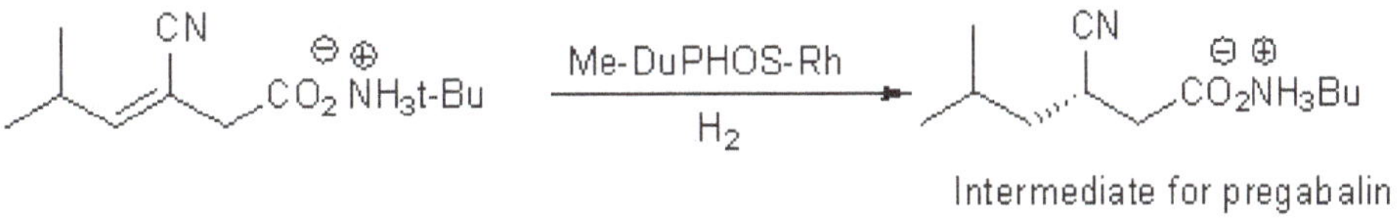

Pfizer Pregabalin Intermediate Synthesis

A more challenging example of an unsaturated nitrile reduction that lacks the carboxylate functional group is the asymmetric reduction of the nitrile shown in Figure. The reduced product is used for the synthesis of chiral 3,3-diarylpropylamine, which is an intermediate for the synthesis of the Arpromidines. The arpromidines analogues are the most potent histamine H_2 receptor agonists known and are promising positive inotropic vasodilators for the treatment of severe congestive heart failure.

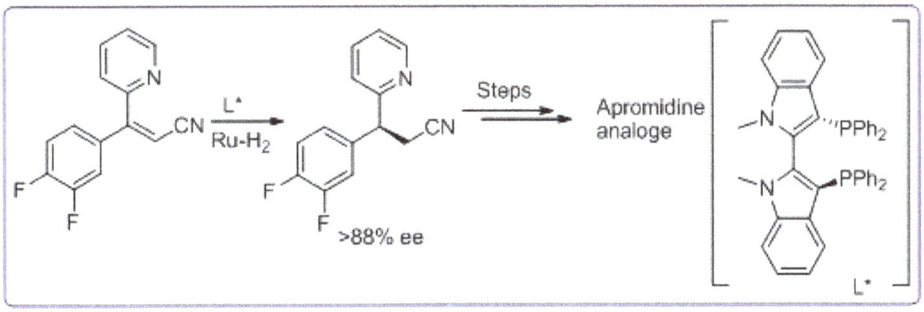

H ydrogenation of Diaryl-substituted α,β-Unsaturated nitriles.

In parallel to Ru, Rh and Ir-based catalytic systems, chiral copper hydride catalysis have been demonstrated for enantioselective 1,4-reductions of 2-alkenyl heteroarenes. Both azoles and azines serve as efficient activating groups for this process.

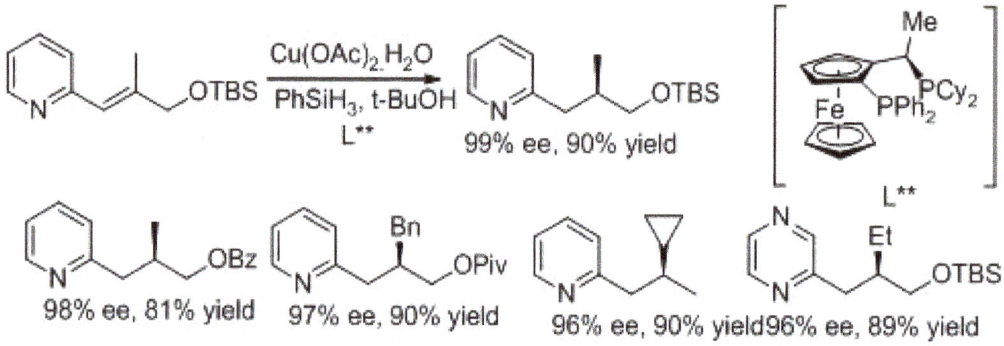

Enantioselective Hydrogenation of Protected Allylic Alcohol

References

- Patel, D. R. (1998). "Hydrogenation of nitrobenzene using polymer anchored Pd(II) complexes as catalyst". Journal of Molecular Catalysis. 130: 57. doi:10.1016/s1381-1169(97)00197-0
- Hudlický, Miloš (1996). Reductions in Organic Chemistry. Washington, D.C.: American Chemical Society. p. 429. ISBN 0-8412-3344-6
- Paul N. Rylander, "Hydrogenation and Dehydrogenation" in Ullmann's Encyclopedia of Industrial Chemistry, Wiley-VCH, Weinheim, 2005. doi:10.1002/14356007.a13 487
- Knowles, W. S. (March 1986). "Application of organometallic catalysis to the commercial production of L-DOPA". Journal of Chemical Education. 63 (3): 222. doi:10.1021/ed063p222
- Moureu, Charles; Mignonac, Georges (1920). "Les Cetimines". Annales de chimie. 9 (13): 322–359. Retrieved 18 June 2014
- Johannes G. de Vries, Cornelis J. Elsevier, eds. The Handbook of Homogeneous Hydrogenation Wiley-VCH, Weinheim, 2007. ISBN 978-3-527-31161-3
- Woodmansee, D. H.; Pfaltz, A. (2011). "Asymmetric hydrogenation of alkenes lacking coordinating groups". Chemical Communications. 47 (28): 7912–7916. PMID 21556431. doi:10.1039/c1cc11430a
- Walling, Cheves.; Bollyky, Laszlo. (1964). "Homogeneous Hydrogenation in the Absence of Transition-Metal Catalysts". Journal of the American Chemical Society. 86 (18): 3750. doi:10.1021/ja01072a028
- Blaser, Hans-Ulrich; Federsel, Hans-Jürgen, eds. (2010). Asymmetric Catalysis on Industrial Scale. Weinheim: Wiley-VCH. pp. 13–16. ISBN 978-3-527-63063-9. doi:10.1002/9783527630639
- Blaser, H. U.; Spindler, F.; Studer, M. (2001). "Enantioselective catalysis in fine chemicals production". Applied Catalysis A: General. 221: 119. doi:10.1016/S0926-860X(01)00801-8
- Sonnenberg, J. F.; Coombs, N.; Dube, P. A.; Morris, R. H. (2012). "Iron Nanoparticles Catalyzing the Asymmetric Transfer Hydrogenation of Ketones". Journal of the American Chemical Society. 134 (13): 5893–5899. PMID 22448656. doi:10.1021/ja211658t
- "F.D.A. Gives Food Industry 3 Years to Eliminate Trans Fats". The New York Times. 2015-06-16. Retrieved 2015-06-16
- Mendham, J.; Denney, R. C.; Barnes, J. D.; Thomas, M. J. K. (2000), Vogel's Quantitative Chemical Analysis (6th ed.), New York: Prentice Hall, ISBN 0-582-22628-7
- Tietze, Lutz F.; Bratz, Matthias (1993). "Dialkyl Mesoxalates by Ozonolysis of Dialkyl Benzalmalonates: Dimethyl Mesoxalate". Organic Syntheses. 71: 214. doi:10.15227/orgsyn.071.0214

- Baeza, A.; Pfaltz, A. (2010). "Iridium-Catalyzed Asymmetric Hydrogenation of N-Protected Indoles". Chemistry: A European Journal. 16 (7): 2036. doi:10.1002/chem.200903105
- March Jerry; (1985). Advanced Organic Chemistry reactions, mechanisms and structure (3rd ed.). New York: John Wiley & Sons, inc. ISBN 0-471-85472-7
- Ouellet; Tuttle, J.; MacMillan, D. (2005). "Enantioselective organocatalytic hydride reduction". Journal of the American Chemical Society. 127 (1): 32–33. PMID 15631434. doi:10.1021/ja043834g
- C. F. H. Allen, F. W. Spangler, and E. R. Webster "Ethyleneimine" Org. Synth. 1950, volume 30, 38. doi:10.15227/orgsyn.030.0038

Permissions

All chapters in this book are published with permission under the Creative Commons Attribution Share Alike License or equivalent. Every chapter published in this book has been scrutinized by our experts. Their significance has been extensively debated. The topics covered herein carry significant information for a comprehensive understanding. They may even be implemented as practical applications or may be referred to as a beginning point for further studies.

We would like to thank the editorial team for lending their expertise to make the book truly unique. They have played a crucial role in the development of this book. Without their invaluable contributions this book wouldn't have been possible. They have made vital efforts to compile up to date information on the varied aspects of this subject to make this book a valuable addition to the collection of many professionals and students.

This book was conceptualized with the vision of imparting up-to-date and integrated information in this field. To ensure the same, a matchless editorial board was set up. Every individual on the board went through rigorous rounds of assessment to prove their worth. After which they invested a large part of their time researching and compiling the most relevant data for our readers.

The editorial board has been involved in producing this book since its inception. They have spent rigorous hours researching and exploring the diverse topics which have resulted in the successful publishing of this book. They have passed on their knowledge of decades through this book. To expedite this challenging task, the publisher supported the team at every step. A small team of assistant editors was also appointed to further simplify the editing procedure and attain best results for the readers.

Apart from the editorial board, the designing team has also invested a significant amount of their time in understanding the subject and creating the most relevant covers. They scrutinized every image to scout for the most suitable representation of the subject and create an appropriate cover for the book.

The publishing team has been an ardent support to the editorial, designing and production team. Their endless efforts to recruit the best for this project, has resulted in the accomplishment of this book. They are a veteran in the field of academics and their pool of knowledge is as vast as their experience in printing. Their expertise and guidance has proved useful at every step. Their uncompromising quality standards have made this book an exceptional effort. Their encouragement from time to time has been an inspiration for everyone.

The publisher and the editorial board hope that this book will prove to be a valuable piece of knowledge for students, practitioners and scholars across the globe.

Index